INTERCONNECTED DYNAMICAL SYSTEMS: STABILITY, DECOMPOSITION AND DECENTRALISATION

NORTH-HOLLAND SYSTEMS AND CONTROL SERIES
VOLUME 5

Series Editors

M. G. SINGH
(Coordinating editor)
University of Manchester
U.K.

P. VARAIYA
University of California at Berkeley
U.S.A.

M. AIZERMAN
USSR Academy of Sciences
Moscow, U.S.S.R.

NORTH-HOLLAND PUBLISHING COMPANY
AMSTERDAM • NEW YORK • OXFORD

INTERCONNECTED DYNAMICAL SYSTEMS: STABILITY, DECOMPOSITION AND DECENTRALISATION

by

J. BERNUSSOU and A. TITLI*

with the collaboration of

G. AUTHIE* and J. L. CALVET

Laboratoire d'Automatique et d'Analyse des Systèmes
Centre National de la Recherche Scientifique
LAAS - CNRS, Toulouse
France

**also with Institut National des Sciences Appliquées Toulouse*

N·H
P·C

1982

NORTH-HOLLAND PUBLISHING COMPANY
AMSTERDAM • NEW YORK • OXFORD

ISBN: 0444 86504 7

Translated by: B. Beeby

Publishers:

NORTH-HOLLAND PUBLISHING COMPANY
AMSTERDAM · NEW YORK · OXFORD

Sole distributors for the U.S.A. and Canada:
ELSEVIER SCIENCE PUBLISHING COMPANY, INC.
52 VANDERBILT AVENUE
NEW YORK, N.Y. 10017

Library of Congress Cataloging in Publication Data
Main entry under title:

Interconnected dynamical systems.

(North-Holland systems and control series ; v. 5)
Translated from the French.
Includes bibliographical references and index.
1. Large scale systems. I. Bernussou, Jacques.
II. Titli, André. III. Centre national de la recherche scientifique (France). Laboratoire d'automatique et d'analyse des systèmes.
IV. Series.
QA402.I5 1982 003 82-14464
ISBN 0-444-86504-7 (U.S.)

PRINTED IN THE NETHERLANDS

TABLE OF CONTENTS

FOREWORDS

The intention of the present work is to examine some topics and problems in the analysis and control of Large Scale Complex Systems of the Interconnected type. It is obviously not an exhaustive survey of the subject which would be a completely unrealistic aim in a field which is continuously changing and where so many approaches are used. The book discusses some works done by the "Decomposition, Interconnected Systems Control" group at the "Laboratoire d'Automatique et d'Analyse des Systèmes" (L.A.A.S.) of the "Centre National de la Recherche Scientifique" (C.N.R.S.) Toulouse, France, in the last few years. The activity covers a number of important aspects such as stability and structural analysis, numerical features, synthesis of decentralized control, networks. An illustrative example taken from the real time management of telecommunication networks is given.

The book is divided into six chapters. The first gives the general setting of some of the main problems encountered in the Large Scale Complex Systems field, notably time and space decomposition, hierarchies, model reduction and multimodelling as well as multicriteria optimisation.

The second chapter is devoted to some numerical issues of interest in the derivation of efficient algorithms to overcome the problems caused by high dimensionality. This is done by means of decomposition, parallel computation and relaxation techniques. Some insight is given into the problem of the adaptation of the control structure and its associated numerical algorithm to the process structure itself.

The third chapter provides an example of suboptimal control for a large scale interconnected system, consisting of finding the suboptimal control in an a priori chosen parametrised class of controls. The method is illustrated on a decentralised regulation example. A numerical algorithm is derived and the stability and suboptimality of the decentralised controlled system discussed.

In the fourth chapter, the stability analysis of interconnected systems by means of the Lyapunov method is examined. The scalar and vectorial approaches are presented through a comparative discussion. The problem of robust control under structural perturbations is dealt with. The potentialities as well as the limitations of the Lyapunov method are indicated.

In the fifth chapter, an example of a large scale complex system is considered with the real time management of telephone networks. This example highlights the crucial link beetwen analysis and modelling with the control aim. A two layer structure for the routing of calls is discussed, this layered structure is based on a time horizon decomposition of the various phenomena to be controlled.

The first part of the final chapter is constituted of a brief exposition of some classical multicommodity flow optimization techniques but the main section consists of a reminder and a presentation of some new results concerning distributed algorithms. In these, the optimal routing in flow and multicommodity flow networks are done in a decentralised coordinated way, the transit modes defining their routing control variables by means of local measurements and information exchanges with the neighbours.

As already stated this book is by no means a complete survey of the field. It is hoped however that the points raised together with the unsolved questions which came up are of sufficient interest to arouse the curiosity of engineers and researchers in associated fields. Although not expressly pointed out in the volume, the authors believe that advances in Large Scale Systems research can only result from a multidisciplinary approach linking structural analysis, modelling, optimisation, data processing ...

This book is the fruit of a collaboration involving several research workers. The contributions of Jean-Louis CALVET and Gérard AUTHIE were particularly important in the second and sixth chapters respectively.

ACKNOWLEDGMENTS - The help of the L.A.A.S. both material and moral, particularly of its Director D. ESTEVE, is gratefully acknowledged. We are also grateful to Mrs E. DUFOUR for the typing and to Mr E. LAPEYRE-MESTRE for the drawing of the diagrams. We thank our colleagues of the "Decomposition, Interconnected Systems Control" group for valuable discussions. Finally, we are indebted to Professor Madan SINGH for his encouragement during the writing of this book.

CHAPTER 1

COMPLEX SYSTEMS, LARGE SCALE SYSTEMS; ASSOCIATED METHODS OF ANALYSIS AND CONTROL

1.1 Notion of large scale systems and complex systems

We shall call large scale systems systems which can satisfactorily be described by normal, traditional mathematical tools but which give large scale (often linear) models, bringing into play, for example in the state representation, a large number of state variables - a hundred or more.

A certain spatial geographical distribution is often associated with this type of large scale system.
Examples of large scale systems are common :
- large production units
- distribution and service networks, etc ...

In this type of system it is the large scale and spatial distribution aspects which give rise to difficulties when problems of management or control of the system are considered.

Complex systems, on the other hand, can be characterised essentially by the difficulty in representing them with traditional mathematical tools, and of obtaining usable models, or by the mathematical complexity of existing models (often obtained from physical and analytical considerations). Examples : representation by systems of equations with partial derivatives, highly non-linear models, qualitative representation, the need to use "fuzzy" concepts ...

Examples of complex systems can easily be found in the Human Sciences, in behaviour, the modeling of social, biological and biotechnological systems.
A large scale system may be complex but it is not necessarily so. Similarly a complex system need not be a large scale system. If complexity and large scale - spatial distribution are common characteristics of a single system, then one can obviously expect to encounter great difficulty in the analysis and control of such a system.

1.2 Problems of control and management of large scale systems and complex systems

Problems to be solved : we shall comment on the following problems :
- help with decision-making : for example, tests of certain scenari in socio-economic systems, through the resolution of problems of optimisation. For the control engineer this means an open loop control calculated for a known initial condition.

- Management : with a more direct contact between the control engineer and the process, this will be a question of defining the best strategies, the best operational policies of the system. This can be obtained through open loop control, repetitive open loop control, closed loop control.

- Control : the problem is to define a structure which, in general, without the action of the decision maker, acts continuously on the process, in a closed loop fashion.

Difficulties encountered : for large scale and/or complex systems, a certain number of difficulties will make the satisfaction of the above objectives difficult to achieve. These difficulties are essentially :
- lengtly calculation times, even taking into account the extremely high rate of technological evolution in the area of computers.
- prohibitive memory space required in traditional approaches, quadratic, cubic or even exponential function of the size of the problem (according to linear and non-linear cases).
- the presence of greatly differing "dynamics" in a single process to be analysed and controlled.
- the impossibility, at the time of definition of management/control objectives, to unite highly different criteria in order to define problems with one single criterion, and the necessity, therefore, of dealing with problems with multiple criteria.
- the consideration of constraints of decentralisation in the control of geographically distributed processes (local control using only local information).
- the use of structure occasionally imposed on less and less costly distributed data processing.

1.3 Introduction of new concepts to overcome these difficulties

Taking into account on the one hand the characteristics of systems to be controlled, and on the other hand the technological evolution of existing data processing systems, new approaches must be developed to overcome the difficulties listed above, namely :
- the decomposition methods
- the techniques of order reduction
- multiple - criteria optimisation

1.3.1. Decomposition methods

These can be classified into three major groups :
- horizontal (or spatial) decomposition in which we shall define, starting from a global problem, as many sub-problems, local problems, as there are sub-systems. So there we make use of the structure of the system (interconnecting sub-systems).
Local control units may or may not see their actions coordinated by higher levels of control (hierarchical or decentralised structure).
Methodologies using this concept of decomposition are, for example, hierarchical control and regular perturbations.
- vertical (or temporal) decomposition giving several levels of control (or management), operating according to different time scales. This type of decomposition works a priori on the com-

plexity of the control function it is hoped to achieve and is found in hierarchical control and in the use of the theory of singular perturbations.

- spatio-temporal decomposition, a combination of the two previous types of decomposition, of interest to geographically distributed systems and giving rise simultaneously to different "dynamics".

1.3.2. Reduction methods

These stem from a different decomposition philosophy. Instead of breaking down one global problem into N sub-problems which are easier to solve (with or without co-ordination) using traditional tools, the reduction methods transform the problem into a reduced scale problem to which traditional methods are then applied.
One method at present stands out amongst other reduction methods : it involves aggregation, developed at the beginning on static economic models and then generalised to dynamic models. Moreover this method makes it possible to bring together a number of reduction methods which were proposed independently several years ago.
The reduced models obtained in this way can be used for the purposes of analysis, simulation, surveillance and also for sub-optimum control and co-ordination in the higher levels of a hierarchical control. Certain structural properties of the original system can also be imposed on them.

1.3.3. Multi-criteria optimisation

In a certain number of situations traditional monocriterion optimisation proves to be powerless :
- pursuit of contradictory, conflicting objectives by one single decider, one single control unit. Example : to maximise production, minimise costs, minimise pollution, maximise a performance, minimise a control structure cost, etc ...
- action of N agents (N deciders) on one single system. Example : an electric network composed of sub-networks, each with their administrative - controlling body.

These problems can be formulated by using a vector criteria giving different concepts of equilibrium examined further.

1.4 Analysis of complex systems

The methods described above require a knowledge of the systems, a certain model and, where applicable, a definition of the sub-systems and their interactions.

This phase is far from obvious for complex and/or large scale systems, and two steps are often necessary to arrive at this stage of knowledge of the system :
- analysis using the data of the complex systems, particularly for the definition of the interconnections and the display of an internal structure of the system;
- structural analysis starting from this structure, to break down the system and define sub-systems.

1.4.1. Analysis by data [RIC - 79]

A complex system may be characterised by a large number of variables and the presence of many links between these variables. Observing such a system on a given horizon we then find ourselves in the presence of a large set of data which it is advisable to structure in order to extract as much information as possible for the purpose of analysis and modeling of the system under consideration.
Corresponding methods of analysis by data which generally require the use of a computer can be classified as :
- statistical methods
- the use of indexes of similarity and dissimilarity
- "informational" approach.

1.4.1.1. Statistical methods

These methods deal with long tables of data without impoverishing them too much, and give the characterisation of the links between variables or groups of variables, or a graphic representation of the influences between variables in the system.
For this we take a vector X with p variables supposed characteristic of the system : $X^T = | x_1, x_2, \dots, x_p |$, variables of which we make N observations (x_i^j therefore represents the j° observation of the variable i^1) providing the data table :

$$\mathbb{X} = \begin{vmatrix} x_1^1 & \dots & x_p^1 \\ \vdots & & \vdots \\ x_1^N & & x_p^N \end{vmatrix}$$

<u>Principal component analysis</u> : from the table we can calculate the vector of the means M and the matrix of co-variance of the variables :

$$M = (\bar{x}_1, \dots, \bar{x}_p)^T ; \; \bar{x}_i = \frac{1}{N} \sum_{j=1}^{N} x_i^j$$

$$C = \| \mathrm{cov}(x_i, x_j) \| \quad , \; \mathrm{cov}\,(x_i, x_j) = \frac{1}{N} \sum_{k=1}^{N} (x_i^k - \bar{x}_i)(x_j^k - \bar{x}_j)$$

In the space R^p of the variables we then proceed as follows :

a) determination of the principal axis of inertia of the cluster of measurements (factorial axis), an axis which minimises the sum of the squares of the distances of the N points to the axis.
b) projection of the cluster of points on R^{p-1} in the preceding direction.
c) repetition with this new cluster to end up with a cluster in R^{p-2} and so on.

The sub-space of dimension k which gives the cluster projection "nearest" to the initial cluster is engendered by the eigenvectors corresponding to the k largest eigenvalues of the matrix of co-variance.

This method makes it possible :
- to separate the proximities between variables, and therefore the dependencies
- to analyse hierarchically the intensity of the coupling

Linear regression : the measurement table is put in the form

$$\mathbb{X} = \begin{bmatrix} y^1 & e_1^1 \ \dots \ e_q^1 \\ \vdots & \\ y^N & e_1^N \quad\ \ e_q^N \end{bmatrix}$$

in which y is a variable to be "explained" from a linear combination of the other variables of the table : $e_1, \dots, e_q$. Working in the space R^{q+1} of the centred variables $y^* = y - \bar{y}$, $e_i^* = e_i - \bar{e}_i$, i = 1 to q, we look in W, sub-space R^q of the variables e_i^*, for the nearest point (in the sense of the least squares) of the variable to be explained y^*. The quality of the representation can be measured by the coefficient of total correlation between y^* and the whole of the e_i^*.

Canonic analysis : this involves analysing the relations between 2 groups of variables E and S in such a way that :

$$X = (x_1 \dots x_p)^T = (E^T\ S^T)\ ;\ E = (x_1, \dots x_{r-1})^T \cdot S = (x_r \dots x_p)^T$$

The matrices of co-variance associated with each of the vectors and between vectors are :

$$C_{EE} \quad C_{ES} \quad C_{SE} \quad C_{SS}$$

We define two canonic variables, a linear combination of the components of E and S as :

$$A = \alpha^T E \qquad\qquad B = \beta^T S$$

The determination of the parameters α and β of the linear combinations is made in such a way that A and B are as closely correlated as possible, with the coefficient of correlation ρ.
The solution is :

$$C_{SS}^{-1}\ C_{SE}\ C_{EE}^{-1} = \rho^2 \beta$$

$$\alpha = \rho^{-1}\ C_{EE}^{-1}\ C_{ES} \cdot \beta$$

from which : - ρ^2 is an eigenvalue of $C_{SS}^{-1}\ C_{SE}\ C_{EE}^{-1}\ C_{ES}$

- β is the corresponding eigenvector.

From spectral analysis of this last matrix, we deduce the two canonic variables (A_i, B_i).

For the purpose of analysis, this technique makes possible the study of the input-output relations of a complex system and the definition of a minimum set characterising the behaviour of the systems.

Factorial analysis of the correspondances : in this method we work with the table of the probabilities of occurence of the jth value of the ith variable :

$$\mathbb{P} = \| p_{ij} \| = \begin{vmatrix} p_{11} & \cdots & p_{pl} \\ \vdots & & \\ p_{1q} & \cdots & p_{pq} \end{vmatrix} \quad . \qquad \begin{array}{l} i = 1 \text{ à } \ell \\ j = 1 \text{ à } q \end{array}$$

We introduce the following probabilities :

$$p_j = \sum_{i=1}^{\ell} \sum_{j=1}^{q} p_{ij} \; ; \; p_i = \sum_{j=1}^{q} p_{ij} \qquad p_j = \sum_{i=1} p_{ij}$$

and the observations are represented in R^p and characterised by $\left(\frac{p_{ij}}{p_j}\right)$ and p_j. To study the relations between variables (ie. the "proximities" between columns of this table), we carry out an analysis in principal components of the cluster of points by using metrics (distributional distance) :

$$d^2\,(i,\ell) = \sum_{j=1}^{q} 1/p_j \left| \frac{p_{ij}}{p_i} - \frac{p_{\ell j}}{p_\ell} \right|^2$$

The graph interpretation is made according to the proximity of the points representing joined variables, or according to the shape of the projected clusters.

Note : The methods of analysis in principal components and factorial analysis of the correspondances with their qualitative aspect associated with a graph representation of the results are often useful for a preliminary analysis of the relations between variables retained for the description of a complex system.
The methods of linear regression, canonic analysis, by their quantitative results, will make it possible to analyse in depth certain relations brought to light in the preceding phase.

1.4.1.2. Use of dissimilarity indexes

The process has two stages :
- construction of dissimilarity indexes
- division of the matrix of dissemblance by a classification process.

Construction of indexes : Starting from the table of measurements already defined (N sampled measurements of period T on q variables), we find the minimum m_i and the maximum M_i of each variable x_i on the horizon of observation NT, which defines the new unit :

$$\text{with} \quad y_i = (x_i - m_i)/(M_i - m_i)$$

The coefficient of coupling between y_i and y_j is expressed from the measurement of the distance between y_i and y_j by construction of the fuzzy images (in the sense of Zadeh) of y_i and y_j and use of the generalised Hamming distance δ_{ij} between these fuzzy images.

Partition of the matrix of dissimilarity (matrix $\delta = \| \delta_{ij} \|$)

We construct a series of matrices D (α_q) representative of the system (α_q being the maximum distance of consideration) such as :

1°) D_{ij} ($\alpha_1 = 0$) = 1 if and only if $\delta_{ij} = 0$

0 elsewhere

⋮

2°) D_{ij} (α_q) = 1 if and only if $0 < \delta_{ij} < \alpha_q$

with $1 > \alpha_q > \alpha_{q-1} \dots \alpha_2 > \alpha_1$

= 0 elsewhere.

The study of the series of these matrices makes it possible to study the links in the system.

Thus :

$$D(\alpha_q) \equiv D(\alpha_{q+1}) \equiv \dots D(\alpha_{q+\ell}) \text{ with } \alpha_q < \alpha_{q+1} \dots < \alpha_{q+\ell}$$

makes it possible to work out a partition of the system between highly uncoupled sub-systems.

1.4.1.3. Informational analysis

Most of the statistical methods previously mentioned make the hypothesis of linearity for the coupling relations between the appropriate variables.

This is a disadvantage and a serious limitation in the analysis of complex systems which raises the use of the concepts of the theory of information, into an "informational" analysis.

Let $u_1 = \{x_{11}, x_{12} \dots x_{1p})\}$ be all the values of a discrete variable x_1 and $n_1(i)$ the number of occurrences of the event $(x_1 = x_{1i})$.

Let $p_1(i) = \frac{n_1(i)}{N'}$ (N' being the total number of observations): the experimental probability of having the event $(x_1 = x_{1i})$, and so the entropy associated with the variable x_1 is :

$$H(x_1) = - \sum_{j=1}^{p} p_1(i) \log p_1(i)$$

Similarly, considering two variables x_1 and x_2 and using the joint and conditional densities of probability, the joint and conditional entropies are defined as :

$$H(x_1, x_2) = - \sum_{j=1}^{p} \sum_{j=1}^{q} p_{12}(i.j) \log p_{12}(i.j)$$

$$H_{x_2}(x_1) = - \sum_{j=1}^{p} \sum_{j=1}^{q} p_{12}(i.j) \log \frac{p_{12}(i.j)}{p_2(j)}$$

The information between variables or "transinformation" is then :

$$I(x_1, x_2) = H(x_1) + H(x_2) - H(x_1, x_2)$$
$$= H(x_1) - H_{x_2}(x_1) = H(x_2) - H_{x_1}(x_2)$$

This transinformation is :
- zero if and only if the variables x_1 and x_2 are statistically independent
- maximum when they are linked.

For N variables, we have :

$$I(x_1,x_2 \ldots x_N) = \sum_{i=1}^{N} H(x_i) - H(x_1,x_2,\ldots x_N)$$

Transinformation appears therefore as a non-negative measurement of the links between the variables of a system and so can be used for analysis of the interactions and the decomposition of complex systems.

<u>Decomposition of static systems</u> : a matrix of the links $M = [(m_{ij})]$ is defined, in which m_{ij}, the measurement of the link between the variables x_i et x_j, is often taken as equal to :

$$m_{ij} = \frac{I(x_1,x_2)}{\sqrt{H(x_1) \cdot H(x_2)}}$$

and therefore has the following properties :

- $O < m_{ij} < 1$
- $m_{ij} = 0$ if, and only if, the variables are statistically independent.
- $m_{ij} = 1$ if, and only if, there is a bi-univocal relationship between these variables.

This linking matrix makes possible a representation of the complex system by means of a weighted graph which will then be partitioned to define the interconnecting sub-systems (see paragraph 1.4.2.).
General results on transinformation and its properties make it possible to study the quality of the partition thus obtained. Indeed, if $S = |x_1,\ldots x_N|$ and if $(S_1, S_2 \ldots S_L)$ is a partition of S, then :

$$I(x_1,x_2 \ldots x_N) = \sum_{j=1}^{L} I(S_i) + I(S_1,S_2 \ldots S_L)$$

This equation states that the measurement of the links between the variables of a system is broken down into the sum of the measurements of the internal links at each sub-system and of the measurement of the links between sub-systems.
Thus, a partition will be valid (the sub-systems $S_1 \ldots S_L$ will be weakly linked) if :

$$\frac{I\ (S_1,\ S_2\ \ldots\ S_L)}{I\ (x_1,\ \ldots\ x_N)} \ll 1$$

Decomposition of dynamic systems

Let x_1 and x_2 be two variables of a dynamic system and let x'_2 be the value of x_2 displaced from T (sampling period) in relation to the measurement of x_1 and x_2.
The dynamic link $x_1 \rightarrow x_2$ is measured by :

$$I\ (x_1,\ x'_2) = H\ (x_1) + H\ (x'_2) - H\ (x_1,\ x'_2)$$

and, defining the linking coefficient by (other variants are possible) :

$$m'_{ij} = \frac{I\ (x_i,\ x'_j)}{\sqrt{H\ (x_i)\ H\ (x_j)}}$$

we can also construct the linking matrix which will be used as a basis for the partition of the system.
Measurement of the quality of the partition is more difficult here. Moreover, in a dynamic case, we are faced with apparent dynamic links (between two variables x_i, x_j for example, comprising a dynamic link $x_i \rightarrow x_h$ followed by a static link $x_h \rightarrow x_j$).
Note : Informational analysis is a tool of analysis complementary of statistical methods and does not require the hypothesis of linearity.
Nevertheless it has limitations of application which should not be underestimated : quantification of the continuous variables, a sufficient number of variables observed and of observations, wise choice of the sampling period.

1.4.2. Structural analysis [RIC - 79b]

The above method of analysis by data has given a coupling matrix or equivalently to a graphic representation of the complex system under consideration, a representation which visualises the interactions, the internal structure of the system.
The next stage is, using this information, to define sub-systems, usually weakly linked, together with their interconnection, with a view to placing hierarchical or decentralised controls, or with a view to studying the stability of the system by examining the stability of the sub-systems and of certain mathematical properties of the interconnections.
Two cases of representation by graph may appear :
1°) a qualitative analysis of the interactions gives a non-valued graph (an interaction either exists or does not exist) which must be broken down into sub-graphs using particularly as a base the properties of connexity of the original graph.
2°) a quantitative analysis of the interactions is made (not only does an interaction exist or not exist, but when it is present its intensity is evaluated) and this time gives a

valued graph which we shall try to partition by minimising the sum of the weights of the interactions.

1.4.2.1. Decomposition of graphs

Structures induced by strong connexity : the reduced graph of a structured graph G (X,U) (X : all the nodes, U : all the arcs) according to the increasing rank of its peaks, causes to appear the hierarchical structure of the graph G. At each level are the nodes of the same rank to which correspond strongly connected components (f-connex) of G, and the movement between the levels is uni-directional in the increasing direction of rank. So we find the components which are simply connected (s-connex), and inside each one analysis of the hierarchy of the links is made through studying the properties of strong connexity. Different algorithms are available for this : Steward's method, Warfield's method algorithm of the reduced graph.

Structures induced through simple connexity : in an s-connected graph G (X,U), all sub-unit C such the sub-graph G_A = (X - C, U_A) may not be s-connected, is called the articulation set.
Consider k sub-sets of points $A_1 \ldots A_k$ of X such that :

$\cup A_i \cap A_j = \emptyset$. $i \neq j$; $i,j = 1$ to k and let a (k+1) partition of X = $(X_1 \ldots X_k, C)$ such that ;
1°) $X_i \supseteq A_i$ $i = 1$ to k
2°) $\cup X_i \cup X_j = \emptyset$ $i \neq j = 1$ to k
3°) $\|C\|$ minimum

then C is called the minimum set of articulation. It is usually non-unique. The partition of X = $(X_1 \ldots X_k, C)$ is such that the sub-graphs G_{X_i} (X_i, U_i) are s-connected.
From the sets of articulation of the sub-graphs successively obtained we define increasingly finer partitions of X which can be represented by a tree.
If G is not s-connected, we join a tree to each of its s-connected components.
Search algorithms of minimum set of articulation are :
- the Malgrange algorithm
- the algorithm of the maximum flow

1.4.2.2. Partition of valuated graphs

a valuated graph is a quadruplet G (X, U, f, g) in which :
X : all the nodes
U : all the arcs
f : weightings of the nodes (f : $X \to R^+$: all the real positives)
g : weightings of the arcs (g : $U \to R$: all the real non-negatives).
indeed, in the problems of decomposition of complex systems there will be no weighting of the nodes and one will be limited to taking into account the triplet G (X, U, g).
The problem will be to determine the partition of X = $(P_1 \ldots P_r)$

which minimises the criterion :

$$J\,(P_1 \ldots P_r) = \sum_{\substack{i,j \\ i \neq j}} \; \sum_{\substack{x \in P_i \\ x' \in P_j}} g\,(x, x')$$

under any constraints which may be specific to the sub-systems. To deal with this problem of optimum partition different methods are available in the appropriate literature :

- method of eigen-vectors
- method of colorations
- method of minimum groups
- method of intergroup exchanges
- method of maximum flow
- method of dynamic clusters

1.5 Horizontal decomposition [SIN - 78]

1.5.1. Introduction. Different types of co-ordination

In order to diminish difficulties of calculation, using the preliminary stage of decomposition of the system mentioned above, we formulate the overall problem from a certain number of smaller separable problems associated with each of the sub-systems defined, in the hope that through the consequent reduction of dimension these sub-problems will be distinctly easier to solve than the overall problem, which is usually the case.
Each sub-system is then, under the responsibility of a local control unit (which solves the corresponding sub-problem) whose actions can be co-ordinated by higher levels of a hierarchy.
This type of decomposition therefore uses the structure of the system in the form of interconnected systems.
Starting from a global problem P, we then try to solve it via a certain number of sub-problems P_i, so as to satisfy :

$$\text{Sol } |P_1 \ldots P_N| \rightarrow \text{Sol } P$$

In fact, such an equation cannot usually be satisfied because of the existence of interactions linking the P_i, and it is necessary to introduce a vector of intervention (or parameter of co-ordination) λ , to replace P_i by $P_i(\lambda)$ so as to satisfy :

$$\text{Sol } \left|P_1(\lambda), P_2\,(\lambda) \ldots P_N(\lambda)\right|_{\lambda=\lambda^*} \longrightarrow \text{Sol } P$$

To choose λ and make it develop from an initial value λ° to a final value λ^* giving the solution to the global problem, is to solve the problem of hierarchical co-ordination in control or calculation when spatial (horizontal) decomposition is used.
Many methods of co-ordination have been proposed. To simplify the situation we shall pick out :

α) co-ordination by action on the objectives of the sub-problems (co-ordination by the criterion, by prices, "non-feasible" method).

β) co-ordination by prediction of interactions for which we must further distinguish between :

β_1) prediction of the outputs of interactions at the sub-system level (method of co-ordination by the model, "feasible" method).

β_2) prediction of the inputs of interactions on the sub-systems. Actually this last method of co-ordination is rarely applicable by itself and recourse must be made at the same time to a prediction of the interactions and to a modification of criteria ("mixed" method).

1.5.2. Hierarchic calculation and control of linear systems with quadratic criteria

We are considering the problem of the optimum control of a group of N dynamic, linear, interconnected sub-systems by minimisation of an additive, separable, quadratic criterion in the form :

$$\min \sum_{i=1}^{N} 1/2 \int_{o}^{T} (x_i^T Q_i x_i + u_i^T R_i u_i)\, dt$$

$$\text{subject to} \begin{cases} \overset{\circ}{x}_i = A_i x_i + B_i u_i + C_i Z_i \;; \quad x_i(0) = x_{i0} \\ Z_i = \sum_{j=1}^{N} (L_{ij} x_j + M_{ij} u_j) \quad i = 1 \text{ to } N \end{cases}$$

$x_i \in R^{mi}$ (state), $u_i \in R^{ni}$ (control), $Z_i \in R^{li}$ (linking variable)

L_{ij}, M_{ii} matrices of interconnection on the states and the controls

$Q_i \geq 0 \quad R_i > 0$

A global formulation of the problem is :

$$\min \; 1/2 \int_{o}^{T} (x^T Q x + u^T Ru)\, dt$$

$$\text{subject to } \overset{\circ}{x} = (A + CL)x + (B+CM)\, u = \tilde{A}x+\tilde{B}u \quad x(0)=x_o$$

in which Q, R, A, B, C are block-diagonal.

If the sub-systems (Z = 0) and the global system are controllable, then there is a stabilising global control of the form :

$$u = - G\, x = - G_d\, x - T\, x$$

in which : G is the state feedback gain obtained by resolving the global matrix Riccati equation

G_d is a block-diagonal gain matrix obtained by resolving the local Riccati equations (without interaction Z=0)

T is a matrix of co-ordination gains obtained by a hierarchical calculation using the structure particular to the system.

<u>Calculation of T</u> : the hamiltonian of the initial problem is :

$$H = 1/2\, (x^TQx + u^TRu) + \lambda^T \,|\, Ax+Bu+CZ \,| + \pi^T \,|\, Z-Lx-Mu \,|$$

(λ : adjoint variable, π : Lagrande multiplier)
or, in composite form :

$$H = \sum_{i=1}^{N} \Big| 1/2(x_i^TQ_ix_i + u_i^TR_iu_i) \Big| + \lambda_i^T \Big| A_ix_i+B_iu_i+C_iZ_i \Big|$$

$$+ \pi_i^T \Big| Z_i - \sum_{j=1}^{N} (L_{ij}x_j+M_{ij}u_j) \Big|$$

The conditions of optimality are written as :

$$H_x = -\overset{\circ}{\lambda} \rightarrow \lambda = -Qx - A^T\lambda + L^T\pi \qquad \lambda(i) = 0$$

$$H_\lambda = \overset{\circ}{x} \rightarrow \overset{\circ}{x} = Ax + Bu + CZ$$

$$H_u = 0 \rightarrow u = + R^{-1} | M^T\pi - B^T\lambda |$$

$$H_Z = 0 \rightarrow C^T\lambda + \pi = 0$$

$$H_M = 0 \rightarrow Z = Lx + Mu$$

if π is fixed (and a fortiori π and Z_i) a separable additive form is obtained for the hamiltonian H :

$$H = \sum_{i=1}^{N} H_i$$

with

$$H_i = 1/2(x_i^TQ_ix_i + u_i^TR_iu_i) + \pi_i^TZ_i - \sum_{j=1}^{N}\pi_j^TL_{ji}x_i - \sum_{j=1}^{N}\pi_j^TM_{ji}u_i +$$

$$+\lambda_i^T | A_ix_i + B_iu_i + C_iZ_i |$$

a separable form giving N independent sub-problems (defined by H_i) at the lower level in a 2-level structure :

Sub-problem no. 1

$$\min_{u_i} \int_0^T | 1/2(x_i^TQ_ix_i + u_i^TR_iu_i) + \pi_i^TZ_i - \sum_{j=1}^{N}\pi_j^T(L_{ji}x_i + M_{ji}u_i) | dt$$

$$\text{sub} \quad \overset{\circ}{x}_i = A_ix_i + B_iu_i + C_iZ_i \qquad x_i(0) = x_{i0}$$

for $\pi_j(t)$ $j = 1$ to N, $Z_i(t)$, $t \in |0,T|$ given.

At the co-ordinating level the variables of co-ordination π and Z will evolve according to the relations (direct iteration scheme - mixed method- Takahara prediction principle) :

$$\pi^{k+1}(t) = -C^T\lambda^k(t); \; Z^{k+1}(t) = L\,x^k(t) + M\,u^k(t)$$

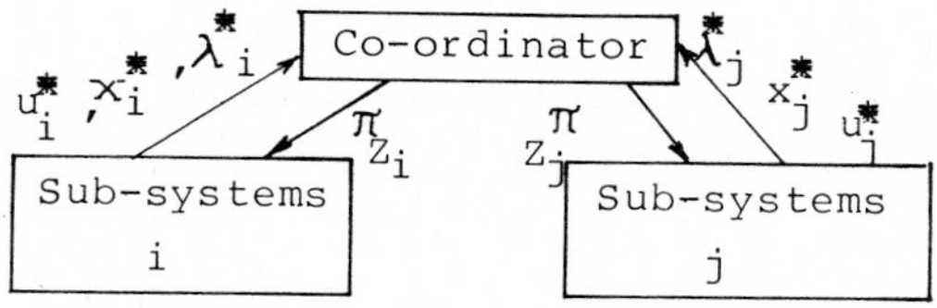

Figure 1 : 2-level structure

To resolve the sub-problems the lower level deals with the equations :

$$\overset{\circ}{\lambda}_i = -Q_ix_i - A_i^T\lambda_i - \sum_{j=1}^{N} L_{ji}^T\pi_j \qquad \lambda_i(T) = 0$$

$$\overset{\circ}{x}_i = A_i x_i - B_i R_i^{-1} B_i^T \lambda_i + B_i R_i^{-1} \sum_{j=1}^{N} M_{ji} \pi_i + C_i Z_i \qquad x_i(0) = x_{i0}$$

for π and Z_i given.

To find a local closed loop control let us put :

$$\lambda_i = K_i x_i + q_i$$

K_i, the local matrix of Riccati, and q_i, defining the open-loop term, satisfy the equations :

$$\overset{\circ}{K}_i + K_i A_i + A_i^T K_i - K_i B_i R_i^{-1} B_i^T K_i + Q_i = 0 \qquad K_i(T) = 0$$

$$\overset{\circ}{q}_i = | K_i B_i R_i^{-1} B_i^T - A_i^T | q_i - K_i C_i Z_i - K_i B_i R_i^{-1} \sum_{j=1}^{N} M_{ji} \pi_j + \sum_{j=1}^{N} L_{ji}^T \pi_j \qquad q_i(T) = 0$$

The control is then :

$$u_i = R_i^{-1} B_i^T K_i x_i + R_i^{-1} | B_i^T q_i - \sum_{j=1}^{N} M_{ji}^T \pi_j |$$

or at the end of the convergence of the structure of calculation at 2 levels (for which we have :

$$\pi_i = - C_i^T \lambda_i = - C_i^T | K_i x_i + q_i | \quad) :$$

$$u_i = R_i^{-1} B_i^T K_i x_i + R_i^{-1} \sum_{j=1}^{N} M_{ji}^T C_j K_j x_j + R_i^{-1} B_i^T q_i + R_i^{-1} \sum_{j=1}^{N} M_{ji}^T C_j q_j$$

that is :

$$u_i = u_i(x) + u_i(q) \rightarrow u = u(x) + u(q)$$

and to obtain a law of control of the form :

$$u = - Gx$$

we must then find U(q) in the form :

$$u(q) = \Gamma_q . x$$

To simplify the presentation of the calculation of Γ_q, let us suppose $T \rightarrow \infty$; then Γ_q is a constant matrix.

After convergence in the 2 - level calculation method, we have x_i^*, u_i^*, K_i (i = 1 to N), therefore $u^*(q)$ can be calculated and the equation :

$$u^*(q) = \Gamma_q x^*$$

can be used.

Let us put : $\mathbb{X} = | x^*(t_o) \; \vdots \; x^*(t_1) \; \ldots \; x^*(t_{m-1}) | \qquad m = \sum_{i=1}^{N} m_i$

$$\mathbb{U} = | u^*(q,(t_o)) \; \vdots \; u^*(q,(t_1)) \; \ldots \; u^*(q,(t_{m-1})) |$$

These matrices are formed by the optimal values x and u considered form different time values for m, during the transient phase of the system.

So then : $\Gamma_q = \mathbb{U}\mathcal{X}^{-1}$

Finally, we have made the synthesis of the following 2 - level control structure :

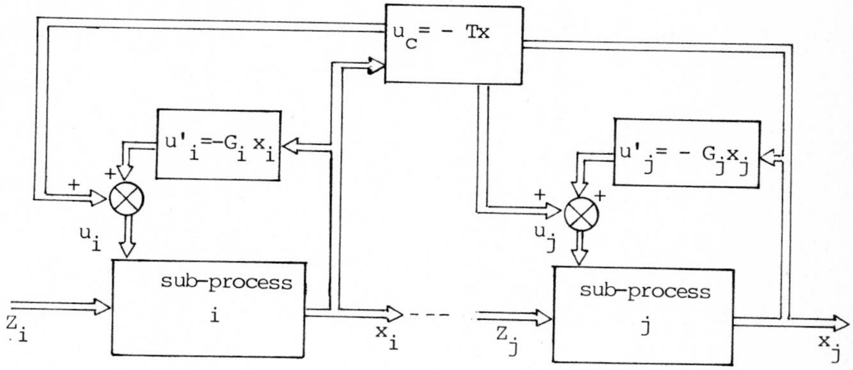

Figure 2 : multi-level control

Notes :

1. This approach can also be considered as the hierarchic calculation of a matrix of global feedback G = Gd + T.
2. If the inversion of $\mathcal{X}$ poses problems, it could be considered possible to choose m initial conditions over x in such a way that $\mathcal{X}$ would be a unit matrix and to carry out m hierarchic calculations off-line (for each initial condition).
3. The approach may be generalised in the case of the finite horizon and in the servomechanism problem.
4. A sub-optimum control easy to calculate is given by :

$$u_i = R_i^{-1} B_i^T K_i x_i + R_i^{-1} \sum_{j=1}^{N} M_{ji}^T C_j K_j x_j$$

This control only brings into play local matrices of Riccati. On the other hand, it has a priori no stabilising property of the system. It must therefore be checked that system thus controlled will be stable.

1.5.3. Generalisation to dynamic interconnections [CAL - 80]

The interconnection system is now considered to be a dynamic system, with its own control variables u', in the following formulation :

$$\min_{u_i, u'} \; 1/2 \int_0^T \left\{ | Z^T Q' Z + u'^T R' u' | + \sum_{i=1}^{N} (x_i^T Q_i x_i + u_i^T R_i u_i) \right\} dt$$

subject to
$$\begin{cases} \overset{\circ}{x}_i = A_i x_i + B_i u_i + C_i V_i ; & x_i(0) = x_i 0 \\ V_i = H_i Z + G_i u' & i = 1 \text{ to } N \end{cases}$$

and subject to
$$\begin{cases} \overset{\circ}{Z} = A'Z + B'u' + C'W & Z(0) = Z_0 \\ W = H'x + G'u = \sum_{i=1}^{N} H'_i x_i + G'_i u_i \end{cases}$$

This mathematical formulation, which may correspond to certain physical realities (ex : power systems) gives the following structure of interconnection (Figure 3).
Let us define ψ_i, λ as the adjoint variables and β_i, β' as the parameters of Lagrange associated with the equations of state and the linking constraints respectively.

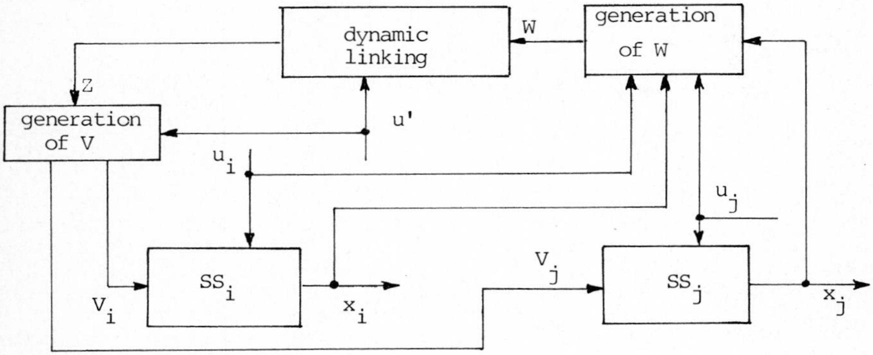

Figure 3 : dynamic interconnection

The Hamiltonian of the problem is then :

$$H = 1/2\ (Z^T Q'Z + u'^T R'u') + \sum_{j=1}^{N} (x_i^T Q_i x_i + u_i^T R_i u_i) + \lambda^T | A'Z + B'u' + C'u |$$

$$+ \beta'^T | W - \sum_i (H'_i x_i + G'_i u_i) | + \sum_i \psi_i^T | A_i x_i + B_i u_i + C_i V_i | + \beta_i^T | V_i - H_i Z - G_i u' |$$

This Hamiltonian assumes a separable additive form by choice of β' and β_i as variables of co-ordination and a fortiori β', W, β_i and V_i (in order to use algorithms of direct prediction - coordination) :

$$H = 1/2 (Z^T Q'Z + u'^T R'u') + \lambda^T | A'Z + B'u' + C'W | + \beta^T | V - HZ - Gu' | + \sum_{i=1}^{N} (1/2 (x_i^T Q_i x_i + u_i^T R_i u_i)$$

$$+ \psi_i^T | A_i x_i + B_i u_i + C_i V_i | \quad - \beta'^T \, | H'_i x_i + G'_i u_i \, |) + \beta'^T W$$

let : $H = \sum_{i=1}^{N} H_i + H_{N+1}$

Each H_j, j = 1 to N+1 defines the sub-problems corresponding to the sub-systems for j = 1 to N and to the system of dynamic linking for j = N+1.

<u>Sub-problem no. i</u> : (i = 1 to N)

$$\min_{u_i} \int_o^T \{ 1/2 | x_i^T Q_i x_i + u_i^T R_i u_i | + \beta'^T W - \beta'^T | H'_i x_i + G'_i u_i | \} dt$$

subject to $\quad \overset{\circ}{x}_i = A_i x_i + B_i u_i + C_i V_i \qquad x_i(0) = x_{i0}$

for β' and V_i, W given.

<u>Sub-problem N+1</u>

$$\min_{u'} \int_o^T \{ 1/2 | Z^T Q' Z + u'^T R u' | + \beta^T | V - HZ - Gu' | \} dt$$

subject to $\overset{\circ}{Z} = A'Z + B'u' + C'W \qquad z(o) = z_o$

for β and V, W given.

The task of co-ordination is to make the variables of co-ordination β, β', V, W, evolve using the information from the solution of the sub-problems and dealing with the following equations :

$$H_{\beta'} = 0 \rightarrow W = H' x + G' u$$
$$H_{V_i} = 0 \rightarrow \beta_i = C_i^T \psi_i$$
$$H_{\beta_i} = 0 \rightarrow V_i = H_i Z + G_i u'$$
$$H_W = 0 \rightarrow \beta' = C'^T \lambda$$

Thus we have obtained the following 2 - level structure :

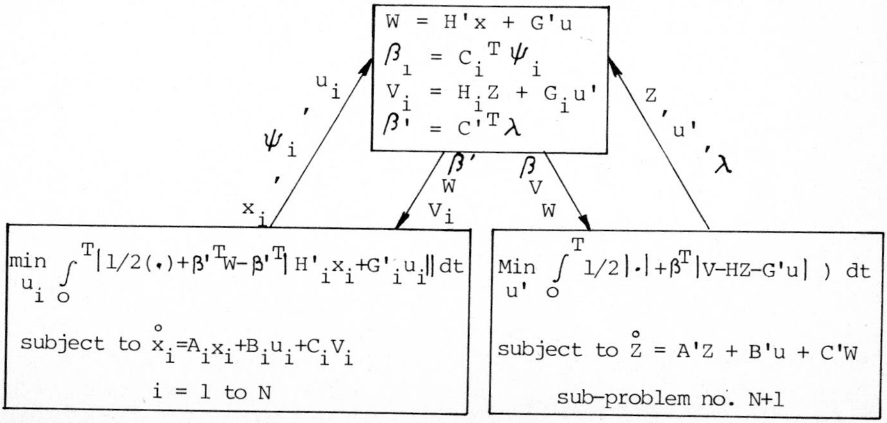

Using the particular structure of the equations of co-ordination and distributing these among the sub-problems, one can thus obtain a different 2 - level structure with a problem of optimisation at each level.
For this, let us note that if the sub-problem N+1 is resolved, is necessarily λ known and β' can be calculated and sent to all the other sub-systems (j = 1 to N).
Conversely, if the sub problem (N+1) receives x and u, directly from the other sub-systems, then W can be calculated at the level of the sub-problem (N+1).

Finally, the term $\beta^T V$ constant at the lower level may be ignored in the criterion.
From another aspect, the same reasoning can be made for the sub-problems j (j= 1 to N) :

- If the sub-problems are resolved, then ψ_j is necessarily known for each one and $\beta_j = C_j^T \psi_j$ can be calculated at this level and transferred to the problem N+1.
- If the sub-problems receive Z and u' directly from sub-problem N+1, they can calculate V_j (at the lower level).
- $\beta'^T W$ may be ignored in the criterion, being constant at the lower level.

All these remarks give the following structure :

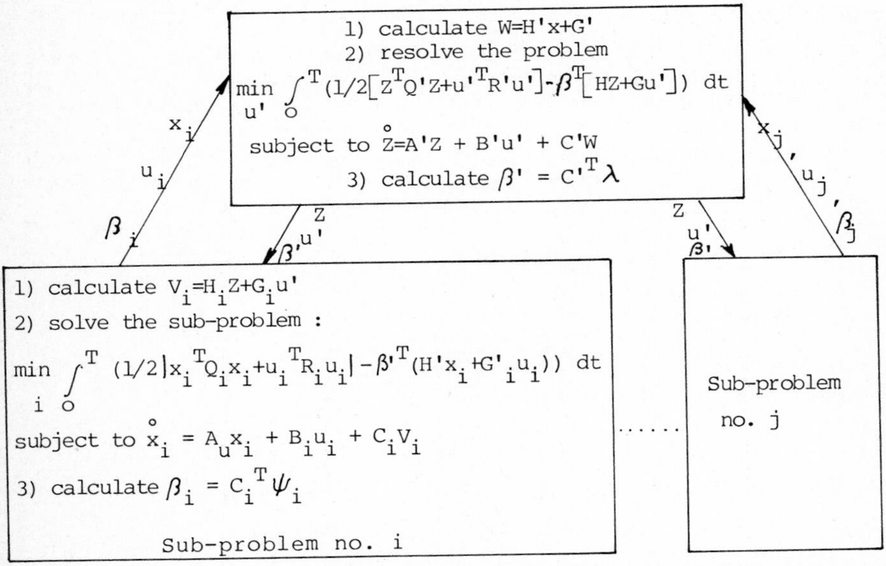

In this procedure the task of co-ordination is distributed between the levels. The task of optimisation is also performed at 2 levels : N independent sub-problems of optimisation are resolved at the lower level whilst one problem of optimisation on the system of dynamic interconnection is resolved at the upper level.

Notes :

1) At each stage of the iteration the algorithm implies a parallel decomposition (see chapter II) into N independent sub-problems at the lower level, combined with a sequential decomposition between the 2 levels.
2) The particular form of the dynamic matrix for the global system is :

A =

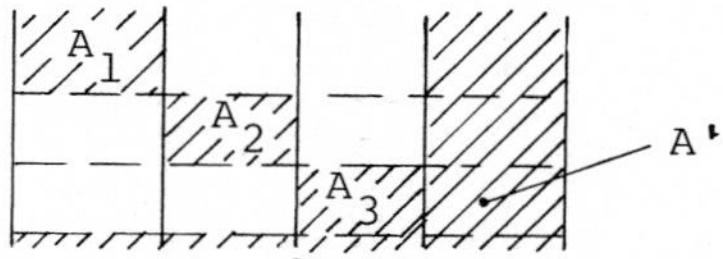

3) By putting $\psi_i = K_i x_i + q_i, \lambda = KZ+q$, and on the traditional hypotheses for each of the "sub-systems", we can make the synthesis of structures with local closed loop controls and with one open loop term which can be determined by co-ordination.

1.5.4. Sub-optimum spatial decomposition : use of regular perturbation techniques (ε-decoupling) [TIT - 79]

We are presenting here a relatively old formulation but in a new light, using the concepts of decomposition co-ordination and decentralisation.
We are considering the problem of the control of N interconnected sub-systems (for reasons of simplicity we are limiting ourselves to 2 sub-systems), weakly linked, which finds expression in the appearance of a small parameter ε ($\varepsilon \in |0,1|$) in the 2nd term of the equation of state :

$$\min_u \quad 1/2 \int_0^\infty (X^T Q x + u^T Ru)\, dt$$

$$\text{subject to } \overset{\circ}{x} = \begin{vmatrix} \overset{\circ}{x}_1 \\ \overset{\circ}{x}_2 \end{vmatrix} = \begin{vmatrix} A_{11} & \varepsilon A_{12} \\ \varepsilon A_{21} & A_{22} \end{vmatrix} \begin{vmatrix} x_1 \\ x_2 \end{vmatrix} + \begin{vmatrix} B_{11} & 0 \\ 0 & B_{22} \end{vmatrix}$$

$$\text{or again} \quad \overset{\circ}{x} = Ax + Bu \qquad x(0) = x_0$$

$$Q = \begin{vmatrix} Q_{11} & 0 \\ 0 & Q_{22} \end{vmatrix} \geqslant 0 \qquad R = \begin{vmatrix} R_{11} & 0 \\ 0 & R_{22} \end{vmatrix} > 0$$

(A,B) , $(A_{ii}, B_{ii}, i=1,2)$ are supposed controllable
Global control is given by :

$$U = - R^{-1} B^T K x$$

in which K is the solution to the equation of Riccati

$$A^T K + K A - K B R^{-1} B^T K + Q = 0$$

In fact, K is a function of ε , $K(\varepsilon)$, and may be approximated by a Taylor series development around $\varepsilon = 0$:

$$K \simeq P = K(\varepsilon)\Big|_{\varepsilon=0} + \frac{\varepsilon}{1!}\frac{\partial K(\varepsilon)}{\partial \varepsilon}\Big|_{\varepsilon=0} + \ldots + \frac{\varepsilon^r}{r!}\frac{\partial^r K(\varepsilon)}{\partial \varepsilon^r}\Big|_{\varepsilon=0}$$

and a sub-optimum control for the global system is given by :

$$u_{s.o} = - R^{-1} B^T P x$$

Note : The choice of r reflects the degree of approximation desired in the use of the matrix P in place of the matrix of Riccati K.
Actually, for practical reasons of execution, r is often limited to 1.

1.5.4.1. Calculation of $K(\varepsilon)\big|_{\varepsilon=0}$

This case corresponds to completely decoupled sub-systems.

$$K(\varepsilon)\Big|_{\varepsilon=0} = \left[\begin{array}{c|c} K_{11} & 0 \\ \hline 0 & K_{22} \end{array}\right],$$ K_{ii} solution to the local equation of Riccati :

$$A_{ii}^T K_{ii} + K_{ii} A_{ii} - K_{ii} B_{ii} R_{ii}^{-1} B_{ii}^T K_{ii} + Q_{ii} = 0$$

If $P = K(\varepsilon)\big|_{\varepsilon=0}$, we obtain a decentralised structure of control (local closed loops, calculation time and memory space limited for the synthesis of this control) whose stabilising effect on the real global system (with interactions) must be verified.

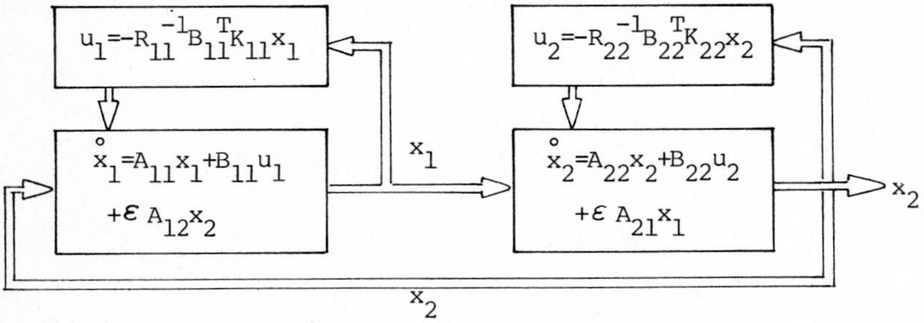

Figure 4 : decentralised control

1.5.4.2. Calculation of $\frac{\partial K(\varepsilon)}{\partial \varepsilon}\Big|_{\varepsilon=0}$

Using the partitioned form of the global equation of Riccati and its derivatives in relation to ε, we get :

with

$$\frac{\partial K(\varepsilon)}{\partial \varepsilon}\bigg|_{\varepsilon=0} = \left|\begin{array}{c|c} 0 & \dfrac{\partial K_{12}}{\partial \varepsilon}\bigg|_{\varepsilon=0} \\ \hline \dfrac{\partial K_{21}}{\partial \varepsilon}\bigg|_{\varepsilon=0} & 0 \end{array}\right| .$$

whith

$$\frac{\partial K_{12}}{\partial \varepsilon}\bigg|_{\varepsilon=0} \left|A_{22}-G_{22}\,K_{22}\right| + \left|A_{11}^{T} - G_{11}\,K_{11}\right| \frac{\partial K_{12}}{\partial \varepsilon}^{T}\bigg|_{\varepsilon=0} + A_{21}^{T}K_{22}\Big|_{\varepsilon=0} + K_{11}\Big|_{\varepsilon=0} A_{12} = 0$$

$$(G_{ii}=B_{ii}R_{ii}^{-1}B_{ii}^{T})$$

This equation (Lyapunov type) makes it possible to calculate

$$\frac{\partial K_{12}}{\partial \varepsilon}\bigg|_{\varepsilon=0} = \frac{\partial K_{21}}{\partial \varepsilon}^{T}\bigg|_{\varepsilon=0} ,$$

knowing $K_{11}\big|_{\varepsilon=0}$, $K_{22}\big|_{\varepsilon=0}$, in terms of the first stage.

This second term of the expansion of the matrix of Riccati induces a crossed co-ordination between the local controls, in accordance with the structure below and so clearly corresponds to a spatial mode of decomposition (Fig. 5).

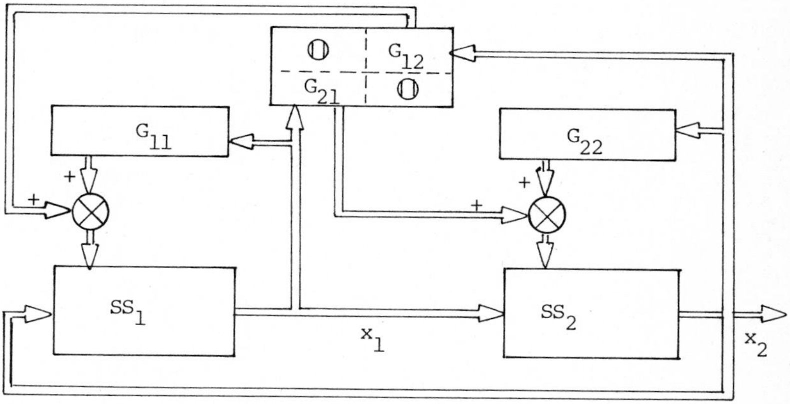

Figure 5 : Crossed co-ordination using techniques of decoupling.

$$G_{12} = -R_{11}^{-1}\,B_{11}^{T}\,\frac{\partial K_{12}}{\partial \varepsilon}\bigg|_{\varepsilon=0}$$

$$G_{21} = -R_{22}^{-1}\,B_{22}^{T}\,\frac{\partial K_{21}}{\partial \varepsilon}\bigg|_{\varepsilon=0}$$

1.5.5. A special case of spatial decomposition : decentralisation

Decentralised control is another aspect of spatial decomposition, it is also a particular case of control limited by structure.

The problem can be formulated as follows :

$$\min_u 1/2 \int_0^\infty (x^T Qx + u^T R u)\, dt$$

subject to
$$\left| \begin{array}{ll} \dot{x} = Ax + Bu & x(0) = x_0 \\ u = K x & \\ K = \text{block diagonal matrix} & \end{array} \right.$$

This formulation signifies that we wish to obtain a local control $u_i = K_i x_i$ at the level of sub-system i uniquely function of the state of this sub-system x_i.

A structure of this type is therefore of great interest in the many cases of control of complex systems geographically distributed over long distances. Example : control/management of distribution and service networks, pollution control of a hydraulic dock, etc.

In such situations the cost of transmission of the information (investment and operation) between the local control level and a co-ordinating level may be prohibitive in relation to the loss of performance entailed in putting into operation a decentralised control (absence of liaison between local control units).

Indeed, such decentralised control is naturally sub-optimum in relation to a control not limited by structure (hierarchical or global). Moreover, it poses two considerable problems :

- dependence in relation to the initial conditions from which it must break free (by considering for example an equivalent stochastic problem or a game theory formulation);
- the solution obtained by optimisation techniques (we are talking about a parametric optimisation problem in relation to limited K) is not necessarily a stabilising control for the system. A posteriori verifications on the stability are necessary.

Given the importance of decentralised control in mastering complex systems, a whole chapter (chapter III) is given over to the subject later in the book.

1.6 Vertical and time decomposition

1.6.1. General aspects of vertical decomposition

Horizontal or spatial decomposition uses the particular structure of the system to be controlled.

As a complement to this, vertical decomposition will take into account the complexity of the control function to be implanted by breaking this down into different control levels.

In the field of automatic control the following levels are currently distinguished (Figure 6) :

- regulation or direct control over the process
- static or dynamic optimisation (determination of the set points or of the input trajectories of the regulators)
- adaptation (adaptation of the model or of the law of control directly)
- organisation (choice of the structures of model, control, policies, in terms of the environment). At this level the deci-

der-computer dialogue will be important.

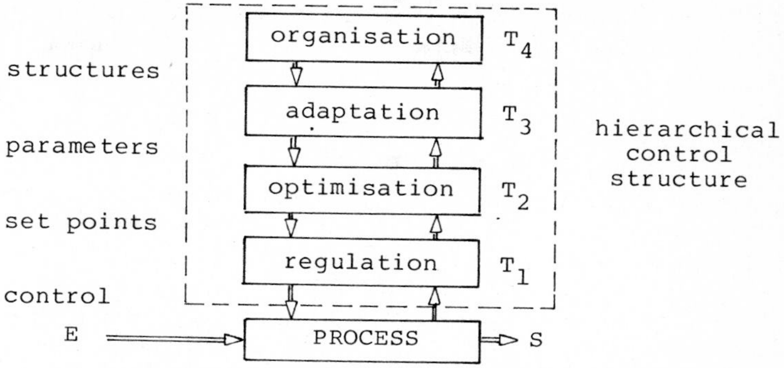

Figure 6 : Hierarchy of the control functions in vertical division.

Let us denote by T_i the period of intervention of level i. We then have the inequalities :

$$T_1 < T_2 < T_3 < T_4$$

The regulation level "follows" the dynamics of the process and the higher levels intervene with periods increasing in length; this explains that in these control hierarchies the models used at each of the levels will be different and one can move from necessarily dynamic models at the bottom of the hierarchy to static models at the upper levels. This control structure therefore appears as a selective filter of the different disturbances affecting the process, the most rapid disturbances being taken over by the lower levels.

This hierarchisation facilitates the synthesis of each of the levels. Indeed, considering the first two levels alone (regulation-optimisation), one has overall a problem of optimal control; using these concepts of division, one can first of all make the synthesis of regulators, based on corresponding traditional techniques. Let us suppose these regulators to be sufficiently "stiff", then seen from the upper level the process thus regulated may appear to be almost always in steady state. We shall therefore at this level have a problem of static optimisation of fixing of the set points of the regulators.

Vertical decomposition may be used together with horizontal division, ie. taking into account the structure of the process if this is complex. This approach will lead to the synthesis of general multi-level multi-objective structures in which, for example, the regulation level will be decentralised in N independent regulators and the optimisation level itself broken down into a 2-level structure (cf. illustrative examples below).

Vertical-temporal breakdown of a function in different levels with different "dynamics" is also used in production management as per the structure below :

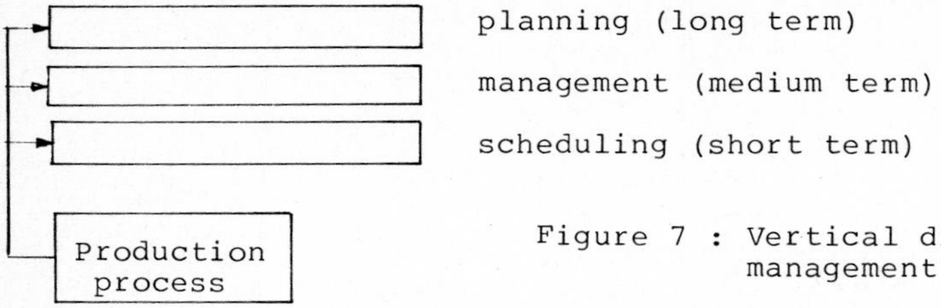

Figure 7 : Vertical division in management production

In this kind of application (contrary to vertical division in control engineering) the definition of the minimal information to be sent to each of the levels remains a great problem. Within each level the problems posed are of different kinds, but techniques of hierarchical optimisation can be used.

1.6.2. Qualitative description of examples and applications of vertical and horizontal divisions in industry.

Vertical division, contrary to horizontal division with its principles of co-ordination and its different breakdown methods, is difficult to formalise on the mathematical level, the great problem being the search for a compromise between the period of intervention of one level T_i and the sophistication of the algorithms working at this level.
Nevertheless, this division has been used for a long time in applications. Therefore we should like to describe here, simply and qualitatively, two significant examples of hierarchical structures being put into operation :
- the hierarchical control of a sugar beet production unit;
- the hierarchical control of a sulphur production complex.

1.6.2.1. Control of a sugar beet production unit [FIN - 69]

In this unit the main disturbances and difficulties are :
- lack of precision with respect to output
- hazardous deterioration in the quality of the raw materials stocked (beet). A hierarchical structure is proposed (Fig. 8) putting into operation from top to bottom :

  . an overall level of dynamic optimisation which fixes the duration of the production season and the level of production,

  . a static optimisation level which, for a fixed production, tends to minimise the losses of the unit by fixation of the set points of the regulators. Taking into account the structure of the process, this optimisation is carried out by a calculation structure which itself has 2 levels.

  . a final level of decentralised regulators acting on the different parts of the process.

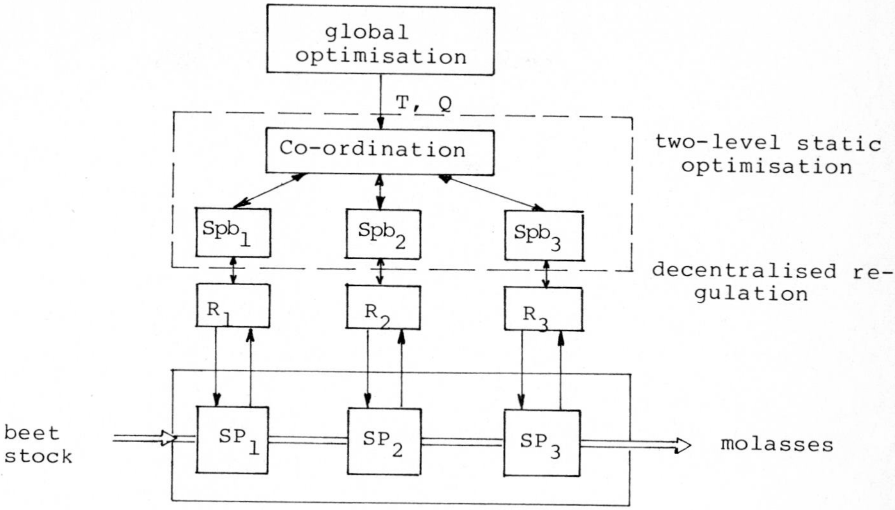

Figure 8 : Hierarchical control of a sugar manufacturing unit

1.6.2.2. Hierarchical control of a sulphur production unit [GRA - 77]

The production complex comprises 3 sulphur factories supplied by the same natural gas and discharging residuary gases in one single treatment unit. Taking into account certain production imperatives, the factories must be able to work alone, with two together, or all three together. Moreover, a number of disturbances affect both the composition and the discharge of the input gas.

The essential objective of the control is to minimise the losses of the whole unit (for the purpose of economic efficiency and also protection of the environment); bearing in mind the limitations above this could no be achieved in a global approach. Consequently, a hierarchical structure was put in place (Fig. 9). It comprises, from bottom to top :

- a level of stochastic decentralised regulation bringing into operation three "Self-tuning regulators"
- a minimisation of losses level by fixation of the set points of the preceding regulators. This optimisation is carried out by a 2 level structure.
- finally, an adaptation level, the aim of which is to follow up the slow development of the process (ageing of the catalyts, clogging-up of the boilers) by fixing parameters in the optimisation and regulation levels.

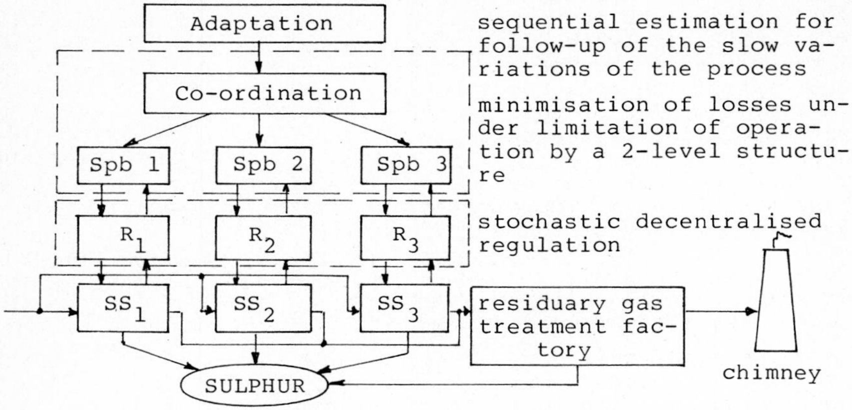

Figure 9 : Hierarchical control of a sulphur production complex

1.6.3. A special case of temporal vertical decomposition using singular perturbations techniques

We have stated previously that there is no general mathematical formalisation associated with vertical decomposition, unlike the situation which exists with horizontal decomposition. However in relation to a more limited problem, the singular perturbations techniques which were developed a long time ago (by Kokotovic et al.) induce a vertical temporal decomposition bringing into play two levels with different dynamics.
The problem has more recently been enlarged [MAG - 81],[FOS - 81] by studies on the analysis and control of systems with several time scales, that is in general systems with several dynamics.

If the dynamics are quite distinct (one slow and one rapid in the case of 2 dynamics, for example) then using certain supplementary conditions (of later in text) the formalism of singular perturbations can be obtained.
Indeed, the results given by [MAG - 81] and [FOS - 81], based on a sometimes temporal, sometimes frequential analysis, apply to arbitrary dynamics and make it possible to obtain accurate or approximate solutions (for the design of observers, the modal control and the optimal control) and according to the same hypotheses, give again the results of the theory of singular perturbations.

For our part, we consider that in a study on complex systems we cannot ignore the aspect of "multiple time scales"; nevertheless, bearing in mind the limitations of time and space, whe shall confine ourselves to older, more standard results although it is worth-while to mention a parallel with the methods of decomposition [BEN - 81].
The singular perturbations method consists of using in a system the property of multiple time scale in order to decompose it, if

possible, into slow or rapid sub-systems.
The sub-problems corresponding to this decomposition are resolved separately, whence the advantages of this method :
- reduction of size of the equations to resolve in order to find the laws of control.
- simplification of the laws of control to be implemented due to the separation of the dynamics of the sub-systems.
Nevertheless the application of this method for the study of complex systems presents serious difficulties namely :
- noting of the multiple time scale property without which the quasi-optimality of the results cannot be guaranteed.
- modelling in the form of a singularly perturbed system which can in fact be a difficult task for physical systems which are of multiple time scale in appearance but whose equations are not in the standardised form of singularly perturbed systems.

In the rest of this section we are going to approach some aspects of these modelling difficulties to deal after with analysis and control of singularly perturbed systems by feedback of state and of output.

1.6.3.1. System with double time scale

Let there be the stationary linear system

$$\overset{\circ}{x} = Ax \qquad\qquad x \in R^{n+m} \tag{1.1}$$

System (1.1) is said to have 2 time scales if it can be decomposed into two sub-systems :

$$\begin{bmatrix} \overset{\circ}{x}_1 \\ \overset{\circ}{x}_r \end{bmatrix} = \begin{bmatrix} A_1 & 0 \\ 0 & A_r \end{bmatrix} \begin{bmatrix} x_1 \\ x_r \end{bmatrix} . \qquad \begin{matrix} x_1 \in R^n \\ x_r \in R^m \end{matrix} \tag{1.2}$$

such as

$$|\lambda_{max} \ (A_1)| \ll |\lambda_{min} \ (A_r)| \tag{1.3}$$

For every square matrix

$$|\lambda_{max} \ (A)| \leq \| A \| = \sqrt{A^T A}$$

$$|\lambda_{min} \ (A) \ |^{-1} \leq \| A^{-1} \| \quad \text{if } A^{-1} \text{ exists}$$

and condition (3) becomes :

$$\| A_r^{-1} \| \ll \| A_1^{-1} \| \tag{1.4}$$

Let us suppose that system (1.1) has effectively n "slow" and m "rapid" variables and let us partition it consequently in the form

$$\begin{bmatrix} \overset{\circ}{x} \\ \overset{\circ}{z} \end{bmatrix} = \begin{bmatrix} A_{11} & A_{12} \\ A_{21} & A_{22} \end{bmatrix} \begin{bmatrix} x \\ z \end{bmatrix} \qquad \begin{matrix} x \in R^n \\ z \in R^m \end{matrix}$$

In order to obtain the diagonal form (1.2) we proceed to change the variables [CHE - 78] :

$$x = x_l + M x_r$$
$$z = x_r + L x_l$$

and we show that L matrix (mxn) and M matrix (nxm) are the real solutions of the Riccati and Lyapunov equations below :

$$A_{22} L - L A_{11} - L A_{12} L + A_{21} = 0 \quad (1.5)$$

$$(A_{11}+A_{12}L) M + M (A_{22} - L A_{12}) + A_{12} = 0 \quad (1.6)$$

If L and M exist, then

$$A_l = A_{11} + A_{12} L \quad (1.7)$$

$$A_r = A_{22} - L A_{12} \quad (1.8)$$

(M will intervene to modify the control matrix B)

One method of resolving these equations (1.5),(1.6) in algebraic manner by considering the A-invariant sub-spaces can be found in [MAG - 81]. Let us give here the results of [KOK - 75]:

If A_{22}^{-1} exists and if $\|A_{22}^{-1}\| < 1/3 \quad (\|A_o\| + \|A_{12}\| + \|L_o\|)^{-1}$

with $L_o = A_{22}^{-1} A_{21}$; $A_o = A_{11} - A_{12} L_o$

then the sequence :

$$L_{k+1} = - A_{22}^{-1} \quad (A_{21} - L_k A_{11} - L_k A_{12} L_k);\ L_o \quad (1.9)$$

converges towards a real bounded root of the equation (1.5). In the same way, and under the same conditions, M solution of (1.6) can be calculated as the solution of asymptotic equilibrium of the difference equation :

$$M_{k+1} = (A_{11}-A_{12}) - M_k - M_k L A_{12} A_{22}^{-1} + A_{12} A_{22}^{-1} \quad (1.10)$$

The condition of convergence $\|A_{22}^{-1}\| < 1/3 \quad (\|A_o\| + \|A_{12}\| + \|L_o\|)^{-1}$ will obviously be satisfied if :

$$\begin{bmatrix} A_{11} & A_{12} \\ A_{21} & A_{22} \end{bmatrix} = \begin{bmatrix} A_{11} & A_{12} \\ \frac{A_{21}^*}{\mu} & \frac{A_{22}^*}{\mu} \end{bmatrix} \quad (1.11)$$

ie. if the system can be written down in the form :

$$\begin{bmatrix} \overset{\circ}{x} \\ \mu \overset{\circ}{z} \end{bmatrix} = \begin{bmatrix} A_{11} & A_{12} \\ A_{21}^* & A_{22}^* \end{bmatrix} \begin{bmatrix} x \\ z \end{bmatrix} \quad (1.12)$$

with μ small parameter, since if μ is small enough one will have :

$$\mu \|A_{22}^{-1}\| \ll (\|A_o\| + \|A_{12}\| + \|L_o\|)^{-1}$$

According to (1.9) $L = L_o + O(\mu)$ and if we keep the approximation $L \simeq L_o$, then :

$$A_1 = A_{11} - A_{12} L \simeq A_{11} - A_{12} L_o = A_o$$

$$A_r = A_{22} + L A_{12} \simeq A_{22} + L_o A_{12} \simeq A_{22} = \frac{A^*_{22}}{\varepsilon}$$

A condition which is sufficient for a system :

$$\begin{bmatrix} \overset{\circ}{x} \\ \overset{\circ}{z} \end{bmatrix} = \begin{bmatrix} A_{11} & A_{12} \\ A_{21} & A_{22} \end{bmatrix} \begin{bmatrix} x \\ z \end{bmatrix} \qquad (1.13)$$

to have the two time scales is therefore

$$\| A_{22}^{-1} \| \ll (\| A_o \| + \| A_{12} \| + \| L_o \|)^{-1} \qquad (1.14)$$

$$(L_o = - A_{22}^{-1} A_{21})$$

This system can be written down in the form (1.12) with :

$$\mu = \| A_{22}^{-1} \| (\| A_o \| + \| A_{12} \| + \| L_o \|)$$

or $$\mu = | \lambda_{max} (A_1) | / | \lambda_{min} (A_r) | \quad \text{or} \quad | \lambda_{max} (A_o) | / | \lambda_{min} (A_{22}) |$$

1.6.3.2. Iterative separation of time scales

The approximation $L \simeq L_o$ makes it possible to define slow and rapid decoupled sub-systems (cf. later in text) bringing into play the dynamic matrices A_1 and A_r.

Indeed these results, which correspond to an approximation ($L \simeq L_o$) can be improved by searching for a more accurate value of L, by recurrence on L_k, or again, following [KOK - 80], by carrying out iterative separation of the time scales by successive application of the transformation matrices, the system obtained at each stage being itself considered as a singularly perturbed system for the following stage.

1.6.3.2.1. System in singularly perturbed form

Let the system be :

$$\overset{\circ}{X} = Ax + Bz \qquad \underline{X}(0) = \underline{X}_o$$

$$\mu \overset{\circ}{Z} = Cx + Dz \qquad \underline{Z}(0) = \underline{Z}_o \qquad (\mu \text{ known physical constant}) \qquad (1.15)$$

we assume that D^{-1} exists and that this system has physical properties of double time scale while not verifying the inequality (1.14)

ie. $$\left\| \frac{D^{-1}}{\mu} \right\| \leq \frac{1}{3 (\| A_o \| + \| B \| . \| D^{-1} C \|)} \quad \text{not verified}$$

with $A_o = A - B D^{-1} C.$

The algorithm (proposed by Kokotovic) which we are working out in this section aims to make the initial system verify this condition, and this by weakening the norm of A_o and above all that of C and by increasing that of D, therefore diminishing that of D^{-1} in the first instance; secondly by weakening the norm of B (and consequently that of A_o) and by increasing that of D.

a) Separation by weakening $\|C\|$.

If we consider that $\mu = 0$ for system (1.15) then

$$0 = Z_1 + D^{-1} C x_1$$

namely $Z_1 = - D^{-1} C x_1$

and let η_1, the difference between Z and Z_1

$$Z - Z_1 = \eta_1 = Z + D^{-1} C X$$

system (1.15) becomes :

$$\overset{\circ}{X} = (A - B D^{-1}) X + B \eta_1 \equiv A_1 X + B \eta_1$$

$$\mu \overset{\circ}{\eta}_1 = \mu D^{-1} C A_1 X + (D + \mu D^{-1} C B) \eta_1 \equiv C_1 X \quad D_1 \eta_1$$

This algorithm therefore consists of distinguishing between a slow part $\dot{Z}$ and a rapid part Z_1 in the vector Z. We repeat this operation, on the hypothesis $\eta_1 = \eta_{11} + \eta_2$ with

$$\eta_2 = \eta_1 + D_2^{-1} C_1 X$$

and after k iterations

$$\eta_k = \eta_{k-1} + D_{k-1}^{-1} C_{k-1} X \qquad \eta_o = Z$$

that is for system (1.15)

$$\begin{cases} \overset{\circ}{X} = A_k x + B \eta_k \\ \mu \overset{\circ}{\eta}_k = C_k x + D_k \eta_k \end{cases} \tag{1.16}$$

with

$$\begin{cases} A_k = A_{k-1} - B D_{k-1}^{-1} C_{k-1} & A_o = A \\ C_k = \mu D_{k-1}^{-1} C_{k-1} A_k & C_o = C \\ D_k = D_{k-1} + \mu D_{k-1}^{-1} C_{k-1} B & D_o = D \end{cases}$$

we note that the matrix C_k is of the order of $O(\mu^k)$, therefore the weakening will be as great as the number of iterations is large.

Furthermore, to find the initial variables again it is sufficient to consider that :

$$\sum_{i=1}^{k} (\eta_i - \eta_{i-i}) = \eta_k - Z = (\sum_{i=1}^{k} D_{i-1}^{-1} \; C_{i-1}) \; x$$

This algorithm consists of isolating in Z, then successively in $\eta_1, \eta_2 \dots \eta_{k-1}$, a slow part and then of making an increasingly clear temporal separation between X and η_k.

For k tending towards infinity, $\|C_k\|$ tends towards O, $\|A_k\|$ decreases and $\|D_k\|$ increases considerably. This has two consequences :

- decouple the rapid modes from the slow modes (triangularisation of the matrix

$$\mathcal{A} = \begin{bmatrix} A & B \\ C & D \end{bmatrix}$$ of the global system)

- separate even more the eigenvalues of A and of D by increasing $\|D\|$ and diminishing $\|A\|$.

b) Separation by weakening of $\|B\|$

The second equation is cut off from the first in system (1.16) and we have :

$$\overset{\circ}{x} = \mu B \, D_k^{-1} \, \eta_k = A_k x + B\eta_k - B \, D^{-1} \, C_k \, x - B \, D^{-1} \, D_k \eta_k$$

$$= (A_k - B \, D_k^{-1} \quad C_k) \quad x$$

we put : $\mathcal{E}_1 = x - \mu B \, D_k^{-1} \, \eta_k$

then : $\overset{\circ}{\mathcal{E}}_1 = A_{k_1} \mathcal{E}_1 + \mu A_{k_1} B \, D_k^{-1} \eta_k \equiv A_{k_1} \mathcal{E}_1 + B_{k_1} \eta_k$

in the same way we define, $\mathcal{E}_2, \mathcal{E}_3 \dots \mathcal{E}_{j+1}$

$$\mathcal{E}_{j+1} = \mathcal{E}_j - \mu B_{kj} \, D_{kj}^{-1} \, \eta_k \qquad \mathcal{E}_o = x$$

where

$$\begin{cases} A_{kj+1} = A_{kj} - B_{kj} \, D_{kj}^{-1} \, C_k & A_{ko} = A_k \\ B_{kj+1} = \mu A_{kj+1} \, B_{kj} \, D_{kj}^{-1} & B_{ko} = B \\ D_{kj+1} = D_{kj} + \mu C_k \, B_{kj} \, D_{kj}^{-1} & D_{ko} = D_k \end{cases}$$

we observe similarly that $\|B_{kj}\|$ is of the order of O (μ^j) and is therefore weaker by as much as the number of iterations j is large.

The new slow and rapid sub-systems show themselves in the form :

$$\overset{\circ}{\mathcal{E}}_j = A_{kj}\,\mathcal{E}_j + B_{kj}\,\eta_k \qquad \mathcal{E}_j(0) = \overset{\circ}{\mathcal{E}}_j$$

$$\mu\overset{\circ}{\eta}_k = C_k\,\mathcal{E}_j + D_{kj}\,\eta_k \qquad \eta_k(0) = \overset{\circ}{\eta}_j \tag{1.17}$$

$A_{kj} = A_1 + 0\,(\mu^{k+j})$ and $A_r = \frac{1}{\mu}\,D_{kj}\left[I - 0\,(\mu^{k+j})\right]$

B_{kj} is of the order $0\,(\mu^{j})$

C_k is of the order $0\,(\mu^{k})$

We can easily see that through this procedure the inequality (1.10) may be verified, if of course we have a system with a double time scale.
We find again the initial variables at the iteration j by :

$$\sum_{i=1}^{j} (\mathcal{E}_i - \mathcal{E}_{i-1}) = \mathcal{E}_j - x = -\mu\,(\sum_{i=1}^{j} \cdot B_{ki-1})\,\eta_k$$

1.6.3.2.2. System in the state space standard form

$$\begin{cases} \overset{\circ}{X} = AX + BZ & X(0) = X_o \\ \overset{\circ}{Z} = CX + DZ & Z(0) = Z_o \end{cases} \tag{1.18}$$

This system is presumed to have double time scale but does not verify the inequality (1.14). Moreover this system (contrary to the one in the previous section) does not have an obvious parameter μ .

a) weakening of $\|C\|$:
We consider that $\overset{\circ}{Z} = \mu\,\overset{\circ}{Z}'$ (fictious μ) and we make μ tend towards zero.

$$0 = C\,X_1 + D\,Z_1 \implies Z_1 = -D^{-1}\,C\,X_1$$

let there be : $\eta_1 = Z - Z_1 = Z + D^{-1}\,C\,X_1$

system (1.18) becomes :

$$\begin{cases} \overset{\circ}{X} = (A - B\,D^{-1}\,C)\,X + B\,\eta_1 = A_1\,X + B\,\eta_1 \\ \overset{\circ}{\eta}_1 = (D^{-1}\,C\,A_1)\,X + (D + D^{-1}\,C\,B)\,\eta_1 = C_1\,X + D_1\,\eta_1 \end{cases}$$

after k iterations we get :

$$\begin{cases} \overset{\circ}{X} = A_k\,X + B\,\eta_k \\ \overset{\circ}{\eta}_k = C_k\,X + D_k\,\eta_k \end{cases} \tag{1.19}$$

with :

$$\begin{cases} A_k = A_{k-1} - B\, D_{k-1}^{-1}\, C_{k-1} \\ C_k = D_{k-1}^{-1}\, C_{k-1}\, A_k\, . \\ D_k = D_{k-1} + D_{k-1}^{-1}\, C_{k-1}\, B \end{cases}$$

the system being assumed to have a double time scale, it follows that :

$$\| D_{k-1} \| \gg \| A_k \|$$

and therefore the product $\| D_{k-1}^{-1} \| \,.\, \| A_k \|$ is very weak. As a result of this the matrix C_k finds itself to be considerably weakened after k iterations. System (1.19) is almost triangular :

$$\Rightarrow \quad \lambda(A) \simeq \lambda(A_k) + \lambda(D_k)$$

the system can be modelled in the form of a singularly perturbed system with :

$$\mu = \frac{\| A_k \|}{\| D_k \|} \quad \text{or} \quad \frac{\lambda_{max}(A_k)}{\lambda_{min}(D_k)}$$

We find Z again by the conversion :

$$Z = \eta_k - \sum_{i=1}^{k} D_i^{-1}\, C_i\, x$$

b) weakening of B :

We use the conversion :

$$\varepsilon_1 = X - B\, D_k^{-1}\, \eta_k$$

At the iteration j :

$$\varepsilon_{j+1} = \varepsilon_j - B_{kj}\, D_{kj}^{-1}\, C_k$$

where

$$\begin{cases} A_{kj+1} = A_{kj} - B_k\, D_{kj}^{-1}\, C_k & A_{ko} = A_k \\ B_{kj+1} = A_{kj+1} \cdot B_{kj}\, D_{kj}^{-1} & B_{ko} = B \\ D_{kj+1} = D_{kj} + C_{kj}\, B_{kj}\, D_{kj}^{-1} & D_{ko} = D_k \end{cases}$$

System (1.19) becomes :

$$\begin{cases} \overset{\circ}{\varepsilon}_j = A_{kj}\, \varepsilon_j + B_{kj}\, \eta_k \\ \overset{\circ}{\eta}_k = C_k\, \varepsilon_k + D_{kj}\, \eta_k \end{cases}$$

with :

$$X = \varepsilon_j \sum_{i=1}^{j} (B_{ki-1}\, D_{ki-1}^{-1})\, \eta_k$$

$\|B_{kj}\|$ diminishes at each iteration in the ratio :

$$\|A_{kj}\| \quad . \quad \|D^{-1}_{kj-1}\|$$

Note :

It is obvious that this algorithm, after a great many iterations (on k, then on j), ends in a perfect separation of the slow and rapid dynamics, and therefore in bloc-triangularisation, then block diagonalisation of the matrix.
However the main difficulty is the accuracy of the calculations and consequently the reliability of the method, which depends on the inversion at each iteration of the matrix D.
We then consider limiting the number of iterations (one or two at the most on k) and applying the singular perturbations method to the system obtained ($\|C\|$ being sufficiently weakened).
Or better still, we can consider making an iteration on k and one on j(weakening of $\|C\|$ and $\|B\|$); the couplings being thus weakened, the regular perturbations method is applied.
Return to the original variables is easily done by multiplications and additions of matrices (in very limited numbers).

1.6.3.3. Analysis of singularly perturbed systems

1.6.3.3.1. Temporal decomposition of the system

Let us consider the following system :

$$\begin{cases} \overset{\circ}{X} = A_{11} X + A_{12} Z + B_1 u & X(0) = X_o \\ \mu \overset{\circ}{Z} = A_{21} X + A_{22} Z + B_2 u & Z(0) = Z_o \\ y = C_1 X + C_2 Z \end{cases} \qquad (1.20)$$

in which dim X = n, dim Z = m, dim u = r, dim y = p.

μ = small positive parameter.

System (1.20) is decomposed into two sub-systems :

- a rapid system which predominates for t approximate to 0.
- a slow system whose behaviour is similar to that of the global system when $t \rightarrow \infty$.

Slow system

It is obtained by putting $\mu = 0$. This comes down to neglecting the rapid modes because their action is only sensitive during the initial phase of evolution of the global system.

$\mu = 0 \rightarrow$ we shall denote all the variables with the index ℓ :

$$\begin{cases} \overset{\circ}{X}_\ell = A_{11} X_\ell + A_{12} Z_\ell + B_1 u_\ell & X(0) = X_o \\ 0 = A_{21} X_\ell + A_{22} Z_\ell + B_2 u_\ell \\ y_\ell = C_1 X_\ell + C_2 Z_\ell \end{cases} \qquad (1.21)$$

If A_{22}^{-1} exists,

$$Z_\ell = - A_{22}^{-1} (A_{21} X_\ell + B_2 u_\ell)$$

then the slow system will be represented by :

$$\begin{cases} \overset{\circ}{X}_\ell = A_o X_\ell + B_o u_\ell & X(0) = X_o \\ y_\ell = C_o X_\ell + D_o u_\ell \end{cases} \tag{1.22}$$

in which

$$\begin{cases} A_o = A_{11} - A_{12} A_{22}^{-1} A_{21} \\ B_o = B_1 - A_{12} A_{22}^{-1} B_2 \\ C_o = C_1 - C_2 A_{22}^{-1} A_{21} \\ D_o = C_2 A_{22}^{-1} B_2 \end{cases}$$

After a very short time it will be possible to identify system (1.20) with system (1.22)

Rapid system

This is obtained by considering that the slow variables of the system vary little during the evolution of the rapid variables. ie. for $t \neq 0$, $\overset{\circ}{Z}_\ell = 0$ and X_ℓ constant

By putting $Z_r = Z - Z_\ell$, $u_r = u - u_\ell$ and $y_r = y - y_\ell$

the rapid system is defined by :

$$\begin{cases} \mu \overset{\circ}{Z}_r = A_{22} Z_r + B_2 u_r & Z_r(0) = Z_o - Z_\ell(0) \\ y_r = C_2 Z_r \end{cases}$$

1.6.3.3.2. Controllability of singularly perturbed systems

Let the system be :

$$\begin{cases} \overset{\circ}{X} = A_{11} X + A_{12} Z + B_1 u \\ \mu \overset{\circ}{Z} = A_{21} X + A_{22} Z + B_2 u \end{cases} \tag{1.23}$$

in which

$$X \in R^n, \quad Z \in R^m$$

by taking again the preceding change of variables

$$\begin{bmatrix} X_\ell \\ Z_r \end{bmatrix} = \begin{bmatrix} I_n + \mu ML & - \mu M \\ - L & I_m \end{bmatrix} \begin{bmatrix} X \\ Z \end{bmatrix}$$

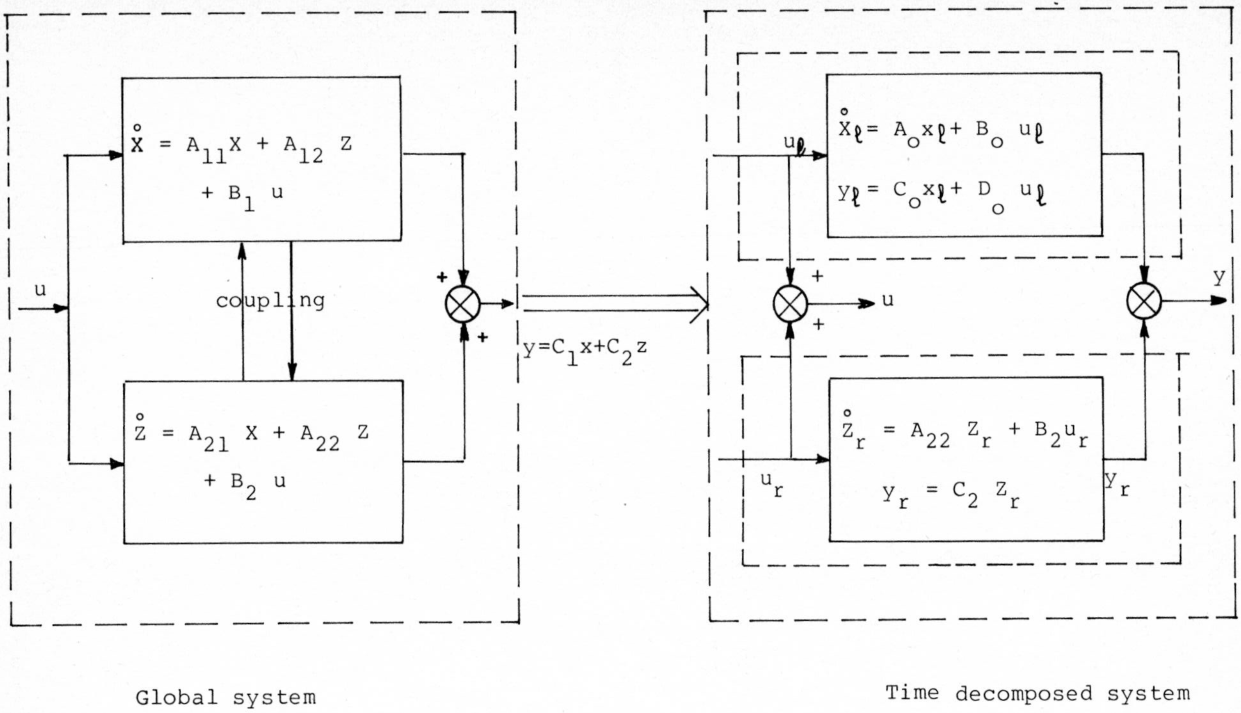

Figure 10 - DECOMPOSITION OF THE SINGULARLY PERTURBED SYSTEM and by neglecting the terms of higher or equal order to $0(\mu^2)$, the system is written as [CHE - 78]:

$$\begin{bmatrix} \overset{\circ}{X}_\ell \\ \mu \overset{\circ}{Z}_r \end{bmatrix} = \begin{bmatrix} A_o + 0(\mu) & 0 \\ 0 & A_{22} + 0(\mu) \end{bmatrix} \begin{bmatrix} X_\ell \\ Z_r \end{bmatrix} + \begin{bmatrix} B_o + 0(\mu) \\ B_2 + 0(\mu) \end{bmatrix} \underline{u} \qquad (1.24)$$

$$= \begin{bmatrix} A_\ell & 0 \\ 0 & A_r \end{bmatrix} \begin{bmatrix} X_\ell \\ Z_r \end{bmatrix} + \begin{bmatrix} B_\ell \\ B_r \end{bmatrix} \underline{u}$$

The controllability properties of system (1.24) are identical to those of system (1.23)

When $\mu \to 0$

$$A \longrightarrow A_o$$
$$A_r \longrightarrow A_{22}$$
$$B \longrightarrow B_o$$
$$B_r \longrightarrow B_2$$

The matrices A_ℓ, B_ℓ, A_r and B_2 are continuous with respect to μ in relation to which enables us to state the following theorem :

Theorem : [POR - 75]

If A_{22} is reversible and if

$$\text{rank}\left[B_o \,\vdots\, A_o B_o \,\vdots\, \text{-------} A_o^{n-1} B_o\right] = n \qquad (1.25)$$

$$\text{rank}\left[B_2 \,\vdots\, A_{22} B_2 \,\vdots\, \text{-------} A_{22}^{m-1} B_2\right] = m \qquad (1.26)$$

then there exists a value $\mu = \mu^*$ such that system (1.23) is controllable whatever may be μ belonging to the interval $[0, \mu^*]$

1.6.3.3.3. Observability

Considering system (1.23)

Theorem

If A_{22}^{-1} exists and if :

$$\text{rank}\left[C_2^T \,\vdots\, A_{22}^T \; C_2^T \,\vdots\, \text{------} (A_{22}^{m-1})^T \; C_2^T\right] = m$$

$$\text{rank}\left[C_o^T \,\vdots\, A_o^T \; C_o^T \,\vdots\, \text{------} (A_o^{n-1})^T \; C_o^T\right] = n$$

then there exists a value $\mu = \mu^*$ such that for every value of μ belonging to the interval $[0, \mu^*]$, system (1.23) is observable.

1.6.3.3.4. Stabilisation

Let the system be

$$\begin{cases} \overset{\circ}{X} = A_{11}\ X + A_{12}\ Z + B_1\ u & \text{(a)} \\ \mu \overset{\circ}{Z} = A_{21}\ X + A_{22}\ Z + B_2\ u & \text{(b)} \end{cases} \quad (1.27)$$

Let us assume that the pair (A_{22}, B_2) is stabilisable. We can then close the loop of system (1.27) by the control.

$$u = k_2 Z + v$$

where k_2 is chosen such that $(A_{22} + B_2\ k_2)$ is stable.
System (1.27) becomes :

$$\begin{cases} \overset{\circ}{X} = A_{11}\ X + (A_{12} + B_1\ k_2)\ Z + B_1 v & \text{(a)} \\ \mu \overset{\circ}{Z} = A_{21}\ X + (A_{22} + B_2\ k_2)\ Z + B_2 v & \text{(b)} \end{cases} \quad (1.28)$$

By putting $\mu = 0$

$$Z = -\ (A_{22} + B_2\ k_2)^{-1}\ \ (A_{21}\ X + B_2\ v)$$

equation (1.28) is in the form :

$$\overset{\circ}{X} = A_o^* X + B_o^*\ v \quad (1.29)$$

where $A_o^* = A_{11} - (A_{12}+B_1k_2)\ (A_{22}+B_2k_2)^{-1}\ A_{21}$

$B_o^* = B_1 - (A_{12}+B_1k_2)\ (A_{22}+B_2k_2)^{-1}\ B_2$

Let us assume that the pair (A_o^*, B_o^*) is also stabilisable, and let us close the loop of system (1.28) by $v = k_1 X$ such that k_1 stabilises system (1.29).

System (1.27) becomes :

$$\begin{cases} \overset{\circ}{X} = (A_{11}+B_1k_1)\ X + (A_{12}+B_1k_2)\ Z \\ \mu \overset{\circ}{Z} = (A_{21}+B_2k_1)\ X + (A_{22}+B_2k_2)\ Z \end{cases} \quad (1.30)$$

The stability property of system (1.30) comes from the following lemma attributable to [KLI - 61].

<u>Lemma</u>

Let the system be

$$\begin{cases} \overset{\circ}{X} = \Gamma_{11}\ X + \Gamma_{12}\ Z & X(o)=X_o \\ \mu \overset{\circ}{Z} = \Gamma_{21}\ X + \Gamma_{22}\ Z & Z(o)=Z_o \end{cases}$$

in which Γ_{22} is a Hurwitz matrix (with eigenvalues having a real part $\leqslant 0$)

If $$R = \Gamma_{11} - \Gamma_{12} \Gamma_{22}^{-1} \Gamma_{21}$$

is a Hurwitz matrix, then there exists a value $\mu = \mu_o$ such that for every μ belonging to the interval $[0, \mu_o]$, the point of equilibrium $X = 0$, $Z = 0$ is asymptotically stable.

For system (1.30)

$$R = (A_{11}+B_1k_1) - (A_{12}+B_1k_2)(A_{22}+B_2k_2)^{-1}(A_{21}+B_2k_1)$$

$$= \left[A_{11} - (A_{12}+B_1k_2)(A_{22}+B_2k_2)^{-1} A_{21}\right] + \left[B_1 - (A_{12}+B_1k_2)(A_{22}+B_2k_2)^{-1} B_2\right] k_1$$

$$R = A_o^* + B_o^* k_1$$

The pair (A_o^*, B_o^*) is assumed to be stabilisable. The gain k_1 is such that system (1.30) is stable, and so R is indeed a Hurwitz matrix.

From this we get the following theorem :

If (A_{22}, B_2) is a stabilisable pair and if there exists a matrix k_2 such that $A_{22}+B_2k_2$ is a Hurwitz matrix and (A_o^*, B_o^*) is a stabilisable pair, then there exists a matrix k_1 and a scalar μ_o such that for every μ belonging to the interval $[0, \mu_o]$, the point of equilibrium $X = 0$, $Z = 0$ of the closed loop system is asymptotically stable.

1.6.3.3.5. Positioning of the poles

Let the system be

$$\begin{cases} \overset{\circ}{X} = A_{11} X + A_{12} Z + B_1 u \\ \mu \overset{\circ}{Z} = A_{21} X + A_{22} Z + B_2 u \end{cases}$$

Assuming that the matrix A_{22} is invertible, we know that the set of eigen values of the matrix :

$$A = \begin{bmatrix} A_{11} & A_{12} \\ A_{21}/\mu & A_{22}/\mu \end{bmatrix}$$

tends, when μ tends towards 0, towards the union, of the following two disjoint sub-sets :

$$\mathcal{L}_o = \{\lambda_1, \lambda_2, \ldots \lambda_n\}$$

and

$$\mathcal{L}_2 = \{\hat{\lambda}_1/\mu, \hat{\lambda}_2/\mu \ldots \hat{\lambda}_m/\mu\}$$

where λ_i, i=1 to n are the eigenvalues of the matrix n x n

$$A_o = A_{11} - A_{12} A_{22}^{-1} A_{21}$$

and $\hat{\lambda}_i$, i=1 to m are the eigenvalues of the matrix m x m A_{22}

Let us apply the following control to the system :

$u = G_1 X + G_2 Z$ where G_2 is calculated so that

$$A_{22} + B_2 G_2 \quad \text{is stable.}$$

The set of eigenvalues of the resulting closed loop matrix :

$$\tilde{A}^* = \begin{bmatrix} A_{11} + B_1 G_1 & A_{12} + B_1 G_2 \\ (A_{21} + B_2 G_1)/\mu & (A_{22}+B_2G_2)/\mu \end{bmatrix}$$

is constituted when μ tends towards 0, of the union of the two disjoint sub-sets :

$$R_o = \{\rho_1, \rho_2 \dots \rho_n\}$$

and

$$R_2(\mu) = \{\hat{\rho}_1/\mu , \hat{\rho}_2/\mu \dots \hat{\rho}_m/\mu\}$$

where $\hat{\rho}_i$, $i=1 \rightarrow m$ are the eigenvalues of the matrix m x m

$$A_{22} + B_2 G_2$$

and ρ_i, $i=1 \rightarrow n$ are the eigenvalues of the matrix n x n

$$\begin{aligned} \tilde{A}_o &= A_{11}+B_1G_1 - (A_{12}+B_1G_2)(A_{22}+B_2G_2)^{-1} (A_{21}+B_2G_1) \\ &= A_o^* + B_o^* G_1 \end{aligned}$$

Notes :

(1) assignation of the poles $\mathcal{L}_2 \rightarrow R_2(\mu)$ is carried out only by the gain G_2 and this provided that the pair (A_{22},B_2) is controllable.

(2) positioning of the poles $\mathcal{L}_o \rightarrow R_o$ is carried out by the gain G_1, provided that the pair (A_o^*, B_o^*) is controllable. Now if A_{22}^{-1} and $(A_{22}+B_2G_2)^{-1}$ exist, then the pair (A_o, B_o) is controllable if (A_o^*,B_o^*) is controllable [CHE - 78]. The gain is then :

$$G_1 = (I + G_2 A_{22}^{-1} B_2) G_o + G_2 A_{22}^{-1} A_{21}$$

where G_o is determined in such a way as to place the poles of the system reduced from $\mathcal{L}_o$ to R_o.

1.6.3.4. Optimisation and control of singularly perturbed systems

In this section we are going to apply the results of the previous section to the quadratic optimisation of systems with two time scales.

1.6.3.4.1. Quasi-optimal control by temporal decomposition

Let the system be :

$$\begin{cases} \mathring{X} = A_{11}X + A_{12} Z + B_1 u & X(o)=X_o \\ \mu \mathring{Z} = A_{21}X + A_{22} Z + B_2 u & Z(o)=Z_o \\ y = C_1 X + C_2 Z \end{cases} \quad (1.31)$$

We are trying to minimise the criterion :

$$J = 1/2 \int_o^{\infty} (y^T y + u^T Ru) \, dt \quad (1.32)$$

<u>Decomposition of the system</u> :

The system is decomposed into :
- a slow sub-system obtained by putting $\mu = 0$ (ie. by neglecting the action of rapid modes which are very quickly damped and have little influence on the steady state, this under certain conditions which we have previously specified.
- a rapid sub-system, by considering that at the time of the transient state the slow modes evolve very little and are therefore quasi-stationary; the preponderant dynamic is therefore that of rapid modes which gives $\mathring{X}_\ell = 0$ in writing.

We then have the two sub-systems :

Slow system : $\mu = 0$

$$\begin{cases} \mathring{X}_\ell = A_o X_\ell + B_o u_\ell & X_\ell(o) = X_o \\ y_\ell = C_o X_\ell + D_o u_\ell \end{cases} \quad (1.33)$$

where

$$\begin{cases} A_o = A_{11} - A_{12} A_{22}^{-1} A_{21} \\ B_o = B_1 - A_{12} A_{22}^{-1} B_2 \\ C_o = C_1 - C_2 A_{22}^{-1} A_{21} \\ D_o = - C_2 A_{22}^{-1} B_2 \end{cases}$$

Rapid system :

$$\mathring{X} = \mathring{Z}_\ell = 0 \quad \text{and} \quad \begin{cases} X_\ell = X \\ Z_\ell = Z - Z_r \end{cases}$$

$$\begin{cases} \mu \mathring{Z}_r = A_{22} Z_r + B_2 u_r & Z_r(o) = Z_o + A_{22}^{-1} A_{21} X_o \\ y_r = C_2 Z_r \end{cases} \quad (1.34)$$

The criterion J will be similarly decomposed into Jr and $J\ell$:

$$J_\ell = 1/2 \int_o^{\infty} (y_\ell^T \, y_\ell + u_\ell^T \, Ru_\ell) \, dt \quad (1.35)$$

and

$$J_r = 1/2 \int_o^\infty (y_r^T\, y_r + u_r^T\, R\, u_r)\, dt \qquad (1.36)$$

The problem of optimisation (1.31),(1.32) is decomposed into two independent reduced problems (1.33),(1.35) and (1.34),(1.36), which leads us to the two solutions u_ℓ^* and u_r^*. The global control u_c will be calculated at a later time from these values u_ℓ^* and u_r^*.

Control of the sub-systems

Rapid sub-system : The optimal control of sub-system (1.34) which minimises the criterion (1.36) is given by :

$$u_r^* = - R^{-1}\, B_2^T\, k_r\, Z_r$$

where k_r is the unique semi-definite positive solution of the Riccati equation :

$$k_2\, A_{22} + A_{22}^T\, k_r - k_r\, B_2^T\, R^{-1}\, B_2^T\, k_r + C_2^T\, C_2 = 0$$

This solution exists on condition that the pair (A_{22}, B_2) is stabilisable and that the pair (A_{22}, C_2) is observable.

Slow sub-system : The slow sub-system is written as :

$$\begin{cases} \mathring{X}_\ell = A_o\, X_\ell + B_o\, u_\ell \\ y_\ell = C_o\, X_\ell + D_o\, u_\ell \end{cases}$$

the criterion to be minimised is :

$$J_\ell = 1/2 \int_o^\infty (y_\ell^T\, y_\ell + u_\ell^T\, R u_\ell) dt \quad \text{with}$$

$y_\ell = C_o\, X_\ell + D_o$

$$J_\ell = 1/2 \int_o^\infty (X_\ell^T\, C_o^T\, C_o\, X_\ell + 2\, u_\ell^T\, D_o^T\, C_o\, X_\ell + u_\ell^T R_o u_\ell) dt$$

$(R_o = R + D_o^T\, D_o)$

The change of control is made :

$$\bar{u} = u_\ell - R_o^{-1}\, D_o^T\, C_o\, X_\ell$$

System (1.33) becomes :

$$\begin{cases} \mathring{X}_\ell = (A_o - B_o\, R_o^{-1}\, D_o^T\, C_o)\, X_\ell + B_o\, \bar{u} \\ \mathring{y}_\ell = (C_o - D_o\, R_o^{-1}\, D_o^T)\, X_\ell + D_o\, \bar{u} \end{cases}$$

with the criterion

$$J_\ell = 1/2 \int_o^\infty \left\{ X_\ell^T\, C_o^T\, (I - D_o\, R_o^{-1} D_o^T)\, C_o\, X_\ell + \bar{u}^T\, R_o \bar{u} \right\} dt$$

The solution of this problem is :

$$u_\ell^* = - R_o^{-1} \; (D_o^T C_o + B_o^T k_\ell \;) X_\ell$$

with k_ℓ the solution of the Riccati equation

$$k_\ell (A_o - B_o R_o^{-1} D_o^T C_o) + (A_o - B_o R_o^{-1} D_o^T C_o)^T k_\ell - k_\ell B_o R_o^{-1} B_o^T k_\ell$$

$$+ C_o^T (I - D_o R_o^{-1} D_o^T) C_o = 0$$

It is shown [CHE - 78] that the solution k_ℓ exists if (A_o, B_o) is controllable, and if the pair (A_o, C_o) is observable.

Composite control

The global control of system (1.31) is :

$$u_c = u_r^* + u_\ell^*$$

$$= - R^{-1} B_2^T k_r z_r - R_o^{-1} (D_o^T C_o + B_o^T k_\ell) X_\ell$$

$$u_c = G_r \; z_r + G_\ell \; x_\ell$$

By using the results of the section on analysis we get

$$u_c = \left[(I + G_r A_{22}^{-1} B_2) G_\ell + G_r A_{22}^{-1} A_{21} \right] x + G_r z$$

$$u_c = k_1 x + k_2 z$$

where

$$k_1 = - \left[(I - R^{-1} B_2^T k_r A_{22}^{-1} B_2) R_o^{-1} (D_o^T C_o + B_o^T k_\ell) + R^{-1} B_2^T k_r A_{22}^{-1} A_{21} \right]$$

and

$$k_2 = - R^{-1} B_2^T k_r .$$

By a transformation (Kokotovic), we go back to

$$k_1 = - R^{-1} B_1^T k_\ell - R^{-1} B_2^T \left\{ (A_{22} - B_2 R^{-1} B_2^T k_r)^T \right\}^{-1} \left\{ (k_r B_2 R^{-1} B_1^T - A_{12}^T) k_\ell - (C_2^T C_1 + k_r A_{21}) \right\}$$

which gives :

$$u_c = - R^{-1} (B_1^T k_\ell x + B_2^T k_m^T x + B_2^T k_r z)$$

$$= - R^{-1} B^T \begin{bmatrix} k_\ell & \mathbb{O} \\ \mu K_m^T & \mu K_r \end{bmatrix} X = - R^{-1} B^T k_c X$$

where $B^T = [B_1^T \;\; B_2^T / \mu]$, $X^T = [x^T \; z^T]$

and $k_m = \left\{ k_\ell (A_{12} - B_1 R^{-1} B_2^T k_r) - C_1^T C_2 + A_{12}^T k_r) \right\} (A_{22} - B_2 R^{-1} B_2^T k_r)^{-1}$

Calculation of the criterion

The optimal criterion is given by the equation :

$$J = \frac{1}{2} X_o^T K X_o$$

where $X_o = \begin{bmatrix} x_o \\ z_o \end{bmatrix}$ and k is the solution of the Riccati equation of the overall problem.

From the composite control calculated by the singular perturbations method, the sub-optimal criterion is written as :

$$J_c = \frac{1}{2} X_o^T P_c X_o$$

P_c being the positive definite solution of the Lyapunov equation.

$$P_c (A - B R^{-1} B^T k_c) + (A-BR^{-1}B^T K_c)^T P_c = - k_c^T B R^{-1} B^T k_c - C^T C$$

$$k_c = \begin{bmatrix} k_\ell & 0 \\ k_m^T & k_r \end{bmatrix}$$ previously defined.

By separating the Riccati equation of the global system from this equation

$$KA + A^T K - K B R^{-1} B^T k + C^T C = 0 \text{ with } C = [C_o \; C_1]$$

and by putting : $W_c = P_c - K$

we get :

$$W_c (A-BR^{-1}B^T k_c) + (A-BR^{-1}B^T k_c)^T W_c + (k-k_c)^T BR^{-1}B^T (k-k_c) \equiv 0$$

By remainding that [CHO - 76]

$$k = k(\mu), \text{ therefore } k - k_c = 0(\mu)$$

the expression $(k-k_c)^T B R^{-1} B^T (k-k_c)$ is a function $0(\mu^2)$

$(A - B R^{-1} B^T k_c)$ is stable, W_c is therefore a function $0(\mu^2)$ or again $P_c = k + 0(\mu^2)$.

From which we get the theorem :

The first two terms of the series expansion of J_c and J^* are the same at $\mu = 0$, ie. $J_c = J^* + 0(\mu^2)$.

Reduced control

When A_{22} is stable we can consider controlling only the slow modes - which comes down to optimising an aggregated system of the initial system - we then arrive at a quasi-optimal control of the order $O(\mu)$.

The problem is defined by :

$$\begin{bmatrix} \dot{x}_\ell = A_o x_\ell + B_o u_\ell & \quad x_\ell(o) = x_o \\ y_\ell = C_o x_\ell + D_o u_\ell & \end{bmatrix}$$

with

$$J_\ell = \frac{1}{2} \int_0^\infty (y_\ell^T y_\ell + u_\ell^T R u_\ell)\, dt$$

The solution of this problem, which has already been dealt with, is :

$$u_\ell = - R_o^{-1} (D_o^T C_o + B_o k)\, x_\ell$$

k_ℓ being the stable solution of the Riccati equation :

$$k_\ell(A_o - B_o^{-1} D_o^T C_o) + (A_o - B_o R_o^{-1} D_o^T C_o)^T k_\ell - k_\ell B_o R_o^{-1} B_o^T k_\ell - C_o^T (I - D_o R_o^{-1} D_o^T)\, C_o = 0$$

Let us put :

$$G_\ell = - R_o^{-1} (D_o^T C_o + B_o^T k_\ell)$$

then

$$J_\ell = \frac{1}{2} x_o^T P x_o$$

where P_ℓ is the positive definite solution of the Lyapunov equation :

$$P_\ell (A + B G_\ell) + (A + B G_\ell)^T P_\ell = - G^T R G - C^T C$$

By putting u_ℓ in the form :

$$U_\ell = - R^{-1} \begin{bmatrix} B_1^T & B_2^T/\mu \end{bmatrix} \begin{bmatrix} \mu k_{11} & 0 \\ \mu k_r & 0 \end{bmatrix}$$

We show, in the same way as in the case of composite control, that :

$$J_\ell = J^* + 0(\mu)$$

1.6.3.4.2. Control by output feedbacks

Control by state feedback gives satisfactory results with respect to performance (stability, optimality of the solution vis à vis the chosen criterion). However, from a practical viewpoint this control has certain disadvantages : in fact the equation $u = - k x$ brings all the states (into play) and therefore requires the calculation of a large number of parameters. Moreover,

if something happens and one of the states becomes inaccessible, the whole control system becomes unusable. One is forced to have recourse to emergency solutions.
Finally, in a large scale system, not all the states are generally measurable. Application of the state control requires the reconstruction of non-measured states. This solution has the major disadvantage of increasing the size of an already large system, and does not simplify the determination of the control.

For these reasons it is useful to study an output feedback solution taking into account the singularly perturbed aspect of the system; the objective of this approach is to decompose the problem of output feedback control into two sub-problems of control resulting from the separation of the output vector into one component with rapid variation and one component with slow variation.

$$y = \begin{bmatrix} y_1 \\ \hline y_2 \end{bmatrix} \begin{matrix} \updownarrow p_1 \\ \updownarrow p_2 \end{matrix} \begin{bmatrix} C_{11} & 0 \\ C_{21} & C_{22} \end{bmatrix} \begin{bmatrix} X \\ \hline Z \end{bmatrix} \begin{matrix} \updownarrow n \\ \updownarrow m \end{matrix}$$

(the sub-matrix C_{12} is zero as we consider that y_1 has slow variation [CHE - 80]).

The problem consists of determining a control in the form :

$$u = G_1 y_1 + G_2 y_2$$

obtained from a temporal decomposition of the global system into two sub-systems, slow and rapid, for which the local controls are in the form : $u_\ell = G_\ell y_{1\ell}$ and $u_r = G_r y_{2r}$
The matrices of gain G_1 and G_2 are determined in terms of G_ℓ and G_r in the following way :

$$\begin{cases} G_2 = G_r \\ G_1 = (I + B_2 C_{22} A_{22}^{-1} B_2)\, G_\ell + G_2\, (C_{22} A_{22}^{-1} A_{21} - C_{21})\, C_{11}^+ \end{cases}$$

with $C_{11}^+ = C_{11}^T \, (C_{11} C_{11}^T)^{-1}$ $\quad P_1 \leqslant n$

The control u will be stable in the following cases [CHE - 80]:

<u>1st case</u> :
C_{11} is of rank n : this means that the control obtained is looped on the rapid outputs and on the slow states. Indeed, the slow outputs then represent only a regular linear transformation of the slow states.

<u>2nd case</u> :
The matrix $(C_{22} A_{22}^{-1} A_{21} - C_{21})$ can be written in the form $C'_{02} C_{11}$. This condition is difficult to verify due to the fact that C_{11} is not square. However, there are cases where this condition is met :

a) $A_{21} = L_{21}\ C_{11}$ and $C_{21} = M_{21}\ C_{11}$

this means that the interaction of the slow system on the rapid system is done by the intermediary of the slow output (on the rapid state and the rapid output). (cf. figures 11a and 11b).

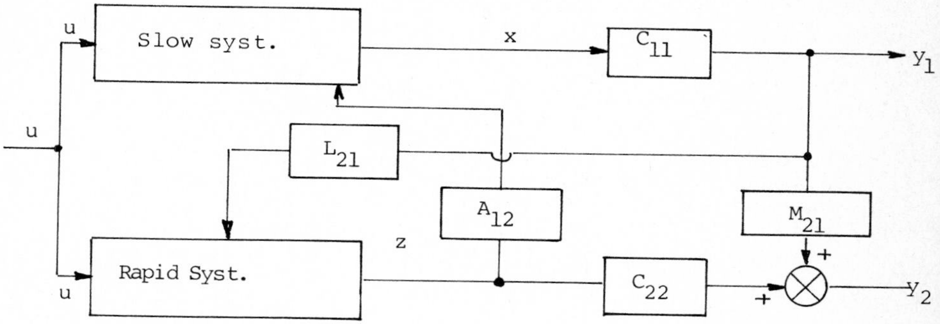

Figure 11a

b) $A_{21} = L_{21}\ C_{11}$ $C_{21} = \mathbb{O}$

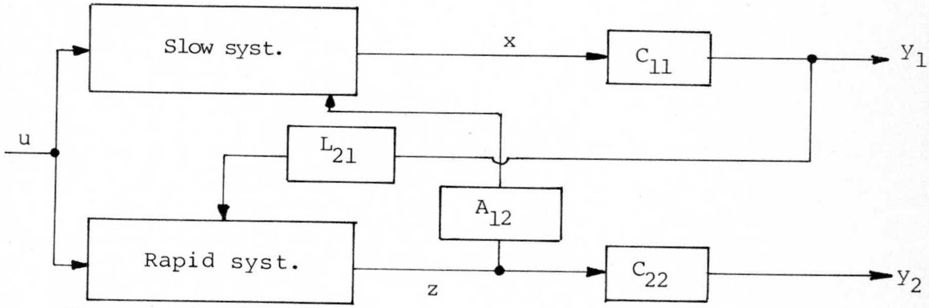

Figure 11b

For these two cases it is suitable to put the system in the necessary form right from modelling and not to look throught the pseudo-inverse of C_{11} for the calculation of the matrix C'_{02}.

3rd case :

The matrix $G_2\ (C_{22}\ A_{22}^{-1}\ A_{21} - C_{21})$ can be written in the form $C''_{02}\ C_{11}$, ie. G_2 is determined so as to stabilise the rapid system and to satisfy the above equation. In the case where an algorithm of optimisation is used for the determination of G_2, this condition is generally not met.

In reality only the 2nd case can be of practical interest. In general, dim y_1 < dim x therefore the 1st case is not often

met. The 3rd case introduces a supplementary constraint for the determination of G_2 which considerably complicates the problem.
The use of an optimisation method makes it possible to determine a control by stable output feedback satisfying certain characteristics defined by the criterion to be minimised under certain conditions of stabilisability.
In the text which follows we are going to work out an approach of this sort for the class of systems mentioned in the 2nd case. Let us note that as there is very little difference between cases 2) a) and 2) b), we shall restrict ourselves to case 2) b) ie. :

- the equation of outputs $y = C_1 X + C_2 Z$ is decomposable into 2 equations which represent respectively the observation of the slow variables and of the rapid variables :

$$y = \begin{bmatrix} y_1 \\ y_2 \end{bmatrix} = \begin{bmatrix} C_1 & 0 \\ 0 & C_2 \end{bmatrix} \begin{bmatrix} X \\ Z \end{bmatrix}$$

- interaction of the slow variables on the rapid system is done by the intermediary of the output variables :

$$\mu \mathring{Z} = L_{21}\, y_1 + A_{22}\, Z + B_2\, u$$

with $\quad A_{21} = L_{21}\, C_1$

<u>Position of the problem</u>

Let the system therefore be :

$$\begin{cases} \mathring{X} = A_{11}\, X + A_{12}\, Z + B_1\, u & X(0) = X_0 \\ \mu \mathring{Z} = A_{21}\, X + A_{22}\, Z + B_2\, u & Z(0) = Z_0 \end{cases}$$

we assume that :

$$* \quad y = \begin{bmatrix} y_1 \\ y_2 \end{bmatrix} = \begin{bmatrix} C_1 & 0 \\ 0 & C_2 \end{bmatrix} \begin{bmatrix} X \\ Z \end{bmatrix}$$

$$* \quad A_{21} = L_{21}\, C_1$$

whence :

$$\begin{cases} \mathring{X} = A_{11}\, X + A_{12}\, Z + B_1\, u \\ \mu \mathring{Z} = L_{21}\, C_1\, X + A_{22}\, Z + B_2\, u \end{cases}$$

$$y = [\tilde{C}_1 \quad \tilde{C}_2] \begin{bmatrix} X \\ Z \end{bmatrix} = \begin{bmatrix} C_1 & 0 \\ 0 & C_2 \end{bmatrix} \begin{bmatrix} X \\ Z \end{bmatrix}$$

we propose to minimise the criterion :

$$J = 1/2 \int_0^{\infty} (y^T y + u^T Ru)\, dt$$

Decomposition of the problem

a) Slow system :

by putting $\mu = 0$, the slow system is defined by :

$$\begin{cases} \dot{x}_\ell = (A_{11}-A_{12}A_{22}^{-1}L_{21}C_1)\, x_\ell + (B_1-A_{12}A_{22}^{-1}B_2)u_\ell \quad x_\ell(0)=x_o \\ y_\ell = \begin{bmatrix} C_1 & 0 \\ 0 & C_2 \end{bmatrix} \begin{bmatrix} x_\ell \\ z_\ell \end{bmatrix} \end{cases}$$

For $\mu = 0$

$$z_\ell = - A_{22}^{-1} (L_{21}\, C_1\, x_\ell + B_2\, u_\ell)$$

that is :

$$\begin{cases} y_{1\ell} = C_1\, x_\ell \\ y_{2\ell} = - C_2\, A_{22}^{-1}\, L_{21}\, C_1\, x_\ell = x_2\, A_{22}^{-1}\, B_2\, u_\ell \end{cases}$$

the slow system is written as :

$$\begin{cases} \dot{x}_\ell = A_o\, x_\ell + B_o\, u_\ell \qquad x_\ell(0) = x_o \\ y_\ell = \tilde{C}_o\, x_\ell + \tilde{D}_o\, u_\ell \end{cases}$$

with :

$$\begin{cases} A_o = A_{11} - A_{12}\, A_{22}^{-1}\, L_{21}\, C_1 \\ B_o = B_1 - A_{12}\, A_{22}^{-1}\, B_2 \\ \tilde{C}_o = \begin{bmatrix} C_1 \\ - C_2\, A_{22}^{-1}\, L_{21}\, C_1 \end{bmatrix} \\ \tilde{D}_o = \begin{bmatrix} 0 \\ - C_2\, A_{22}^{-1}\, B_2 \end{bmatrix} \end{cases}$$

Rapid system

$$\dot{z}_\ell = 0, \qquad x_\ell = \text{constant}$$

$$z_r = z - z_\ell\,, \quad y_r = y_2 - y_{2\ell}\,, \quad u_r = u - u_\ell$$

we get the system :

$$\begin{cases} \mu \overset{\circ}{Z}_r = A_{22} Z_r + B_2 u_r \\ y_r = \begin{bmatrix} 0 \\ C_2 \end{bmatrix} Z_r \end{cases}$$

the criterion J is decomposed into two independent criteria :

$$\begin{cases} J_\ell = \frac{1}{2} \int_0^\infty (y_\ell^T \; y_\ell + u_\ell^T R u_\ell) \, dt \\ J_r = \frac{1}{2} \int_0^\infty (y_r^T \; y_r + u_r^T R u_r) \, dt \end{cases}$$

Control of the rapid sub-system

We have the system :

$$\begin{cases} \mu \overset{\circ}{Z}_r = A_{22} Z_r + B_2 u_r & Z_r(0) = Z(0) - Z_\ell(0) \\ y_r = \begin{bmatrix} 0 \\ C_2 \end{bmatrix} Z_r \end{cases}$$

we are looking for the control $u_r = k_r y_r$ which minimises J_r. As the rapid system is not interconnectected with y_1, it is therefore sufficient to loop on y_{2r} (y_{1r} does not exist).

whence $u_r = k_2 y_{2r} = k_2 C_2 Z_r$

It follows that :

$$\mu \overset{\circ}{Z}_r = (A_{22} + B_2 K_2 C_2) Z_r \qquad Z_r(0) = Z(0) - Z_\ell(0)$$

and

$$J_r = \frac{1}{2} \int_0^\infty (Z_r^T C_2^T C_2 Z_r + Z_r^T C_2^T k_2^T R K_2 C_2 Z_r) \, dt$$

$$= \frac{1}{2} \int_0^\infty \left[(Z_r^T C_2^T) (I + K_2^T R K_2) (C_2 Z_r) \right] dt$$

K_2 being the solution of a parametric optimisation problem associated with the rapid system.

Control of the slow sub-system

This system is defined by :

$$\begin{cases} \overset{\circ}{X}_\ell = A_o X_\ell + B_o u_\ell & X_\ell(0) = X_o \\ y_\ell = C_o X_\ell + D_o u_\ell = \begin{bmatrix} y_{1\ell} \\ y_{2\ell} \end{bmatrix} \end{cases}$$

We know, (hypothesis), that y has two components : $y = \begin{bmatrix} y_1 \\ y_2 \end{bmatrix}$;

the rapid system is not interconnected with y_1; on the other

hand the slow system is linked to the two components of y because of the slow part of the vector Z.
Nevertheless the 1st hypothesis formulated at the beginning of the section leads us to look, for the slow sub-system, for a control of the form :[CHE - 81] :

$$u_\ell = K_\ell \, y_{1\ell}$$

The closed loop system becomes :

$$\overset{\circ}{X}_\ell = (A_o + B_o K \; C_1) \, X_\ell$$

the criterion to be minimised :

$$J_\ell = \frac{1}{2} \int_0^\infty (y_\ell^T \; y_\ell + u_\ell^T \; R u_\ell \;) \, dt$$

being :

$$J_\ell = \frac{1}{2} \int_0^\infty (X_\ell^T \; C_1^T \, Q_\ell \, C_1 \; X_\ell \,) \, dt$$

with :

$$Q_\ell = \left[(\; I + (L_{21} + B_2 k_\ell \;)^T (A_{22}^{-1})^T C_2^T \; A_{22}^{-1} \; (L_{21} + B_2 k_\ell) \; + k^T_\ell \; R \; k_\ell \right]$$

This is again a parametric optimisation problem; subject to the stability of the system, there exists a positive definite matrix P_ℓ , solution of the Lyapunov equation :

$$P_\ell \; (A_o + B_o \, k_\ell \, C_1) + (A_o + B_o \, k_\ell \, C_1)^T \; P_\ell + Q_\ell \equiv 0$$

the value of the criterion is then :

$$J_\ell = \frac{1}{2} \; X_\ell^T \; (0) \; P_\ell \; X_\ell \; (0) \qquad \text{(For a reduced control)}$$

<u>Composite control</u>

This is defined by :

$$\begin{aligned} u_c &= u_r + u_\ell \\ &= k_2 \; y_{2r} + k_\ell \; y_{1\ell} \\ &= k_2 \; C_2 Z_r + k_\ell \; C_1 \; X_\ell \\ &= G_2 \; Z_r + G \;\; X_\ell \end{aligned}$$

according to the results of the section on analysis

$$\begin{aligned} u_c &= \left[(\; I + G_2 \; A_{22}^{-1} \; B_2) \;\; G_\ell + G_2 \; A_{22}^{-1} \; L_{21} \; G \right] \; X + G_2 \; Z \\ &= \left[(\; I + K_2 \; C_2 \; A_{22}^{-1} \; B_2) \; K_\ell + K_2 \; C_2 A_{22}^{-1} \; L_{21} \right] \; C_1 \; X + K_2 \; C_2 \; Z \\ &= \left[(\; I + K_2 \; C_2 \; A_{22}^{-1} \; B_2) \; K_\ell + K_2 \; C_2 \; A_{22}^{-1} \; L_{21} \right] y_1 + K_2 \; y_2 \end{aligned}$$

the composite control is then put in the form :

$$u_c = \left[(I + K_2 C_2 A_{22}^{-1} B_2) K_\ell + K_2 C_2 A_{22}^{-1} L_{21} \;\vdots\; K_2 \right] y$$

1.6.3.5. Spatio temporal decomposition [TIT - 79]

We are combining here the singular perturbations method with techniques of spatial decomposition to deal with interconnected systems for which we would have demonstrated the double time scale property at the level of each of the sub-systems.

Position of the problem

We consider the set of interconnected sub-systems represented by the equations :

$$\overset{\circ}{X}_i = A_i x_i + B_i u_i + C_i z_i + \sum_{\substack{j \neq i \\ j=1}}^{N} (L_{ij} x_j + M_{ij} u_j)$$

$$\mu_i \overset{\circ}{Z}_i = D_i X_i + E_i Z_i + F_i u_i \qquad \left| \begin{array}{l} X_i(0) = X_{io} \\ Z_i(0) = Z_{io} \end{array} \right. \qquad (1.37)$$

$$i = 1, N$$

$$X_i \in R^{n_{i1}}, \quad Z_i \in R^{n_{i2}}, \quad u_i \in R^{m_i} \qquad (0 < \mu_i \ll 1)$$

We note that in this formulation the interconnections between the sub-systems are made by the intermediary of the slow modes of each sub-systems. In other words, the rapid parts are decoupled from each other.
We propose to control this system by minimising the separable quadratic criterion.

$$\min_{u_i} \sum_{i=1}^{N} \frac{1}{2} \int_0^\infty \left[X_i^T Q_{i1} X_i + Z_i^T Q_{i2} Z_i + u_i^T R_i u_i \right] dt$$

Temporal decomposition of the problem

System with slow dynamic : this is obtained by making $\mu_i = 0$ in (1.37). We then have, by generalisation of the results of section 1.6.3.4. and on the hypothesis of the existence of E_i^{-1}

$$\forall i = 1 \text{ to } N$$

$$\overset{\circ}{X}_{i\ell} = A_{i\ell} X_{i\ell} + B_{i\ell} u_{i\ell} + \sum_{\substack{j \neq i \\ j=1}}^{N} (L_{ij\ell} X_{j\ell} + M_{ij\ell} u_{j\ell}).$$

$$X_{i\ell}(0) = X_{io} \qquad i = 1, N$$

with the criterion

$$\min_{u_{i\ell}} \sum_{i=1}^{N} \frac{1}{2} \int_0^\infty (X_{i\ell}^T Q_{i\ell} X_{i\ell} + u_{i\ell}^T R_{i\ell} u_{i\ell} + 2 X_{i\ell}^T S_{i\ell} u_{i\ell}) \, dt$$

and

$$A_{i\ell} = \left[A_i - C_i E_i^{-1} D_i \right], \quad B_i = \left[B_{i\ell} - C_i E_i^{-1} F_i \right]$$

$$\begin{cases} Q_{i\ell} = Q_{i1} + D_i^T (E_i^T)^{-1} Q_{i2} E_i^{-1} D_i \\ R_{i\ell} = R_i + F_i^T (E_i^T)^{-1} Q_{i2} E_i^{-1} F_i \\ S_{i\ell} = D_i^T E_i^{T-1} Q_{i2} E_i^{-1} F_i \\ L_{ij\ell} = L_{ij} - N_{ij} E_j^{-1} D_i \\ M_{ij\ell} = M_{ij} - N_{ij} E_j^{-1} F_i \end{cases}$$

This is conventional optimisation problem of a set of N interconnected sub-problems for which decomposition co-ordination methods can be applied.

System with rapid dynamics : to obtain the formulation of this problem we assume that the variables with slow dynamics are constant during the transitory states.

$\overline{X}_i = X_{i\ell} = \text{cte}, \quad \overline{u}_i = u_{i\ell} = \text{cte}, \quad \overline{Z}_i = \text{cste}$

$$\text{and } \overset{\circ}{\overline{Z}}_i = 0 \qquad Z_i = \overline{Z}_i + Z_{ir}$$

we can write :

$$\mu_i \overset{\circ}{\overline{Z}}_i = D_i X_i + E_i Z_i + F_i u_i$$

and

$$\mu_i \overset{\circ}{\overline{Z}}_i = 0 = D_i \overline{X}_i + E_i \overline{Z}_i + F_i \overline{u}_i$$

by splitting up these two equations, we get :

$$\mu_i (\overset{\circ}{Z}_i - \overset{\circ}{\overline{Z}}_i) = E_i (Z_i - \overline{Z}_i) + F_i (u_i - u_{i\ell})$$

that is

$$\mu_i \overset{\circ}{Z}_{ir} = E_i Z_{ir} + F_i u_{ir} \qquad Z_{ir}(0) = Z_i(0 - \overline{Z}_i(0)$$

$$i=1, \text{ to } N$$

with the criterion

$$\min_{u_{ir}} \frac{1}{2} \sum_{i=1}^{N} \int_o^\infty (Z_{ir}^T Q_{i2} Z_{ir} + u_{ir}^T R_i u_{ir}) \, dt$$

which corresponds to the resolution of N totally independent sub-problems, whence the control structure outlined on the following figure 12.

1.7 Methods of size reduction [BER - 79]

We have just devoted several paragraphs to a natural step - decomposition - to overcome the difficulties linked with the control of complex systems.
Another step, which is gaining renewed interest, is reduction of size which makes it possible to return to the application of traditional methods.

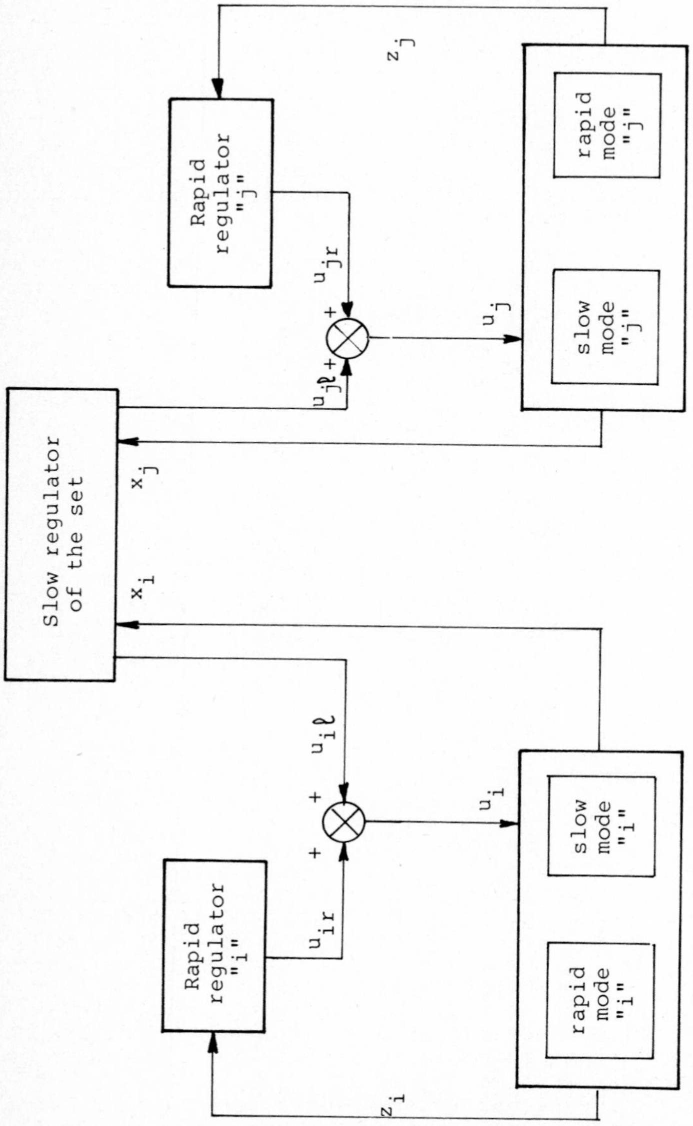

Figure 12 - REGULATION OF THE OVERALL SYSTEM

1.7.1. General comments on methods of reduction

Let the initial problem be :

$$\min \int_0^\infty 1/2 \ (x^T Q x + u^T R u) \ dt$$

subject to $\overset{o}{x} = Ax + Bu \qquad \qquad x(0) = x_0$

$$x \in R^n, \ u \in R^r$$

for which we are looking for a solution in the form $u = Kx$
Let us consider a reduced model of the initial system :

$$\overset{o}{Z} = FZ + Gu$$

$$Z \in R^m \qquad m < n, \qquad u \in R^r$$

bringing in a state $Z(t)$ linked to the initial state $x(t)$ by the equation :

$$Z(t) = \emptyset \ (x(t))$$

Let us join to this reduced model the criterion :

$$J_e = \int_0^\infty 1/2 \ (Z^T Q_e Z + u^T Ru) \ dt$$

which makes it possible to obtain an optimum control (in the sense of J_e) :

$$u^* = K^* \ Z(t)$$

and a sub-optimum control for the real system :

$$u_{s.0} = K^* \ \emptyset \ (x(t))$$

The problem of control with reduced model will therefore be to determine simultaneously the accepted mode of size reduction (determination of $\emptyset$) and the equivalent criterion of performance (determination of Q_e) in such a way as to have satisfactory behaviour of the system thus controlled.
The methods of reduction applicable to a state space formulation have given 2 large families of models :
- the reduced models retaining the dominant modes of the system, i.e. the modes with the strongest influence on the behavior of the system;
- the optimum reduced models minimising, for a given class of entry, an error of representation of the system by the reduced model.
Among these general methods of reduction the technique of size reduction by linear aggregation plays an increasingly important role. Moreover, it makes it possible to rediscover a considerable number of reduced models (retaining the dominant modes of the real system) already available in the literature. Therefore we shall present in the following pages only aggregated models and their use in optimum control.

1.7.2. The concept of linear aggregation

Let the linear dynamic system be

$$\overset{o}{x} = Ax + B\overset{\cdot\cdot}{u}$$

$$y = Cx \qquad \qquad x \in R^n \ \text{(state)} \qquad u \in R^r \ \text{(control)}$$

Let there be a linear relationship of aggregation between the state of the initial system x and a reduced state Z :

$$Z = L x$$

L : matrix of aggregation of size (m x n) $Z \in R^m$ ($m \ll n$, $r < m$).
A simplified representation of the initial system (aggregated model) is :

$$\mathring{Z} = FZ + Gu$$

$$\hat{y} = HZ$$

if (and only if) the following equations are verified :

$$FL = LA$$
$$G = LB$$
$$Z(0) = L\,x\,(0)$$

Let u_i be a eigenvector of A and λ_i the associated eigenvalue :

$$A\,u_i = \lambda_i\,u_i$$

from which $L\,A\,u_i = F\,L\,u_i = \lambda_i\,L\,u_i$

So for all of $L\,u_i \neq 0$, $L\,u_i$ is an eigenvector of F connected to the eigenvalue of λ_i.

The dynamic matrix F of the aggregated model therefore retains m eigenvalues of the initial model and its corresponding vectors are the aggregates of the eigenvectors of the matrix A corresponding to these eigenvalues.

Here again, determination of the matrix of aggregation can be made :

- by choice of the dominant modes according to structural properties of A and C, or to energy considerations;
- by minimisation of a criterion of error (initial model - aggregated model) of the form :

$$\sum_{i=1}^{r} \int_{o}^{\infty} \| y_i - \hat{y}_i \|^2 \, dt$$ in which y_i, $\hat{y}_i$ are the responses of the initial model and of the aggregated model when only the component i of the control is applied (the other components being nil) for given classes of control (deterministic or stochastic, general or standard inputs).

Notes : This concept of aggregation can be generalised to interconnected systems when one wishes to conserve the structure of the initial system after order reduction.

1.7.3. Sub-optimum control by aggregated model

We are considering the problem of initial control :

$$\min \int_{o}^{\infty} 1/2\,(x^T Qx + u^T R\,u)\,dt$$

$$\text{subject to } \mathring{x} = Ax + B\,u$$

for which we presume that we have an aggregated model, through the equation $Z = L\,x$:

$$\mathring{Z} = F\,x + G\,u$$

If we also have an equivalent criterion :

$$J_e = 1/2 \int_0^{\infty} (Z^T\ Q_e\ Z + u^T\ R\ u)\ dt$$

we can deduce from it a law of optimum control for the aggregated model :

$$u = K^* Z(t)$$

and sub-optimum for the real system :

$$u_{S.O.} = K^* L\ x(t)$$

This law of control will stabilise the real system if K^* stabilises the aggregated model and if the aggregated model retains all the unstable modes of A, because the corresponding closed loop dynamic basis :

$$\tilde{A} = A + B\ K^* L$$

possesses :
- the (n-m) eigenvalues of A not retained in F,
- the m eigenvalues of $F + G\ K^*$

The definition of an equivalent criterion J_e, through the intermediary of Q_e, plays an important role in the determination of sub-optimum control. This equivalent criterion must be chosen in such a way that its minimum value is close to the minimum value of J with the optimum control u = K x.
There are at least two ways to determine Q_e :
- by comparison of the equations of Riccati associated with the initial problem and with the aggregated problem, which gives [AOK -68].

$$Q_e = (L^T)^+\ Q\ L^+$$

(in which L^+ designates the Moore-Penrose pseudo-inverse of L);
- by direct optimisation of J in relation to $Q_e = S^T S$, using gradient procedures; the problem is :

$$\min_S J = \text{trace} \left| Q + L^T\ P\ G\ R^{-1}\ G^T\ P\ L \right| W$$

$$\text{subject to} \quad P\ F + F^T\ P - P\ G\ R^{-1}\ G^T\ P + S^T\ S = 0$$

$$\tilde{A}\ W + W\ \tilde{A} + x_o\ x_o^T = 0$$

By using the matrix maximum principle, calculation of the gradient $\partial J / \partial S$ is made by resolution of a Riccati equation (of reduced size m) and of 3 Lyapunov equations of dimensions (nxm), (nxn), (mxm).
This approach [SIR-79] which is more complex than that of Aoki, gives better results and can be further simplified by taking into account the particular structure of A, if any, for a large dimensional problem.
Note : The sub-optimality of this control may be evaluated by lower and upper limits on the value of the criterion.
Generalisations have been developed for sub-optimum control by state feedback or with observers.

1.8 Multi-criteria formulation for problems associated with complex systems (*) [BAP - 80]

1.8.1. Introduction

The multi-criteria formulation is becoming more and more a necessity in decision-making, not only in socio-economic systems but also in the more restricted field of automatic control of complex systems.
Two obvious examples are the search for the cost/efficiency compromise of a complex control structure and simultaneous identification/control.
The consideration of different criteria rules out the traditional and convenient notion of optimal solution because it gives rise to a space of optimality which is partially ordered and comparison between two solutions becomes difficult, if not impossible.
We are going to consider the process of decision making for a system (or the control of a system) by N agents or deciders (by N local control units) and examine the different concepts of optimality and equilibrium which may be presented.

1.8.2. Notion of optimality - Points of equilibrium

Taking into account the interactions between deciders via the system, two extreme cases can occur :
- each decider tries to optimise its own criterion without regard to the other deciders and associated criteria;
- each decider negotiates with the others and seeks a final co-operative equilibrium. If the deciders decide to have a single common criterion, the problem becomes a team problem.

Between non-co-operation and the joint strategy, we can envisage the formation of sub-groups with common interests. Thus we arrive at the following table :

equilibrium
- non-co-operative
  - Nash point (selfish strategy)
  - min max (pessimistic strategy)
  - Stackelberg point (passive strategy)
- ↕ coalitions
- co-operative (negotiation)
  - Pareto point (compromise)
  - team (single criterion)

To describe these strategies we suppose that the N deciders have corresponding decision variables u_i, $i=1,\ldots,N$, with $u_i \in U_i$ (U_i normed vectorial space) and the different criteria will be designated by $J_1(u_1,\ldots,u_N),\ldots, J_N(u_1,\ldots,u_N)$, or, to simplify the notations, $J_1(u),\ldots,J_N(u)$, putting

(*) In this section the authors make frequent reference to the doctoral thesis of L.F.B. BAPTISTELLA.

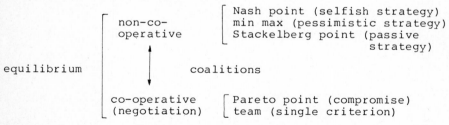

$u = \{u_1, \ldots, u_N\}$. We therefore suppose that the different criteria may depend on the decision variables of the other deciders.

1.8.2.1. Nash point

This is the equilibrium with the perfect information hypothesis. Each decider takes as data the decision criteria of the others and then tries to minimise its own criterion. The strategies are announced simultaneously and before beginning the optimisation procedure (or the game) see [STA - 69], [HO - 70] and [NAS - 51].

Definition : A point $u = (u_1^N, \ldots, u_N^N)$ is a Nash point if :

$$u_i^N \in U_i \qquad i = 1 \ldots N \tag{1.38}$$

$$J_i(u_1^N, \ldots, u_i^N, \ldots u_N^N) \leq J_i(u_1^N, \ldots, u_i, \ldots, u_N^N) \quad \forall u_i \in U_i$$

The Nash strategy therefore protects each decider from an attempt at improvement made by another decider.

If we consider a dynamic problem, with :

$$\mathring{x} = f(x, u_1, \ldots, u_N) \tag{1.39}$$

$$J_i(u) = K_i(x(T)) + \int_0^T L_i(x; u_1, \ldots, u_N)\, dt \qquad i=1 \ldots N$$

in which

$x(t) \in R^n$, $u(t) \in R^m$, $f : R^n \times R^m \rightarrow R^n$ is C^1 and $L_i : R^n \times R^m \rightarrow R$ is C^1, and with the associated Hamiltonians :

$$H_i = L_i(x, u_1 \ldots u_N) + p_i^T . (f(x, u_1 \ldots u_N)) \qquad i=1 \ldots N \tag{1.40}$$

in which $p_i \in R^n$ are the respective adjoint variables, two cases may occur [PIN - 77] :

Open loop optimisation

In this case each decider seeks an optimum trajectory (from its own criterion) and tries to follow it during the optimisation. period. The necessary optimality conditions associated with (1.39) and (1.40) are given by :

$$u_i^N \text{ minimises } H_i(x^N, u_1^N, \ldots, u_i, \ldots, u_N^N, p^N) \quad \forall u_i \in U_i \tag{1.41}$$

in which x^N and p^N are solutions to the equations

$$\mathring{x} = \frac{\partial H_i}{\partial p_i} \qquad x(0) = \mathring{x} \qquad i=1 \ldots N \tag{1.42}$$

$$\mathring{p}_i = - \frac{\partial H_i}{\partial x} \qquad i=1 \ldots N \tag{1.43}$$

$$p_i(T) = \frac{\partial K_i}{\partial x(T)} \qquad i=1 \ldots N \tag{1.44}$$

Interpretation of these conditions is no different from the mono-criterion case, except for the new adjoint variables which now represent the marginal variational costs in relation to the criterion of each decider in particular. Application of these conditions to a quadratic linear problem introduces a system of coupled Riccati equations [PIN - 77] .

Closed loop optimisation

In this case, each decider calculates a law of control before commencing the planning procedure and uses this law revising its policy all the time from observation of the state of the system. If we consider that the decisions are dependent on the state :

$$u_i = w_i(x) \qquad i=1...N \qquad (1.45)$$

then the conditions of optimality (1.41), (1.47) remain valid, except for equation (1.43), which changes into [CLE - 79] :

$$\overset{\circ}{p}_i = -\left[\frac{\partial H_i}{\partial x} + \sum_{j=1}^{N} \frac{\partial H_i}{\partial u_j} \cdot \frac{\partial u_j}{\partial x} \right] \quad i=1...N \qquad (1.46)$$

an equation in which the crossed terms show that each decider changes its policy all the time in response to the evolution of the strategy of the other deciders (and of the state of x of the resulting system). Thus we can note that the philosophy of this closed loop deterministic strategy is different from that used in the traditional mono-criterion case. Here the decider recognises that the variations in its own policy will change the state of the system and the other deciders will change their policy in consequence, even in the deterministic case. For this reason, open loop and closed loop behaviour are usually different.

1.8.2.2. Min-max pessimistic strategy

The ideal situation of perfect information of the Nash equilibrium is obviously not always met, for example when the deciders do not know the criteria of the others. So this strategy is no longer calculable and alternative behaviours must be adopted.
One possibility is the choice of a pessimistic strategy, in which each decider considers that its "rivals" will try to harm it as much as possible. So then, knowing only its own criterion, each decider tries to minimise it in the case corresponding to the worst decision for itself that the other decider can make. In a system with two deciders, this takes us to :

$$\begin{aligned} &\text{decider 1} : \quad \min_{u_1} \max_{u_2} J_1(u_1,u_2) \\ &\text{decider 2} : \quad \min_{u_2} \max_{u_1} J_2(u_1,u_2) \end{aligned} \qquad (1.47)$$

The conditions of existence of a point of equilibrium which satisfies (1.47) must obviously be verified [ISA - 65] , chap. 12;

[BLA - 73] . The most noteworthy results of this approach, which have been known for a number of years([NEU - 28] see also [ISA - 65], for the differential games) are associated with the systems with two deciders with zero sum of the criteria, ie. :

$$J_1 + J_2 = 0 \qquad (1.48)$$

In this case, from (1.47) we get

$$\min_{u_1} \max_{u_2} J_1 (u_1, u_2) \qquad (1.49)$$

and

$$\min_{u_2} \max_{u_1} (-J_1(u_1, u_2)) = \max_{u_2} \min_{u_1} J_1(u_1, u_2) \qquad (1.50)$$

According to the fundamental theorem of [NEU - 28], (p. 307), if J_1 satisfies some supplementary hypotheses (see also [STA - 79]), equations (1.49) and (1.50) are identical and show that any set with two deciders and with zero sum has a min max solution.
On the other hand, since (1.48) implies a knowledge of the behavior of the deciders, we observe that, according to the definition of the Nash point, with $J_1 = - J_2$

$$J_1(u_1^N, u_2^N) \leq J_1(u_1, u_2^N)$$

$$-J_1 (u_1^N, u_2^N) \leq - J_1 (u_1^N, u_2)$$

$$\Longrightarrow J_1(u_1^N, u_2) \leq J_1(u_1^N, u_2^N) \leq J_1 (u_1, u_2^N) \qquad (1.51)$$

which coincides with the definition of a saddle point. So in problems with a saddle point, the strategies of Nash and max-min coincide.

1.8.2.3. Passive Stackelberg strategy

If the deciders do not wish to follow a passive strategy as above, another possibility is to adopt a passive behaviour, ie. the decider waits for the announcement of the other participants strategy and optimises its situation from that. In the same way, this situation may occur in systems where different means of information, more or less rapid, enable one decider to announce its strategy before the others. This sequential strategy (due to [STA - 52]) corresponds to the case where one of the deciders assumes the role of "leader" and the others of "followers"([SIM - 73 a, 73 b]). This situation was at first proposed for a dipole : generalisation in the case of several deciders was proposed by [LEI - 78] . If several deciders wish to assume the role of leader, an imbalence may occur, and similarly when no-one wishes to assume this role [BAS - 73] ; [OKU - 76] .
In this strategy, the leader knows the follower's criterion, whereas the follower does not know that of the leader, but only

its own strategy. The leader fixes its strategy beforehand and so encourages the follower to take favourable action, since the latter will react in an optimal manner to the trajectory fixed by the former.

Definition for two deciders

If there is a transformation $T : U_1 \rightarrow U_2$ such that, for u_1 fixed by decider 1, decider 2 chooses $u_2 = T(u_1)$ such that :

$$J_2(u_1, T(u_1)) \leq J_2(u_1, u_2) \qquad \forall\, u_2 \in U_2 \tag{1.52}$$

then (u_1^{S1}, u_2^{S1}) is a Stackelberg strategy if there exists $u_1^{S1} \in U_1$ and $u_2^{S1} = T(u_1^{S1})$, $u_2^{S1} \in U_2$ such that :

$$J_1(u_1^{S1}, T(u_1^{S1})) \leq J_1(u_1, T(u_1)) \qquad \forall u_1 \in U_1 \tag{1.53}$$

In such a case, decider 1 is the leader and decider 2 is the follower (whence the index S1). The definition is similar when the roles are reversed. According to this definition, we observe that an important characteristic is contributed by the whole set of rational reactions of the follower, defined, in the case where decider 2 is the follower, by :

$$D_2 = \left\{ (u_1, u_2) \in U_1 \times U_2 \ / \ u_2 = T(u_1) \right\} \tag{1.54}$$

We shall make the hypothesis that the follower is "rational". Similarly

$$D_1 = \left\{ (u_1, u_2) \in U_1 \times U_2 \ / \ u_1 = T(u_2) \right\} \tag{1.55}$$

is the whole set of rational reactions when decider 1 is the follower. In accordance with these definitions we can observe that :

Proposal 1 -

(u_1^{S1}, u_2^{S1}) is a Stackelberg point if and only if

$$(u_1^{S1}, u_2^{S1}) \in D_2 \tag{1.56}$$

$$J_1(u_1^{S1}, u_2^{S1}) \leq J_1(u_1, u_2) \qquad \forall u_1, u_2 \in D_2$$

Proposal 2 -

(u_1^N, u_2^N) is a Nash point if and only if

$$(u_1^N, u_2^N) \in D_1 \cap D_2 \tag{1.57}$$

The demonstrations are direct from the definitions of the Stackelberg point and the Nash point.

In addition, let us note that, according to (19) :

$$J_1\ (u_1^{S1},\ u_2^{S1}) \leq J_1\ (u_1,\ u_2) \qquad \forall u_1,\ u_2 \in D_2 \qquad (1.58)$$

in particular $(u_1, u_2) \in D_1 \cap D_2$. So then, according to (1.57)

$$J_1\ (u_1^{S1},\ u_2^{S1}) \leqslant J_1\ (u_1^{N},\ u_2^{N}) \qquad (1.59)$$

and for the leader, the Stackelberg strategy is at least as good as the Nash strategy. For the follower we cannot say anything a priori. Nevertheless it is possible to see if :

$$(u_1^{S1},\ u_2^{S1}) = (u_1^{S2},\ u_2^{S2}) \qquad (1.60)$$

ie. if the leader-follower permutation does not change the solution, that the Nash and Stackelberg solutions coincide according to the proposals 1 and 2.

Let us take the example in Figure 13 to illustrate these ideas [SIM - 73 a] :

The functions J_1 and J_2 are the criteria of the deciders 1 and 2 respectively, and D_1 and D_2 are the sets of rational reactions associated with decider 1 when it is the follower, and decider 2 when it is the follower, respectively. It is easy to verify that S1 is a Stackelberg point when decider 1 is the leader, in the same way that S2 is a Stackelberg point when decider 2 is the leader. N is a Nash point.

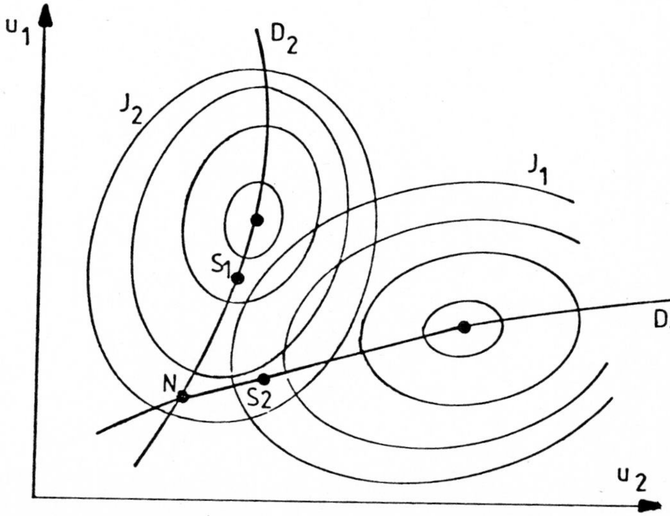

Figure 13 - A problem with two deciders - The Nash and Stackelberg solutions.

It is interesting to note here that the Stackelberg strategy can also bring benefits to the follower, compared with the Nash solution. In addition, the existence of the Stackelberg point does not imply the existence of the Nash point (it is sufficient that $D_1 \cap D_2 = \emptyset$) and vice-versa (see a counter-example in [SIM - 73 a], together with the conditions of existence of these strategies).
If we now consider the dynamic problem (1.39) two cases can also occur [CRU - 78] :

Open loop optimisation

In this case, the deciders know only the initial state of the system and will try to find an optimum trajectory. As the leader's trajectory is assumed to be known and fixed, the problem of optimisation of the follower becomes standard.

$$u_2^{S1} \text{ minimises } H_2(.),\ \forall u_2 \in U_2 \text{ with } \begin{cases} \overset{\circ}{x} = \dfrac{\partial H_2}{\partial p}, & x(o) = x^\circ \\ \overset{\circ}{p} = -\dfrac{\partial H_2}{\partial x}, & p(T) = \dfrac{\partial K_2}{\partial x(T)} \end{cases} \tag{1.61}$$

with

$$H_2\ (x,u_1,u_2,p) = L_2\ (x,u_1,u_2) + p^T \cdot f\ (x,u_1,u_2) \tag{1.62}$$

On the other hand, the conditions of optimality of the leader must include the knowledge that the latter has from the reaction of the follower to its trajectories. Thus, for the leader equations (1.61) are the constraints and :

u_i^{S1} minimises $H_1(.)$, $\forall\ u_i \in U_i$, $i = 1,2$ with

$$\overset{\circ}{x} = \frac{\partial H_1}{\lambda_1},\quad \overset{\circ}{\lambda}_1 = -\frac{\partial H_1}{\partial x},\quad x(o) = x^\circ \tag{1.63}$$

$$\lambda_1(T) = \frac{\partial K_1}{\partial x(T)} - \lambda^T{}_2(T) \cdot \frac{\partial^2 K_2}{\partial x(T)^2}$$

$$\overset{\circ}{\lambda}_2 = -\frac{\partial H_1}{\partial p},\quad \lambda_2(o) = 0$$

where H_1 is defined by

$$H_1\ (x,u_1,u_2,p,\lambda_1,\lambda_2,\beta_2) = L_1(x,u_1,u_2) + \lambda^T{}_1 . f\ (x,u_1,u_2) =$$

$$+ \lambda^T{}_2 \cdot \left(-\frac{\partial H_2}{\partial x}\right) + \beta^T{}_2 \cdot \left(\frac{\partial H_2}{\partial u_2}\right) \tag{1.64}$$

Explicit solutions for the linear-quadratic problem by resolution of Riccati equations, are given by [SIM - 73 a] .

Closed loop optimisation

Although the conditions of optimality for the Stackelberg trajectory with a fixed interval and initial known conditions x^o were obtained in the open loop case from standard variational techniques, the same is not true for the closed loop case, which leads us on to a non-standard problem of control [CRU - 78 , PAP - 79] .
When the strategies at each moment of time t are functions of the state x(t), ie. $u_i = w_i\ (x(t))$, i = 1...N, we have a non-standard control problem because $\partial u/\partial x$ (see eq. 1.46) appears in the conditions necessary to the follower. As these conditions are considered by the leader as differential equations, the presence of $\partial u/\partial x$ prevents direct use of variational techniques to resolve the problem of the leader.
Therefore, taking into account the difficulties of characterisation of the necessary conditions in closed loop until recently [PAP - 79] , simplifications have been sought, for example from linear laws of control and random initial conditions [MED - 77] .
Note : A numerical disadvantage of the Stackelberg strategy, even in the open loop case, is the non-validity of the principle of optimality of dynamic programming for the problem of the leader [SIM - 73 b] . Therefore a sub-optimum strategy has been developed, based on dynamic programming (see also [CRU - 78] and recently [PAP - 79] have shown that a condition of application of the leader is, in fact, a team decision problem in which all the deciders optimise a single objective function.

1.8.2.4. Coalitions

If the attitude of competition of the deciders seems to be natural, it is nevertheless not rational. The notion of non-co-operative equilibrium can be criticised from the point of view of decisions at group level.
Usually, by co-operating, the N deciders can improve the situation of each one among them in relation to the non-co-operative equilibirum.
Thus, if $u = (u_1, \dots, u_N)$ is a non-co-operative point of equilibrium, another set of decisions may exist

$$u^+ = (u_1^+, \dots, u_N^+) \text{ such that } u_i^+ \in \mathcal{U}_i \text{ and } J_i\ (u^+) \leq J_i\ (u),$$

$\forall i = 1 \dots N$. It is clear that the N deciders may agree to choose u^+. Obviously the simplest situation is obtained when $J_1 = J_2 = \dots = J_N = J$, which signifies that the N deciders collaborate completely to optimise one single criterion. This introduces the notion of team studied, in the case of dynamic problems by [HO - 72] and [CHU - 72].
Thus, as [BEN 72] we can conceive, between non-co-operation and teamwork, that some sub-groups form a coalition against the other deciders.
Let us take, for example, the set $u = (u_1, \dots, u_N) = (u_C, U_{\bar{C}})$. in which C is a sub-set of the indices (I...N). The deciders

$i \in C$ form a <u>coalition</u> ($\bar{C}$ is the complement to C).
Let us suppose that there exists u_C^+ such that $(u_C^+, u_{\bar{C}})$ is compatible with the limitations and such that :

$$J_i\ (u_C^+,\ u_{\bar{C}}) \leq J_i\ (u_C,\ u_{\bar{C}})\ .\quad \forall\ i \in C \qquad (1.65)$$

It is clear that the coalition C will prefer $(u_C^+ , u_{\bar{C}})$ to $(u_C, u_{\bar{C}})$. If it can, it will try to oppose the decision $(u_C, u_{\bar{C}})$. We can say that it <u>blocks</u> $(u_C, u_{\bar{C}})$. This leads us on to the following definition :

<u>Definition</u> : A set $u = (u_1, \ldots u_N)$ cannot be "blocked" by a coalition C of size K if, for any coalition C with K members, the following property is verified :

$(u_C^+, u_{\bar{C}})$ verifying the constraints and

$$\text{if } J_i\ (u_C^+ , u_{\bar{C}}) \leq J_i\ (u_C , u_{\bar{C}}) \quad \forall i \in C \qquad (1.66)$$

$$\Rightarrow J_i\ (u_C^+ , u_{\bar{C}}) = J_i\ (u_C , u_{\bar{C}})$$

ie. there is not set $(u_C^+ , u_{\bar{C}})$ which improves the situation of all the deciders $i \in C$ at the same time, in relation to $(u_C , u_{\bar{C}})$. If it did exist, coalition C would block $(u_C, u_{\bar{C}})$.
Thus we can see that a Nash point is a set of decisions which cannot be blocked by a coalition of size 1. In the same way, a decision will be as co-operative at a group level as K will be large and as the coalition will protect a greater number of deciders.

1.8.2.5. Pareto point

According to the above definitions it appears that if the N deciders come to an agreement and form a coalition of size N, they will obtain a set of decisions which cannot be blocked by anyone. This defines <u>a co-operative equilibrium</u> known as <u>Pareto point</u>, or <u>non-inferior</u> solution, or again <u>effective</u> solution [STA - 79].

<u>Definition</u> :
A point $u = (u_1, \ldots, u_N)$ is a Pareto point if there is no other point $u^+ = (u_1^+, \ldots, u_N^+)$, compatible with the constraints, such that :

$$J_i\ (u^+) \leq J_i\ (u) \quad \forall i\ ,\ i = 1 \ldots N \qquad (1.67)$$

This is therefore a point such that it is not possible to find another set of decisions compatible with the constraints and improving the situation of <u>everyone</u> at the same time. This concept leads to a compromise solution between the deciders and is the equivalent of the team problem in which the N deciders are

in agreement as to the choice of a common criterion [BEN - 72]. We can also note that this co-operative equilibrium may be associated with the problem of a single decider with several criteria to optimise. This problem is equivalent, from the mathematical point of view, to the problem of several deciders, each with a single criterion.
If we take up the example in figure 14 we can compare the different equilibriums and notice that the Pareto points, defined on the curve AB, improve the situation of the two deciders at the same time, in comparison with the non-co-operative equilibriums.

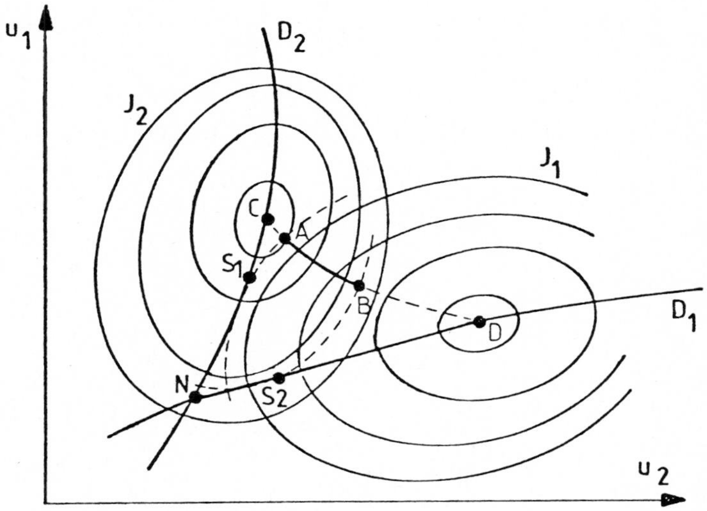

Figure 14 - A problem with two deciders. The Stackelberg, Nash and Pareto solutions.

We can also observe that all the points on the curve CD are Pareto points, ie. they are points where it is no longer possible to improve the situation of a decider without harming the situation of the other. Nevertheless, there are points on this curve, apart from AB, where the situation of one of the deciders is worse in relation to the non-co-operative equilibrium. Thus we can differentiate the set of non-inferior solutions CD, where the deciders co-operate, from the set of non-inferior solutions AB, where the deciders co-operate but also establish a negotiation.
Whatever the case may be, we can observe that the problems with continuous differentiable criteria have an <u>infinite</u> number of co-operative solutions.
Thus, a negotiation between deciders must define a final point of co-operation, or a sub-set of the curve of non-inferior points, which will be called <u>preferred solution.</u>

How to obtain a Pareto point

According to the above definition, we can see that a Pareto point $u^P = (u_1^P, \dots u_N^P)$ is the solution to the set of the N following problems of optimisation [BEN - 72] :

$$\text{Problem no. } i \quad \left| \begin{array}{l} \min J_i(u) \\ u_j \in U_j \quad j = 1 \dots N \\ J_j(u) \leq J_j(u^P) \ \forall j \neq i \end{array} \right. \tag{1.68}$$

In fact, if u meets the limitations of problem (1.68) we must then have

$$J_i(u) \geq J_i(u^P)$$

for otherwise we would have $J_i(u) < J_i(u^P)$, which would contradict the fact that u^P is a Pareto point. Reciprocally, if u^P is the solution to the N problems (1.68) u^P is a Pareto point. Indeed, if not, for every u compatible with the limitations and verifying $J_i(u) \geq J_i(u^P)$ there would exist i_o such that :

$$J_{i_o}(u) < J_{i_o}(u^P)$$

but then u^P would not be the solution to problem (1.68) no. i_o.

However, characterisation of the Pareto points enables us, in fact, to return to a single problem of optimisation. For this, it is sufficient to introduce the notion of ideal or utopic solution point.

It is evident that, for each isolated decider, the best possible solution would be :

$$\bar{J}_i = \min J_i(u) \qquad i = 1 \dots N$$
$$u \in U_1 \times U_2 \times \dots \times U_N \tag{1.69}$$

and the point $\bar{J} = (\bar{J}_1, \dots, \bar{J}_N)$ can be defined as the ideal solution point. As this point is normally non admissible, taking into account the limitations of the overall problem, the deciders will try to draw as near to it as possible, in the negotiation process. Thus, to give a geometric interpretation to this process, we can introduce the notion of distance from a point to the utopic point, given by the following metrics family L_P :

$$L_P = \left[\sum_{i=1}^{N} (J_i - \bar{J}_i)^P \right]^{1/P} \qquad 1 \leq P < \infty \tag{1.70}$$

It is not difficult to show that, if $u^P = (u_1^P, \dots, u_N^P)$ is the solution of $\min_u L_P$, for a p given ($1 \leq P < \infty$), then u^P is a Pareto point. Indeed, let us suppose that u^P is the solution of $\min_u L_P$.

As $J_i \geq \bar{J}_i$, $\forall$ i, L_P is strictly increasing and u^P also minimises

$$\bar{L}_P = \sum_{i=1}^{N} (J_i - \bar{J}_i)^P \tag{1.71}$$

by providing :

$$\sum_{i=1}^{N} (J_i(u) - \bar{J}_i)^P \geq \sum_{i=1}^{N} (J_i(u^P) - \bar{J}_i)^P \tag{1.72}$$

$\forall$ u verifying the constraints.

Let there be u satisfying the constraints and, moreover

$$J_j(u) \leq J_j(u^P) \qquad \forall\, j \neq i \tag{1.73}$$

and as $\quad J_j \geq \bar{J}_j$, $\forall$ j,

$$(J_j(u) - \bar{J}_j)^P \leq (J_j(u^P) - \bar{J}_j)^P \qquad \forall j \neq i \tag{1.74}$$

From (1.72) there comes the immediate result :

$$(J_i(u^P) - \bar{J}_i)^P \leq (J_i(u) - \bar{J}_i)^P + \sum_{\substack{j=1\\ j\neq i}}^{N} \left[(J_j(u)-\bar{J}_j)^P - (J_j(u^P) - \bar{J}_j)^P \right] \tag{1.75}$$

and according to (36)

$$(J_i(u^P) - \bar{J}_i)^P \leq (J_i(u) - \bar{J}_i)^P \tag{1.76}$$

and as $\quad J_i \geq \bar{J}_i$, $\forall_i$, we have :

$$J_i(u^P) \leq J_i(u) \tag{1.77}$$

which proves that u^P is the solution to problem (1.68) no. i. As i is arbitrary, u^P is consequently a Pareto point. We observe that this requires no hypothesis on the functions J_i.

Examples of metrics :

$$P = 1 \quad \left[\min_u \bar{L}_P = \min_u \sum_{i=1}^{N} (J_i - \bar{J}_i) \Longrightarrow \min_u \sum_{i=1}^{N} J_i(u) \right.$$

(linear combination of criteria)

$$P = 2 \quad \left[\begin{array}{l} \min \bar{L}_p = \min\limits_u \sum\limits_{i=1}^{N} (J_i - \bar{J}_i)^2 \\ \\ \text{(Euclidean norm)} \end{array} \right.$$

If we now do :

$$\min_u \bar{L}_p = \min_u \sum_{i=1}^{N} (J_i - \bar{J}_i)^{p-1} \cdot (J_i - \bar{J}_i)$$

$$= \min_u \sum_{i=1}^{N} W_i^{p-1} \cdot (J_i - \bar{J}_i) \tag{1.78}$$

we note that the deviations in relation to $\bar{J}$ have weighting coefficients which are function of p, and the greatest deviations will have the greatest weight.
Thus for $p = \infty$, we arrive at :

$$\min_u \bar{L}_\infty = \min_u \max_i (J_i - \bar{J}_i) \tag{1.79}$$

Let us take the following example to illustrate these ideas.

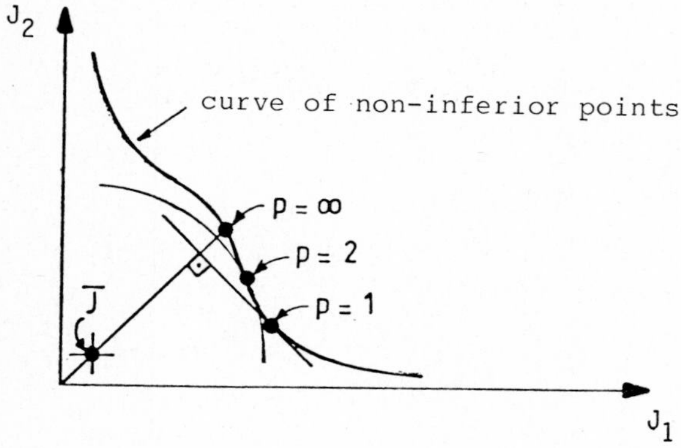

Figure 15 - The different metrics and the solutions on the curve of non-inferior points.

In Figure 15, we represent the curve of non-inferior points in the space of the criteria, with J_i, i = 1,2 continuous and non-convex. The different solutions obtained with the different metrics show us that if the deciders have equal importance for all the criteria, the choice of a particular p already implies an implicit allocation of the weights to the deviations.
Nevertheless, the choice of a p may be in direct opposition

with the classification of the criteria made by the deciders. Thus, they can choose a set of weights $a = (a_1, \dots, a_N)$, $a_i \geq 0$, such that :

$$\bar{L}_p(a,u) = \sum_{i=1}^{N} a_i . (J_i(u) - \bar{J}_i)^{\,p} = \sum_{i=1}^{N} a_i . W_i^{p-1} . (J_i(u) - \bar{J}_i) \qquad (1.80)$$

where the weights give the relative importance among the criteria. We note that an appropriate variation of the weights ($\sum a_i = 1$, $a_i \geq 0$, we can always return to this case) and of the p's can provide the whole set of non-inferior points. From the mathematical point of view, co-operative equilibrium does not, therefore, present a new problem, since everything comes back to a problem of optimisation with one single criterion. However, this presupposes that the N deciders have come to an agreement on the coefficients of ponderation a_i of their influence and on the metric P.

Conditions of optimality of a Pareto point

According to the aforementioned considerations, we observe that when we consider a dynamic problem like (1.39), obtaining the conditions of optimality of a Pareto point takes us back to the problem of optimal control with a single criterion. According to (1.68) we also note that this problem of optimum control is subject to isoperimetric limitations. Thus, according to the necessary conditions of optimality for this type of problem [SCH - 76], we have the following theorem [SCH - 74] :

Theorem : If $u_N^P = (u_1^P, \dots, u_N^P)$ is a Pareto point, then there exists $a \in R^N$, $a \geq 0$, a continuous vector $p \in R^n$, with $(p(t), a) \neq 0$ in $[0,T]$, a vector $\gamma \in R^n$ and a function $H(.) : R^n \times R^m \times R^n \times R^N \rightarrow R$, where :

$$H(x, u, p, a) = a^T . L(x,u) + p^T . f(x,u) \qquad (1.81)$$

with $L(x,u) = \left[L_1(x,u), \dots, L_N(x,u) \right]$, such that :

a) $p^P(.)$ is the solution of :

$$\overset{\circ}{p}(t) = -\frac{\partial H(.)}{\partial x}, \quad p(T) = \sum_{i=1}^{N} \gamma_i . \frac{\partial K_i}{\partial x(T)}$$

and $x^P(t)$ is the solution of :

$$\overset{\circ}{x}(t) = \frac{\partial H(.)}{\partial p}, \quad x(o) = x^{\circ}$$

b) $H(x^P(t), u^P(t), p^P(t), a) \leq H(x^P(t), u(t), p^P(t), a)$

$$\forall u \in U$$

c) $H(x^P(t), u^P(t), p^P(t), a) = \text{constant}.$

1.8.3. Multi-criteria methods of optimisation [HWA - 80]

For decision-making in complex systems, the terms multi-objectives, multi-attributes, multi-criteria or multi-dimensional are used indifferently to describe the decision situations, ie. the process of selection of one action from among all the possible alternatives. If the term "decision with several criteria" is normally accepted to represent these methodologies as a whole, we can nevertheless define two categories of methods :
- methods of decision with multiple attributes. In this case we select an alternative from an explicit, finite number (usually small) of alternatives. These problems are considered rather as problems of choice and are reviewed by [MAC - 73 , COL - 76]
- methods of decision with multiple objectives. In this case we select an alternative (or a sub-set) from an infinite number of alternatives, implicitly defined by the structural constraints of the system. These methods are associated more with project or planning problems.

The concept of non-inferior solution or of Pareto is imposed by the structure of the system, whereas the concept of preferred solution is introduced by the decider, who must provide as it were information on its preferences in the decision-making process. Thus, we find that decision-making in a complex system implies the action of two agents : the decider and the analyst. The latter's task is to make the preliminary analysis of the problem, considering the technical, physical and economical limitations so as to be able to reduce the initial set of alternatives by elimination of inferior solutions (with an optimisation process, for example). It is on the resulting set that the decider must act to select the final solution or supply information on its preferences. The interaction of the decider and the analyst, the type of information given, play a fundamental part in the classification of methods. According to the articulation of the information on the preferences we can classify these principal methods from the following table :

Decision with multiple objectives	1. no articulation of the information on preferences	
	2. a priori articulation of the information on preferences	2.1 Cardinal information
		2.2 Mixed information cardinal and ordinal
	3. progressive articulation of the information on preferences (interactive methods	3.1 Explicit rates of substitution
		3.2 Implicit rates of substitution
	4. a posteriori articulation of the information on preferences	4.1 Implicit rates of substitution

1.8.3.1. Methods with no articulation of preferences

These methods do not require information on the preferences of the decider because it "knows" the overall criterion which is suitable, for example by fixing parameters p and a of (1.80) [SAL - 72 , HUA - 72] . The problem reverts to a mono-criterion problem.
The advantage of these methods is the non-interference of the analyst in the action of the decider, a situation which is always preferable for the latter in a decision process : the analyst resolves the problem defined by the decider. Moreover, this problem is resolved only once. The obvious disadvantage is the often unrealistic hypothesis of the knowledge of the overall criterion, and the decider risks having to be content with a final unsatisfactory solution.

1.8.3.2. Methods with a priori articulation of preferences

The difference of this class of method in comparison with the previous class is that here the analyst resolves the overall problem after obtaining from the decider the structure of preferences from a process of comparison and evaluation of the preferences between different alternatives, the systematic use of which gives an approximation of the global criterion. If the set of alternatives of the overall problem is explicit and finite, we revert to the multi-attributes methods [MAC - 73] .
We can distinguish two types of methods in this category :
- those which use only cardinal information such as methods obtaining the function [FIS - 70 , KEE - 76] and curves of indifference [BRI - 66 ; MAC - 75] where the decider must judge specific levels of preferences or rates of substitution between the criteria;
- those which use mixed information, cardinal and ordinal, ie. the decider also classifies the criteria by order of importance (lexicographic methods [FIS - 74] , the programming methods by objectives [CHAR - 77] , the method of obtaining goals [GEM - 74]) .
The greatest advantage of these methods is that if the approximation of the overall criterion has been well made and well used, the solution of the problem will inevitably be the decider's preferred solution. They are well adapted to the multi-attributes problems with some qualitative and independent criteria [FIS - 75] .
The disadvantage always arises from the necessity to obtain the global criterion in a sure manner, an almost impossible task, principally in multi-objective project problems. We observe that the questions of evaluation of the preferences are developed in the space of the criteria, without connection with the internal structure of the system since the information is aggregated in the attributes. Thus it is difficult to ensure the admissibility and the non-inferiority of the alternatives tested, which may lead the decider into false judgements.

1.8.3.3. Methods with progressive articulation of preferences

This class of methods turns on the progressive definition of the decider's preferences by exploration of the space of criteria. This progressive definition is obtained with a decider-analyst dialogue at each stage of calculation, where the decider is questioned on substitution rates (or its preferences) based on the solution available (or the set of solutions available) with a view to determining a new solution. These methods recognise that the decider is incapable of defining a priori its global preferences because of the complexity of the problem, but assume that it is capable of giving local information of its preferences on a particular solution. As the process develops the decider indicates its preferences, but also increases its knowledge of the problem and may modify its aspirations as a consequence.

We can distinguish two types of methods in this category :

- Those which require explicit information of the substitution rates desired by the decider between the levels reached by the criteria at each stage: This explicit information is given by direct evaluation [DYE - 72 , GEO - 72 , ZIO - 76] or by comparison [HAI - 75 , CHAN - 77] .
- Those which accept implicit information of the substitution rates desired by the decider, which are obtained by evaluation of the acceptability of the levels reached by the criteria at each stage [BEN - 71 , MON - 76 , ZEL - 74 , STE - 77] .

The principal advantages of these methods are :

- the a priori definition of the global structure of the preferences of the decider is not necessary. Only local information on the preferences is necessary and there are fewer restrictive hypotheses in comparison with the aforementioned methods.
- It is an apprenticeship process for the decider, who learns of the behaviour of the system. As the decider is part of the process, the final solution has more chance of being accepted as the preferred one.
- The decider does not need to take account of the internal structure of the system, as this is done by the analyst. Thus, resolution of the problem is accomplished at two distinct, interactive levels : the decision and definition of preferences level, associated with the decider, and the analytical level at which the structural limitations of the problem are considered.

Let us note a further advantage if the method accepts implicit information of the substitution rates, firstly because the decider normally prefers to indicate levels of acceptability of the criteria and not these rates directly. Moreover, the rates are not valid only for a particular value of the criteria, but usually over a wider set of values which are not presented to (or considered by) the decider in the methods with explicit valuation. This may lead to erroneous selection of the rates of substitution.

The principal disadvantages are associated with the dependence of the final solution on the accuracy of estimations of the local preferences of the decider. Moreover, several methods do not guarantee the preferred solution is obtained after a finite number of interactive stages. This may also represent a very great effort of calculation. Finally, the decider is called on to make a greater effort than in the preceding methods.

1.8.3.4. Methods with a posteriori articulation of preferences

This class of method is known as generator of non-inferior solutions, since they determine a sub-set of the complete set of non-inferior solutions without the decider's participation. This work is centred on the analyst, who presents the resulting sub-set to the decider, who chooses the preferred solution from some implicit rates of substitution between the criteria, based on some unidentified criterion. Whatever the case, this information is obtained after use of the method, and after generation of the sub-set of the non-inferior solutions.
So these methods do not use information on the decider's function of preference. They are characterised by two stages : transformation of the multi-criteria problem into an equivalent mono-criterion problem and resolution of the latter, the solution to which is non-inferior for the initial problem under certain hypotheses. The two principal approaches are : the parametric method [GAL - 72, ZEL - 74] and the method of the ε -limitation [HAI - 75]. The great majority of methods specialised in multi-criteria linear programming also belong to this class [EVA - 73, ISE - 77, YU - 76].
It is interesting to return to the definitions of non-inferior solution or Pareto point, represented by the resolution of problem (1.68) or by the minimisation of metric (1.80) and to note that :
- the parametric method uses (1.80) with parameter p fixed at p=1, which brings us to the minimisation of the balanced sum of the criteria (equivalent to the hypothesis of linear, additive utility for the decider). In this case the weights a do not reflect the relative importance between the criteria, but are only parameters which are varied to obtain non-inferior solutions. If the admissible region of the criteria is convex, the systematic variation of the a supplies the complete set of non-inferior solution. If admissible region is not convex, there are certain non-inferior solutions which cannot be reached with any arbitrary a [HAI - 75].
- the method of the ε-constraint exploits the definition of Pareto point from problem (1.68), ie. one of the criteria is considered in the optimisation (the reference criterion) and the others are introduced as limitations of inequality to be satisfied. These limitations are restricted by the ε levels tolerated for each criterion, and the parametric variation of ε determines the influence on the criterion of reference and changes the associated Pareto point. This method does not suffer from the problems of the previous method and can achieve all the non-inferior solutions.
One disadvantage which severely limits the use of these methods is the compromise necessary between obtaining a sufficient number representative of the set of non-inferior solutions (which represents a very great, if not impossible, effort of calculation in complex systems) and the obtaining of a number of non-inferior solutions which is sufficiently small for the decider to be able to choose the preferred solution. Thus, these methods are usually incorporated as a part of the decision process in other interactive methods [ZIO - 76 , HAI - 75 , CHAN - 77].

Note

According to the foregoing considerations, methods with progressive articulation of the information of preferences are better adaptated to the study of complex systems, firstly because the introduction of two interactive levels divides the work and the structural limitations of the system are considered independently of the decider, complex though they may be. The decider can therefore concentrate on the decision process, with more limited and usable information as a consequence. On the other hand, the information transmitted to the decider is always admissible for the system under consideration. Secondly, these methods do not require a priori definitions of the global structure of the decider's preferences and the local estimations, being inserted in an apprenticeship process, may provide a solution with more chance of being accepted as the preferred one.

1.8.4. Pareto equilibrium and complex systems

Here we are interested in the case of a single decider (one single control structure) having to take into account, on a complex system, different criteria.
Therefore the concepts of co-operative equilibrium and Pareto point apply.

Let the problem be : $\min_{x \in X} \begin{vmatrix} f_1(x) \\ \vdots \\ f_n(x) \end{vmatrix}$

where $f_i(x)$, $x \in R^m$ are known, differentiable function criteria. If the decider knows its preferred function, its utility function $u\ (f_1(x) \ldots f_n(x))$, then the problem is written as :

$$\min_{x \in X \subset R^m} u(f_1(x)\ \ldots\ f_n(x)\ )$$

a mono-criterion problem to which can be applied standard optimisation techniques, for example, of gradient projection with the following stages.

a) choice of an initial admissible condition x
b) test of the optimality of solution x. If this is not optimal, evaluate in x the direction which permets a decrease of the function u locally. This direction is usually the gradient of the criterion function in x.
c) If the gradient is an admissible direction, ie. if it makes it possible to calculate a new $x \in X$ which improves the criterion then move on to d). If not, define an admissible direction by projection of the gradient on the set of constraints.
d) Take, in this admissible direction, the step which minimises the criterion.
e) Return to b).

Unfortunately, for the problem posed (multi-criterion optimisation of a complex system), the above procedure is not directly applicable for at least two reasons :

- the decider does not explicitly know u, which makes stages b) and d) impossible;
- the projection stage may be a difficult task for complex problems.

1.8.4.1. Difficulties linked with the multi-criterion aspect

There are two possibilities for overcoming these difficulties :
- the decider does not explicitly know u (.), but can express local preferences. In other words, he knows u (.) approximately, at all admissible points, and can estimate the gradient of u in x. These estimations have to be repeated at each iteration until there is convergence of the procedure.

If the decider is coherent, the final solution will be the preferred non-inferior solution [GEO - 72].
- the decider estimates a utility function u (.) (often a linear function of the functions $f_i(.)$, i = 1 to n) and tries to modify u (.) in terms of the result obtained and of local knowledge. This step, which must be used cautiously [HAI - 75], accelerates the resolution of the problem.

1.8.4.2. Adaptation of the gradient projection to large scale multi-criteria problems

The gradient necessary to this algorithm is expressed as follows :

$$\nabla_x u(f_1(x) \ldots f_n(x)) = \sum_{i=1}^{n} \frac{\partial u}{\partial f_i} \cdot \nabla_x f_i(x)$$

where : $\frac{\partial u}{\partial f_i}$ is the partial derivative of U in relation to f_i

$\nabla_x f_i(x)$ is the gradient of $f_i(x)$ in relation to x.

The important information of this gradient in the algorithm is not its amplitude but its non-variant direction in any multiplication made by a positive quantity :

$$\tilde{\nabla}_x u(f_1(x) \ldots f_n(x)) = \sum_{i=1}^{n} \frac{\partial u/\partial f_i}{\partial u/\partial f_j} \nabla_x f_i(x) = \sum_{i=1}^{n} W_{ij} \nabla_x f_i(x)$$

In this expression, $\partial u/\partial f_j > 0$ is the partial derivative of the utility function in relation to the criterion j which plays a role of reference; the quantity W_{ij}, well known in the field of techniques of aids to decision-making, is a marginal substitution rate between criteria i and j.

Let us now suppose that we wish to apply a gradient technique to the following problem :

$$\min_{x \in X} u \; (f_1(x), \ldots f_n(x))$$

subject to $x \in X = \{ x \,/\, Ax = b, \underline{x} \leqslant x \leqslant \bar{x} \}$

where equality (of model) and inequality (of limit) constraints

appear. If $x \in R^m$, m large, the gradient projection technique may become cumbersome.
So then, to find an admissible direction S such that :

$$u(f_1(x-\alpha S),\dots f_n(x-\alpha S)) < u(f_1(x),\dots f_n(x))$$

(α : step in direction S)

we define :

$$I(x) = \{i/x_i = \underline{x}_i, \quad i = 1 \dots m\}$$
$$J(x) = \{j/x_j = \bar{x}_j, \quad j = 1 \dots m\}$$

The search for S then returns to the following quadratic problem :

$$\min_{S \in S(x)} \{1/2 \|\tilde{\nabla}_x u - S\|^2 \ / \ A S = 0\}$$

with $$S(x) = \{S/S_i < 0,\ i \in I(x);\ S_j > 0\ ,\ j \in J(x)\}.$$

Knowing $\tilde{\nabla}_x u$, this quadratic problem is very easy to solve even for very large dimensions for which decomposition techniques may be introduced (dual methods in particular).
Generalisation in the dynamic case is direct and the study of convergence was carried out in [BAP - 80].

1.8.4.3. Estimation of the W_{ij} [BAP - 80]

The difficulty in this approach is still the failure to know U and, consequently, W_{ij}.
However, under hypotheses of local knowledge, the decider will be capable of estimating W_{ij} and of improving his estimation at each step.
For this, fuzzy interactive methodologies can contribute a lot :

a) the decider expresses its preferences numerically, but approximately by proposing W_{ij} rates between criteria i and j as follows :
"W_{ij} must be in the vicinity of $\bar{W}_{ij}$".
b) the substitution rates are given by the decider in the form :
"W_{ij} must be a approximate real number between W_{ij}^+, W_{ij}^-, with an average $\bar{W}_{ij}$".
c) the decider expresses not quantitative but qualitative linguistic preferences.

Notes : Associated with the algorithm of the gradient projection these 3 procedures have been developed and applied to a certain number of concrete problems : management of hydroelectric systems, joint identification-optimisation in socio-economic systems represented by econometric models [BAP - 80]

Conclusion of chapter I

In this chapter we have presented a broad panorama of the problems linked with complex systems and large scale systems, together with associated methods of analysis and control. Later in the book, some of these methods will be taken up in detail; others will only appear in this introductory chapter, which will have presented a sample of the problems and methods. It is not possible to broad the whole question at length in one single book.

REFERENCES

AOK - 68 AOKI M., "Control of Large Scale Dynamic systems by aggregation", IEEE Trans. Aut. Control, AC-13, June 1978.

BAP - 80 BAPTISTELLA L.F.B., "Contribution à l'optimisation multicritère de systèmes dynamiques", Ph. D., University of Toulouse, 1980.

BAS - 73 BASAR T., "On the relative leadership property of stackelberg strategies", Journal of optimization, theory and applications, vol. 11, pp. 655-661, 1973.

BEN - 71 BENAYOUN R., DE MONGOLFIER J., TERNY J. and LARICHEV O., "Linear programming with multiple objective functions : step method (STEM)", Mathematical Programming, vol. 1, pp. 366-375, 1971.

BEN - 72 BENSOUSSAN A., J.L. LIONS and R. TEMAN, "Sur les méthodes de décomposition, de décentralisation, de coordination et applications", Cahier IRIA, n° 11, tome 2, 1972.

BEN - 81 BENNIS O., P. CHEMOUIL, A. TITLI, "Systèmes dynamiques à échelles de temps multiples : perturbations singulières", Technical Report, LAAS - DCSI, 81.I.48, Novembre 1981.

BER - 79 BERTRAND P., J.M. SIRET, G. MICHAILESCO, "On the use of aggregation techniques", in M.G. SINGH and A. TITLI (Editors). Handbook of Large Scale Engineering applications, North Holland, Amsterdam, 1979.

BLA - 73 BLAQUIERE A., (Ed.), Topics in differential games, North Holland, Amsterdam. 1973.

BRI - 66 BRISKIN L.E., "A method of unifying multiple objective functions", Management Science, vol. 12, n° 10, 1966

CAL - 80 CALVET J.L., A. TITLI, "Hierarchical optimization and control of large scale system with dynamical interconnection system", 2nd IFAC Symposium on Large Scale System Theory and Applications, Toulouse, 26-28 June, 1980.

CHAN - 77 CHANKONG V., "Multiobjective decision-making analysis : the interactive surrogate worth trade-off method", Ph. D. Thesis, Case Western Reserve University, 1977.

CHAR - 77 CHARNES A. and W.W. COOPER, "Goal programming and multiple objective optimization : part I", European Journal on Operations Research, vol. 1, pp. 39-54, 1977.

CHE - 78 CHEMOUIL P., "Analyse et commande des systèmes dynamiques à plusieurs échelles de temps". Docteur Ingénieur Thesis, University of Nantes 1978.

CHE - 80 CHEMOUIL P., A.M. WHADAN, "Output feedback control of systems with slow and fast modes", Large Scale Systems Journal, vol. 1, N° 4, 1980.

CHO - 76 CHOW J.H., P.V. KOKOTOVIC, "A decomposition of near optimum regulators for systems with slow and fast modes", IEEE Trans. A.C., vol. 21, n° 5, 1976.

CHU - 72 CHU K., "Team decision theory and information structures in optimal control problems : part II", IEEE Trans. on Autom. Control, vol. AC-17, pp. 22-28, 1972.

CLE - 79 CLEMHOUT S. and H.Y. WAN Jr, "Interactive economic dynamics and differential games", Journal of Optimization Theory and Applications, vol. 27, pp. 7-30, 1979.

COL - 76 COLE J.D. and A.P. SAGE, "On assessment of utility and worth of multi-attributed consequences in large scale systems", Information Sciences, vol. 10, pp. 31-57, 1976.

CRU - 78 CRUZ J.B., Jr. ,"Leader follower strategies for multilevel systems", IEEE Trans. Aut. Control, vol. AC-23, pp. 244-255, 1978.

DYE - 72 DYER J.S., "Interactive goal programming, Management Science", Vol. 19, pp. 62-70, 1972.

EVA - 73 EVANS J.P. and R.E. STEVER, "A revised simplex method for linear multiple objective programs", Mathematical programming, vol. 5, pp. 54-72, 1973.

FIN - 69 FINDEISEN W., A. MANITIUS , J. PUTACZEWSKI, "Multilevel optimization and dynamic coordination of mass flows for a beet sugar plant", 4th IFAC World Congress Warsaw, 1969.

FIS - 70 FISHBURN P.C., "Utility theory for decision making", John Wiley, New York, 1970.

FIS - 74 FISHBURN P.C., "Lexicographic orders, utilities and decision rules : a survey", Management Science, vol. 20, pp. 1142-1471, 1974.

FIS - 75 FISHBURN P.C. and R.O. KEENEY, "Generalized utility independence and some iplications", Operations Research, vol. 23, pp. 928-940.

FOS - 81 FOSSARD A.J., J.F. MAGNI, "Modélisation et commande des systèmes à échelles de temps multiple (survey), Congrès AFCET, Automatique, Nantes 1981.

GAL - 72 GAL T. and J. NEDOMA, "Multiparametric linear programming", Management Science, vol. 18, pp. 406-421, 1972.

GEM - 74 GEMBICKI F., "Performance and sensitivity optimization : a vector index approach", Ph. D. Dissertation, Case Western Reserve University, 1974.

GEO - 72 GEOFFRION A.M., DYER J.S. and A. FEINBERG, "An interactive approach for multi-criterion optimization, with an application to the operation of an academic department", Management Science, vol. 19, pp. 357-368, 1972.

GRA - 77 GRATELOUP G., P. BLANDIN, A. TITLI, "La commande hiérarchisée : ses principes généraux et son application à un complexe de production de soufre", Colloque Modélisation et optimisation des procédés chimiques, Mai 1977, Toulouse.

HAI - 75 HAIMES Y.Y., W.A. HALL and H.T. FREEDMAN, "Multiobjective optimization in water resources systems : the surrogate worth trade-off method", Elsevier, New-York, 1975.

HO - 70 HO Y.C., "Differential games, dynamic optimization and generalized control theory", Journal of Optimization Theory and Applications, vol. 6, pp. 179-209, 1970

HO - 72 HO Y.C. and K. CHU, "Team decision theory and information structures in optimal control problems : part I", IEEE Transactions on Automatic Control, vol. AC-17, pp. 15-22, 1972.

HUA - 72 HUANG S.C., "Note on the mean square strategy for vector valued objective functions", Journal of Optimization Theory and Applications, vol. 9, pp. 364-366, 1972

HWA - 80 HWANG C.L., S.R. PAIDY, K. YOON and A.S.M. MASUD, "Mathematical programming with multiple objectives : a tutorial", Computers & Operations Research, vol. 7 pp. 5-31, 1980

ISA - 65 ISAACS R., "Differential games", John Wiley, New-York, 1965.

ISE - 77 ISERMANN H., "The enumeration of the set of all efficient solutions for a linear multiple objective program", Operations Research Quarterly, vol. 28, pp. 711-725, 1977.

KEE - 76 KEENEY R.L. and H. RAIFFA, "Decision with multiple objectives : preferences and value tradeoffs", John Wiley, New York, 1976.

KLI - 61 KLIMUSHEV A.I., N.N. KRASOVSKII, "Uniform asymptotic stability of systems of differential equations with small parameter in the derivative terms", Pritel. Math. Meth. vol. 25, n° 4, 1961

KOK - 75 KOKOTOVIC P.V., "A Riccati equation for block-diagonalisation of ill-conditionned systems", IEEE Trans. on Autom. Control. 20 n° 6, 1975

KOK - 80 KOKOTOVIC P.V., J. ALLEMONG, J.R. WINKELMANS, J.H. CHOW "Singular perturbation and iterative separation of time scales', Automatica, vol. 16, pp. 23-33, 1980

LEI - 78 LEITMANN G., "On generalized Stackelberg strategies", Journal of Optimization Theory and Applications, vol. 26, pp. 637-644, 1978.

MAC - 73 MAC-CRIMMON K.R., "An overview of multiple objective decision making", in J.L. COCHRANE and M. ZELENY (Ed.). Multiple criteria Decision Making, University of South Carolina Press, Columbia, pp. 18-44, 1973.

MAC - 75 MAC-CRIMMON K.R. and D.A. WEHRUNG, "Trade off analysis : indifference and preferred proportion", Proceedings of a Workshop on Decision Making with Multiple Conflicting Objectives, IIASA, Laxenburg, Austria, vol. 3, 1975.

MAG - 81 MAGNI J.F., "Analyse et commande des systèmes à plusieurs échelles de temps", Docteur-Ingénieur thesis, ENSAE, Toulouse, 1981.

MED - 77 MEDANIC J., "Closed loop Stackelberg strategies in linear quadratic problems", in Proceedings 1977 JACC, pp. 1324-1329, San Francisco, Ca.

MON - 76 MONARCHI D.E., J.E. WEBER and DUCKSTEIN L., "An interactive multiple objective decision making aid using nonlinear goal programmin", in M. ZELENY (Ed.), Multiple criteria decision making : Kyoto, Springer Verlag, New York, pp. 235-253, 1976.

NAS - 51 NASH J.F., "Non cooperative games", Annals of Mathematics, vol. 54, n° 2, 1951.

NEU - 28 VON NEUMANN J., "Zur theorie der Gesellsch aftsspiele", Mathematische Annalen, vol. 100, pp. 295-320, 1928.

OKU - 76 OKUGUCHI K., "Expectations and stability in oligopoly models", in Lecture Economics and Mathematical Systems, Mathematical Economics, vol. 138, Springer Verlag, New-York.

PAP - 79 PAPAVASSILOPOULOS G.P. and J.B. CRUZ Jr., "Non classical control problems and Stackelberg games", IEEE Transactions on Automatic Control, vol. AC-24, pp. 155-165, 1979.

PIN - 77 PINDYCK R.S., "Optimal economic stabilization policies under decentralized control and conflicting objectives", IEEE Transactions on Automatic Control, vol. AC-22, pp. 517-530, 1977.

POR - 75 PORTER B., "Dynamical characteristics of multivariable linear systems with slow and fast modes", Proc. of the 4th world congress on the theory of machines and mechanisms, New Castle, pp. 987-990, 1975.

RIC - 79a RICHETIN M., J. DUFOUR, "Analyse structurale des systèmes complexes par l'analyse des données", chap. 3, in A. TITLI et al "Analyse et commande des systèmes complexes", Cepadues Editions, Toulouse, 1979.

RIC - 79b RICHETIN M., M. MILGRAN, "Analyse structurale et partition des systèmes complexes par les graphes", chap. 5 in A. TITLI et al "Analyse et Commande des systèmes complexes", Cepadues Editions, Toulouse, 1979.

SAL - 72 SALUKVADZE M.E., "On optimization of control systems according to vector valued performance criteria", Proceedings of the 5th World Congress IFAC, Paris, section 40.5, 1972.

SCH - 74 SCHMITENDORF W.E. and G. LEITMANN, "A simple derivation of necessary conditions for Pareto optimality", IEEE Transactions on Automatic Control, vol. AC-19, pp. 601-602, 1974.

SCH - 76 SCHMITENDORF W.E., "Pontryagin's principle for problems with isoperimetric constraints and with inequality terminal constraints", Journal of Optimization Theory and Applications, vol. 18, pp. 561-567, 1976.

SIM - 73a SIMAAN M. and J.B. CRUZ Jr, "On the Stackelberg strategy in nonzero-sum games", Journal of Optimization, Theory and Applications, vol. 11, pp. 533-555, 1973.

SIM - 73b SIMAAN M. and J.B. CRUZ Jr, "Additional aspects of the Stackelberg strategy in nonzero-sum games", Journal of Optimization Theory and Applications, vol. 11, pp. 613-626, 1973.

SIR - 79 SIRET J.M., G. MICHAILESCO, P. BERTRAND, "On the use of aggregation techniques", pp. 20-32 in M.G. SINGH, A. TITLI, Handbook of Large Scale Systems Engineering Applications, North Holland, 1979.

STA - 52 VON STACKELBERG H., "The theory of the market economy", Oxford University Press, Oxford, 1952.

STA - 79 STADLER W., "A survey of multicriteria optimization or the vector maximum problem", Part I : 1776-1960. Journal of Optimization Theory and Applications, vol. 29, pp. 1-52, 1979

STA - 69 STARR A.W. and Y.C. HO, "Nonzero-sum differential games", Journal of Optimization, Theory and Applications, vol. 3, n° 3 and n° 4, 1969

STE - 77 STEUER R.E., "An interactive multiple objective linear programming procédure" in K. STARR and M. ZELENY (Ed.), Multiple criteria decision making, North Holland, Amsterdam, pp. 225-239, 1977.

TIT - 79a TITLI A., R. HURTEAU, "Decomposition spatio-temporelle des systèmes interconnectés; commande hiérarchisée, décentralisée, Application aux systèmes de puissance" Rapport technique E. Polytechnique de Montréal, EP 80. R.2., 1979.

TIT - 79b TITLI A., R. HURTEAU, "Partitioning and time decomposition for the control of interconnected systems". 2nd IFAC Symposium on optimization method, application, aspects. Varna, Bulgarie, 1979.

YU - 76 YU P.L., and M. ZELENY, "Linear multiparametric programming by multicriteria simplex method", Management Science, vol. 23, pp. 159-170, 1976.

ZEL - 74 ZELENY M., "Linear multiobjective programming", Springer-Verlag, New-Yor, 1974.

ZIO - 76 ZIONTS S. and J. WALLENIUS, "An interactive programming method for solving the multiple criteria problem" Management Science, vol. 22, pp. 652-663.

CHAPTER 2

THE CONCEPT OF PARTITION APPLIED TO CONTROL PROBLEMS

2.1 Introduction

This chapter is directed towards the control of large scale dynamic systems and in particular towards what is generally called the hierarchical approach to problems of control. This approach does not correspond to a new mathematical theory but rather to a combination of methods, the foundations of which were elaborated by Mesarovic et al. [MES - 70] and whose basic principle resides in the notion of decomposition-co-ordination. The main interest is linked with the fact that one is finding the solution of the global problem by calculations of reduced scale, generally by a process of successive approximations of localised problems.
May we point out that the framework of the numerous principal contributions to be found in the literature is convex Mathematical Programming and that most of the iterative methods proposed use the minimisation of generally quadratic functionals [BER - 76] [COH - 78][SIN - 78][FIN - 80].
In this chapter we have decided to present the decomposition methods of problems of control using "splitting methods" issued from numerical analysis. The aim is not to compare these two approaches, both of which have their use and which have, moreover, numerous interfaces, and it would certainly be deceptive to try to disassociate them in any fundamental way. For example, it is quite obvious that there is an equivalence between the resolution of the linear system :

$$Ax = b \quad x \in \mathbb{R}^n; \quad A \in \mathbb{R}n \times n ; \qquad A \text{ non singular}$$

and the minimisation of the quadratic functional

$$\min_x \langle Q\,(Ax-b) \; , \; Ax-b \rangle$$

where Q is a positive definite symmetrical matrix.

In making this choice we wanted to highlight this type of method which, moreover, does not present itself in terms of optimisation. This aspect may be of great interest in complex systems where we do not elaborate a policy of control based on optimisation strategies; the search for an admissible policy may already appear as a very satisfactory strategy. Here again let us be careful not to be too categorical since the approach by way of problems of artificial optimisations associated with other

objectives may prove efficient, see for example [SUR - 78]. Furthermore, we believe that the methods of decomposition and partitioning of numerical analysis which have been specially examined for the resolution of algebraic equations systems, particularly at the level of the study of convergence, of acceleration of convergence, of the partitions based on structural properties of the systems (irreducible matrices, matrices with diagonal dominance, triangular and symmetric matrices, M-matrices, S-matrices, block-diagonal matrices, cyclic matrices,...) may constitute a useful guide for the approach to problems of control.
Of course, as we shall point out when dealing with dynamic optimisation, the theoretical study of the convergence of iterative schemes of decomposition-co-ordination does not give such easily exploitable results. However, the extensions of methods of numerical analysis retain all their interest if only on a heuristic level.
Furthermore, transference to the dynamic case leads to problems which are specific to the problem of control such as "on-line" control, closed loop aspects, which it seems to us could be approached in quite an interesting way using decomposition methods.

2.2 Iterative methods of decomposition in numerical analysis

2.2.1. General definitions

These methods involve the double concept of approximation and iteration and are often designated by phrases such as "iterative process" or better "process of successive approximations". The approximation can be carried out by means of a Taylor series expansion (quasi-linearisation method, for example) or in the more general form of a polynomial expansion.
This leads to the construction of an approximating sequence : $\{x^{(k)}\}$: $x^{(k+1)} = T[x^{(k)}]$ (where k represents the index of iteration) which converges towards a fixed point of the map T, ie. towards a solution of $x^* = T[x^*]$.

We write down :

$$\lim_{k \to \infty} x^{(k)} = x^*$$

or in terms of the Euclidean norm :

$$\|x^{(k)} - x^*\| \to 0 \quad \text{when } k \to \infty.$$

We shall limit ourselves here to linear mappings in order to recall some elementary definitions :

Definition 1 : Stationary linear methods
These apply the iterative scheme :

$$x^{(k+1)} = T\,x^{(k)} + t; \quad k = 0,1,\dots ; \quad x^{(0)} = x_0 \text{ given}$$

We shall call T the associated iteration matrix.

Definition 2 : Non-stationary linear methods

$$x^{(k+1)} = T_{k+1}\, x^{(k)} + t_{k+1}; \quad k = 0,1,\ldots ;$$

$$x^{(0)} = x_0 \text{ given}$$

Definition 3 : Spectral radius

The spectral radius of a matrix T n x n with eigen values $\lambda_1,\ldots,\lambda_n$ is defined as being the real number $\geqslant 0$; ρ such that :

$$\rho(T) = \max_i |\lambda_i|$$

Note : It is generally difficult to calculate and determine accurately the spectral radius of a matrix and one is then led to finding upper limits [VAR - 62] .

For example, for any complex matrix T_{nxn} we have :

$$\rho(T) \leq \max_i \sum_{j=1}^{n} |t_{ij}|$$

or again :

$$\rho(T) \leq \max_j \sum_{i=1}^{n} |t_{ij}|$$

Definition 4 : Spectral norm

The spectral norm of the matrix T is a matrix norm defined using the Euclidean norm $\| x \| = (\bar{x}' x)^{1/2}$ as follows :

$$\| T \| = \sup_{x \neq o} \frac{\| Tx \|}{\| x \|}$$

Let us observe that in addition this norm possesses the multiplicative property :

$$\| T_1 T_2 \| \leq \| T_1 \| \, \| T_2 \|$$

Note : It can be proved that $\| T \| = [\rho(\bar{T}' T)]^{1/2}$

A corollary of this is that for an arbitrary complex matrix T nxn we have $\| T \| \geqslant \rho(T)$ and $\| T \| = \rho(T)$ for a hermitian matrix $(T = \bar{T}')$

* $\bar{x}'$ represents the complex conjugate and transpose x.

2.2.2. Study of convergence

We give a few definitions and some results found in [VAR - 62].

Definition 1 : Error of convergence at the kth step.
We shall call it :

$$\varepsilon^{(k)} = x^{(k)} - x^*$$

For the stationary linear methods we shall have :

$$\varepsilon^{(k)} = T\,\varepsilon^{(k-1)} = \ldots = T^k\,\varepsilon^{(0)}$$

which leads to the Euclidean norm inequality :

$$\|\varepsilon^{(k)}\| \leq \|T^k\| \quad \|\varepsilon^{(0)}\|$$

We find that $\|T^k\|$ can act as a basis of comparison between different methods. Furthermore the condition $\|T\| < 1$ appears as a sufficient condition of convergence towards a unique solution $\forall\,\varepsilon^{(0)}$.

Theorem 1 : A complex matrix T nxn is said to be convergent if and only if

$$\rho(T) < 1$$

or in an equivalent way :

$$\lim_{k \to \infty} T^k = \mathbb{O}$$

Definition 2 : Average rate of convergence for k iterations.

This is the number r_k defined as follows for every $k > 0$ such that $\|T^k\| < 1$:

$$r_k(T) = -\,\mathrm{Log}\,[\,\|T^k\|\,]^{1/k}$$

Notes :

- if $r_k\,(T_1) < r_k\,(T_2)$, then T_2 is iteratively faster than T_1 for k iterations.
- $\frac{1}{r_k(T)}$ measures the number of iterations required in order that $\|\varepsilon(k)\| \leqslant \frac{1}{e}\,\|\varepsilon(0)\|$

where e is the base of the natural logarithms

Definition 3 : Asymptotic rate of convergence

This is defined as follows :

$$r_\infty\,(T) = \lim_{k \to \infty} r_k\,(T) = -\,\mathrm{Log}\,\rho\,(T)$$

Note : This is the simplest number which makes it possible to appreciate the speed of convergence of an iterative method. An iterative method will be considered better by as much as the asymptotic rate of its matrix of iterations is great, or in the same way, by as much as its spectral radius (< 1) is small.

We shall end this section by expressing the necessary and sufficient conditions of convergence of linear methods :

1) Stationary linear methods

$$x^{(k+1)} = T\,x^{(k)} + t$$

Necessary and sufficient condition of convergence $\forall x^{(0)}$

$$\rho(T) < 1 \quad \text{or} \quad \lim_{k \to \infty} T^k = \mathbb{O}$$

2) Non-stationary linear methods

$$x^{(k+1)} = T_{k+1}\, x^{(k)} + t_{k+1}$$

let

$$x^{(k+1)} = \prod_{i=1}^{k+1} T_i\, x^{(0)} + \hat{t}_{k+1}$$

Necessary and sufficient condition of convergence

$$\lim_{k \to \infty} \prod_{i=1}^{k+1} T_i = \mathbb{O} \text{ and the sequence } \{\hat{t}_{k+1}\} \text{ converges.}$$

Of course the structural characteristics of the matrices of iteration play a very important role in the convergence properties of such methods. We shall try to point this out in the presentation of different methods in the following sections where we shall consider the improvement of the properties of convergence.

2.2.3. Contraction principle

We have seen that a sufficient condition of convergence towards a unique solution of stationary linear methods was $\|T\| < 1$. This can be extended not only to the non-linear operators in R^n, but also to the operators of derivation, integration,... by the expedient of the "contraction mapping theorem".
The use of the contraction condition represents in fact a standard way, in functional analysis, to prove the convergence of methods of successive approximations used for the solution of algebraic, differential, integral etc. equations [LIU - 65].
We give below a few definitions and theorems which have mostly come from [ORT - 70].

Definition 1 : Contraction mapping

A mapping $T = D \subset R^n \to R^n$ is contractive on a set $D_0 \subset D$ if there exists a number α : $0 \leqslant \alpha \leqslant 1$ such that $\|T[x] - T[y]\| \leqslant \alpha \|x-y\|$ $\forall x, y \in D_0$.

Definition 2 : Fixed points

The fixed points of T are the solutions of $x^* = T[x^*]$

Theorem 1 : Let us suppose that $T : D \subset R^n \to R^n$ is contractive on a closed set $D_0 \subset D$ and that $T D_0 \subset D_0$. Then T has a unique fixed point in D_0.

Corollary : We can construct an approximating sequence $\{x^{(k)}\}$.

$$x^{(k+1)} = T[x^{(k)}] \qquad k = 0, 1, \ldots$$

which converges towards the fixed point x^* from any initial approximation $x^{(0)} \in D_0$.

Theorem 2 : Contraction principle

Let us consider a mapping $T : D \subset R^n \to R^n$ of a closed set $D_0 \subset D$ into itself and such that :

$$\| T[x] - T[y]\| \leq \alpha \|x-y\| \quad \forall x, y \in D_0 \qquad 0 \leq \alpha < 1$$

Then, $\forall x^{(0)} \in D_0$, the sequence $x^{(k+1)} = T[x^{(k)}]$, $k = 0,1,...$ converges towards the unique fixed point x^* of T in D_0. Furthermore, if the constant of contraction α is known, we have the following estimate of the error of approximation :

$$\| x^{(k)} - x^* \| \leq \frac{\alpha}{1-\alpha} \| x^{(k)} - x^{(k-1)} \| \qquad k = 1,2,...$$

Let us note that in the special important case where $D_0 = D = R^n$, (T is a contraction on all R^n), then theorem 2 provides a condition of global convergence, ie. the approximating sequence converges towards the unique fixed point of T in R^n, $\forall x^{(0)} \in R^n$.

Let us note that the preceding definitions and theorems apply in a more general way to the complete metric spaces.
Let us also quote the following theorem from [KOL - 57].

Theorem 3 : Generalised contraction principle.

Let T be a continuous mapping of the complete space X into itself, of which a power T^k is contractive; then the equation $Tx = x$ has one solution and one only, ie. that every fixed point for the mapping T is fixed for the contracting mapping T^k which can have only one fixed point.

In order to obtain a measure of the convergence behaviour in terms of the individual components of x, rather than of a norm in R^n, it may be useful to use the absolute values of the vectors on R^n :

$$|x| = (|x_1|, ..., |x_n|)^T$$

This gives the following definition of a contraction under a partial ordering in R^n :

Definition 3 : P contraction

An operator $T : D \subset R^n \rightarrow R^n$ is a P contraction on a set $D_0 \subset D$ if there exists a linear operator $P \subset L(R^n)$ with the following properties :

$$\rho(P) < 1; \text{ P is non-negative i.e. } p_{ij} \geq 0, \forall i,j$$

and such that :

$$|T[x] - T[y]| \leq P |x-y| \; ; \quad \forall x, y \in D_0$$

Let us note that theorem 2 applies naturally to the P-contractions and that the estimate of the error becomes :

$$|x^{(k)} - x^*| \leq ((\mathbb{1} - P)^{-1} P) |x^{(k)} - x^{(k-1)}|$$

This makes it possible in particular to determine conditions of global convergence for certain non-linear operators [ORT - 70].
In the same way we can hope to obtain a measurement of conver-

gence in terms of sub-vector norms $\|x_i\|$ corresponding, for example, to a decomposition of the vector $x \in \mathbb{R}^n$ in N sub-vectors x_i. It may then be useful to use the norm termed canonic vector norm of x [MIE - 75].

$$|||x||| = (\|x_1\|, \ldots, \|x_N\|)^T$$

and then consider a P-contraction such that :

$$|||T(x) - T(y)||| \leqslant P\ |||x-y|||.$$

Let us note that this makes it possible to join to the principal system x = T [x] a reduced scale system of comparison z = Pz.

2.3 Relaxation schemes for the resolution of a system of algebraic equations

We have chosen to recall a certain classification of iterative methods of decomposition-co-ordination for the solution of a system of algebraic equations :

$$Ax = b \ ; \quad A = (a_{ij}) \quad ; \ b = (b_i) \tag{2.1}$$

plus the hypothesis det $A \neq 0$ guaranteeing the uniqueness of the solution.
Indeed we believe that these methods, which have been the subject of in depth studies [VAR - 62],[YOU - 71],[ORT - 70],[LIO - 74],[GOL - 81],[MIE - 75], constitute an essential guide for the definition of methods of decomposition of control problems. The study of basic particular decompositions, of their variants (methods of under, over-relaxation, semi-iterative methods), of their extensions (iterative methods by group, block, chaotic...) bring numerous elements of response to the problem of the management of large scale systems, whether it be at the level of calculation structures (sequential, parallel, aperiodic ...) of their processing implementation and of information structures (centralised, hierarchical, distributed) and of the structures of the systems themselves (serial, parallel, tree, cyclic...).
Of course, we can consider extending this type of method to the problems of dynamic optimisation, as we do in section 2.4.
It is clear, moreover, that the system of equations (2.1) may be directly deduced from the solution of discretised dynamic problems.
To finish this introduction, may we emphasize on the one hand the fact that these methods originate from a purely algorithmic preoccupation transforming one problem into sub-problems of reduced scale or complexity. On the other hand, the methods are based on a principle which can be described as relaxation, the definition of which we have borrowed from [LHO - 79] :

<u>Definition</u> : Relaxation strategy

To relax the SSi consists of decoupling it from the others, considering the coupling variables as known and of resolving it separately on this basis.

2.3.1. Point iterative methods

These correspond to an "ultimate decomposition" into n sub-systems of the system (2.1) which we note down again :

$$\sum_{j=1}^{n} a_{ij} \ x_j = b_i \qquad i = 1,\ldots,n \tag{2.2}$$

and to which we add the further hypothesis $a_{ii} \neq 0 \quad \forall i = 1,\ldots,n$

2.3.1.1. Stationary methods

$$x^{(k+1)} = T\, x^{(k)} + t \qquad k = 0,1,\ldots$$

a) Jacobi method

This applies the following iterative scheme from an initial approximation of x : $x(0) = x_0$

$$a_{ii} x_i^{(k+1)} = - \sum_{\substack{j=1 \\ j \neq i}}^{n} a_{ij} \ x_j^{(k)} + b_i \qquad i = 1,\ldots,n; \quad k \geq 0$$

Definition : Point Jacobi matrix

We define thus this matrix associated to the point algorithm of Jacobi :

$$J = D^{-1} \ (L + U)$$

where $\quad D = \text{Diag}\ (A) = \{a_{ii}\}$

L and U are respectively the (nxn) strictly triangular lower and triangular upper of - A

$$L = \{-a_{ij} \quad ; \ j < i\} \ ; \qquad U = \{- a_{ij} \quad ; \quad j > i\}$$

It enables a parallelisation of the calculation scheme by making possible the simultaneous resolution of the n equations (2.2). In a general way, we note this type of mapping as :

$$x_i^{(k+1)} = f_i \ (x_j^{(k)}) \quad j \neq i \qquad i = 1,\ldots,n$$

In terms of information, let us note that this method necessitates global storage at a sort of an upper level of the n values $x_i^{(k)}$, $i = 1,\ldots,n$.

Figure (2.1) represents the centralisation of the information involved in this method.

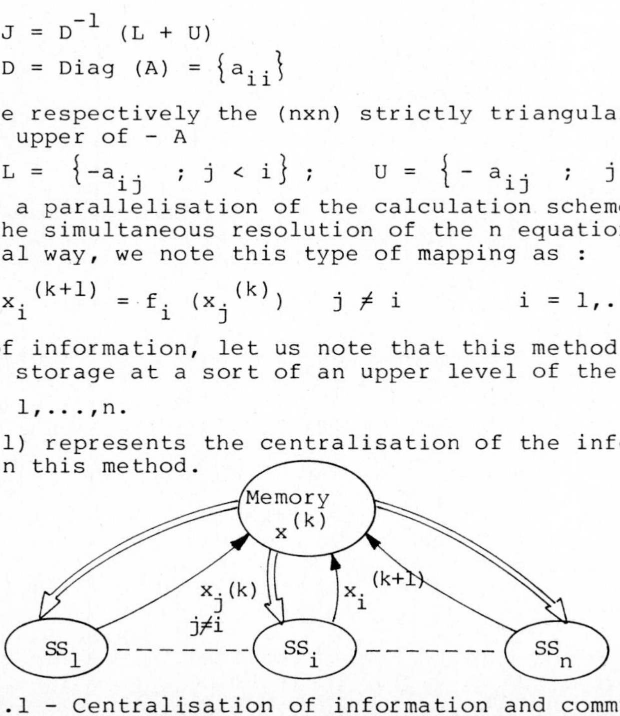

Figure 2.1 - Centralisation of information and communications in a process of parallel calculation

b) <u>Gauss-Seidel method</u>

This applies an iterative scheme of the type :

$$a_{ii}x_i^{(k+1)} = - \sum_{j=1}^{i-1} a_{ij}\, x_j^{(k+1)} - \sum_{j=i+1}^{n} a_{ij}x_j^{(k)} + b_i \qquad i=1,\ldots,n$$

or again :

$$a_{ii}x_i^{(k+1)} = - \sum_{j=1}^{i-1} a_{ij}\, x_j^{(k)} - \sum_{j=i+1}^{n} a_{ij}\, x_j^{(k+1)} + b_i$$

These schemes translate a sequentialisation of the calculation scheme, the first in a direct sense, the second in a sense which we describe as retrograde.
In general terms, we note :

$$x_i^{(k+1)} = f_i\ (x_j^{(k+1)},\ x_l^{(k)}) \quad ; \quad j < i \quad , \quad l > i$$

or vice-versa for the retrograde sense.
We can consider that this method involves :

- on the one hand, a decentralisation of information, to the extent that one part of this information is recorded at a level at which the n values $x_i^{(k)}$, are memorised, another part being processed sequentially : $x_j^{(k+1)}$, $j < i$ in the direct sense.
- on the other hand, a hierarchisation of this second part of the information induced by the relationship of order, natural ordering in this case, which governs the processing of the subsystems.

Figure 2.2 represents information transfers in the direct sense.

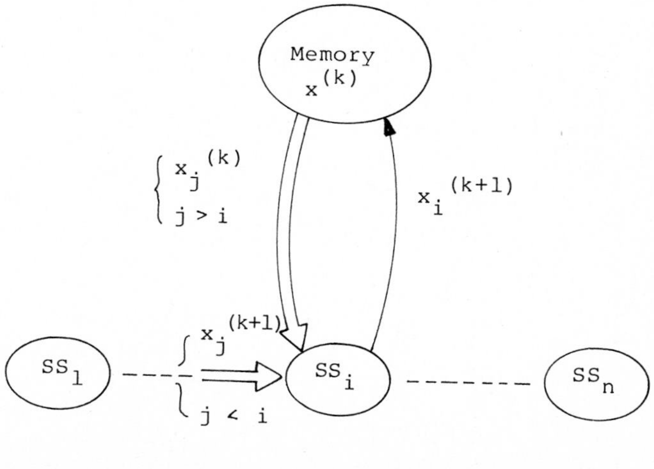

<u>Figure 2.2</u> - Decentralisation and scheduling of information in a process of sequential calculation.

Let us also note that one of the advantages of the sequential method concerns the quantity of information to be memorised, reduced by half in relation to the parallel method which necessitates the storage of $x^{(k)}$ and $x^{(k+1)}$ namely 2n variables, whereas here we need only to know $x_j^{(k+1)}$ for $j < i$ and $x_\ell^{(k)}$ for $\ell > i$, namely n variables (if using a mono-processor computer). One important question is linked to the comparison of the convergence behaviour of the Jacobi method and the Gauss-Seidel method. Although neither method can in general be favoured over the other, elements of comparison may be deduced from certain structural properties of the matrix A. We refer to [VAR - 62] for these comparisons. However, we shall resume the theorem of STEIN-ROSENBERG as follows :

Theorem :

If the matrix of Jacobi J (whose diagonal elements are zero) has coefficients $j_{ij} \geq 0$, then the particular methods of Jacobi and Gauss-Seidel are both either convergent or divergent. Furthermore, if they converge, the Gauss-Seidel method is asymptotically faster.

Finally, let us state that the particular methods of Jacobi and Gauss-Seidel are always convergent for the strictly diagonally dominant matrices A

$$\left(|a_{ii}| > \sum_{\substack{j=1 \\ j \neq i}}^{n} |a_{ij}| \right)$$

Similarly, let us remember that if A is a tridiagonal matrix the result given above remains, but in addition we know that $R_\infty(GS) = 2\, R_\infty(J)$, ie. that the Gauss-Seidel method is asymptotically twice as rapid. We shall return to this theorem in paragraph 2.3.2.

c) Method of over (under)-relaxation

A third basic type of stationary methods concerns the point iterative methods of over-relaxation.

- Method of simultaneous over-relaxation

$$\begin{cases} \tilde{x}_i^{(k+1)} = f_i\,(x_j)^{(k)} & j \neq i \qquad i = 1,\dots,n \\ x_i^{(k+1)} = (1-\omega)\, x_i^{(k)} + \omega\, \tilde{x}_i^{(k+1)} & \end{cases}$$

- Method of successive over-relaxation

$$\begin{cases} \tilde{x}_i^{(k+1)} = f_i\,(x_j^{(k+1)}, x_l^{(k)}) & j < i; \quad l > i; \quad i=1,\dots,n \\ x_i^{(k+1)} = (1-\omega)\, x_i^{(k)} + \omega\, \tilde{x}_i^{(k+1)} & \end{cases}$$

ω is called the relaxation factor,
$\omega < 1$ characterises a method of under relaxation,
$\omega > 1$ characterises a method of over-relaxation,
$\omega = 1$ corresponds to the Jacobi and Gauss-Seidel methods respectively.

Let us note that these methods, which appear to be extrapolations of the preceding methods, have the advantage of involving few supplementary calculations.
For a great majority of systems treated in practice, and as Young, has shown for a wide class of matrices, it proves to be the Gauss-Seidel method with $\omega > 1$ which gives the best improvements in convergence. Therefore it is often called the over-relaxation method. [YOU - 71].
Here again studies of convergence concern particular structural cases. Let us remember, however, a theorem from [LIO - 74], stating the NSC of convergence of the Gauss-Seidel over-relaxed method, in the case where A is a symmetric matrix and where D is positive definite :

Theorem :

The Gauss-Seidel over-relaxed method converges if and only if A is positive semi-definite and $\omega \in]0,2[$, or else -A is positive semi-definite and $\omega \notin [0,2]$, ω must be such that $(D - \omega L)$ is non-singular.
In a general non-linear case it is more useful to limit oneself to under-relaxation (see ref. Comte in [LAN - 77]).

2.3.1.2. Semi-iterative methods

This concept, introduced by VARGA, defines non-stationary methods associated with linear stationary methods with a view to improving the conditions of applicability or of convergence. We present two of the most popular of these :

a) RICHARDSON method type

This applies the following iterative scheme :

$$x_i^{(k+1)} = x_i^{(k)} + \alpha_{k+1} \left(\sum_{j=1}^{n} a_{ij} x_j^{(k)} - b_i \right) \quad i=1,\dots,n \quad k \geq 0.$$

Let us note that for this method it is not necessary to have $a_{ii} \neq 0$ and that for $\alpha_{k+1} = -\frac{1}{a_{ii}}$, we go back to the Jacobi method.
With respect to the necessary and sufficient condition of convergence, it corresponds to :

$$\lim_{k \to \infty} \prod_{j=1}^{k} (\mathbb{1} + \alpha_j A) = \mathbb{O}$$

Let us also point out the so-called RICHARDSON second-order type method, which however, requires storage in memory of twice as much information as before :

$$x^{(k+1)} = x^{(k)} + \alpha_{k+1} (Ax^{(k)} - b) + \beta_{k+1} (x^{(k)} - x^{(k-1)}); \; k \geq 0$$

b) Semi-iterative method using the CHEBYSHEV's polynomials

This is a variant of the Richardson method, which moreover gives a way for choosing the acceleration parameters ω_{k+1} defined from the Ck polynomials of Chebyshev [VAR - 62] .

This method consists in finding a sequence :

$$y^{(k)} = \sum_{j=o}^{k} \gamma_j(k) \quad x^{(j)} \text{ with } \sum_{j=o}^{k} \gamma_j(k)=1 \qquad k \geqslant 0$$

which converges faster to the solution x^*, than the sequence :

$$x^{(k+1)} = T\, x^{(k)} + t \qquad k \geqslant 0 \qquad \forall x^{(o)}$$

When considering the coordination errors :

$$\tilde{\varepsilon}^{(k)} \triangleq y^{(k)} - x^* \quad \text{and} \quad \varepsilon^{(k)} \triangleq x^{(k)} - x^*$$

it is readily found that :

$$\tilde{\varepsilon}^{k} = Pk\ (T) \quad \varepsilon^{(o)} \quad k \geqslant 0 \text{ with } Pk\ (T) = \sum_{j=o}^{k} \gamma_j^{(k)}\ T^j ;$$

$$Pk\ (\mathbb{1}) = \mathbb{1}$$

In the case when only the spectral radius ρ (T) is known (T being assumed hermitian and convergent matrix), a way to improve the convergence can be given by considering the optimization problem :

$$\min_{-1 < -\rho(T) \leqslant x \leqslant \rho(T) < 1} \left\{ \max \ | pk(x) | \right\} \qquad pk(1) = 1$$

Its solution expressed by means of the Chebyshev's polynomials Ck $(1/\rho)$ leads to the following algorithm :

$$y^{(k+1)} = \omega_{k+1} \ [Ty^{(k)}+t - y^{(k-1)}] + y^{(k-1)} \quad ; \quad k \geqslant 1$$

$$y^{(o)} \text{ arbitrary;} \quad y^{(1)} = Ty^{(o)} + t$$

$$\text{with } \omega_{k+1} = \frac{2Ck\ (1/\rho)}{\rho Ck+1\ (1/\rho)} = \frac{1}{1 - \dfrac{\rho^2 \omega_k}{4}} \quad ; \ k \geqslant 1; \quad \omega_1 = 1$$

Note 1. In the special case of a Jacobi type mapping for example, let us recall that T = J and t = D^{-1} b.

Note 2. Let us point out that this method may also apply for a wider class of matrices T [GOL - 81] .

2.3.2. Group iterative methods

We are borrowing this definition from [YOU - 71] ; its title is certainly not ideal and corresponds in fact to an application of the principle of relaxation given at the beginning of Section 2.3 where the notion of sub-system assimilated to an equation, in the case of point mapping, is here joined to sets or groups of equations.

Definition : Groupe iterative methods

In these methods equations are assigned to groups of equations, each equation belonging to only one groupe. Each group will the-

refore calculate the vectorial solutions x_i, the other vectors
$xj : j = 1,...,N$ being considered as known (relaxed).
A particular case corresponds to the partitioning of the matrix A; the iterative methods based on the principle of partitioning are called partitioned methods or block methods.
We shall try in the rest of this section to emphasize the interest of these partitioned methods on an algorithmic point of
view (improvement of the conditions of application and convergence), and also on the means of calculation plane (improvement
of the conditions of implementation).
Let us first go back to the interest which this generalisation
of the principle of relaxation may represent to group iterative
methods.
Indeed the concept of group introduces, in addition to the possibility of association or co-operation between sub-systems,
the possibility of arranging the different sub-systems in a way
which is not necessarily based on the natural order (case of
partitioning). To illustrate this let us quote the different
ordered groupings envisaged in [YOU - 71] (natural ordering,
orderings by rows, columns diagonal, red-black ordering...)
for the solution of linear systems corresponding to the derivation of approximations to the finite differences of partial differential equations.
In this sense let us draw attention to the intensive studies of
consistently ordered matrices to which we shall return at the
end of this section.
Let us finally note that taking into account the notion of ordering for the resolution of sub-systems may lead to schemes of
decomposition which are not necessarily periodical. This generalisation corresponds to the concept of chaotic iterations
[MIE - 75] . Let us also draw attention to the concept of chaotic iterations with distributed lags which introduces the possibility of taking into account delays in the different phases
of calculation. This naturally opens a very useful way towards
asynchronous methods of decomposition. [COUR - 80].

2.3.2.1. Stationary methods

1) Block methods

The problem of ordering being considered solved, system (2.1)
can be rearranged in such a way as to make the following partitioning of matrix A and vectors x and b.

$$A = \begin{bmatrix} A_{11} & \cdots & A_{1N} \\ \vdots & & \vdots \\ A_{N1} & \cdots & A_{NN} \end{bmatrix}, \quad x = \begin{bmatrix} x_1 \\ \vdots \\ x_N \end{bmatrix}; \quad b = \begin{bmatrix} b_1 \\ \vdots \\ b_N \end{bmatrix}$$

where the matrices A_{ii} are square, of dimensions $n_i \times n_i$

the vectors x_i and b_i of dimension n_i

such that $\sum_{i=1}^{N} n_i = n$,

which makes it possible to re-write system (2.1) in the form :

$$\sum_{j=1}^{N} A_{ij}\, x_j = b_i \qquad i = 1,\dots,N$$

We can then consider applying Jacobi, Gauss-Seidel type methods, over-relaxation not on the components of x, but on several components simultaneously, thus defining methods which we describe as block methods of relaxation.
For example, we shall define the method of successive block over-relaxation as follows :

$$A_{ii}\, x_i^{(k+1)} = A_{ii} x_i^{(k)} + \omega \Big\{ - \sum_{j=1}^{i-1} A_{ij}\, x_j^{(k+1)}$$

$$- \sum_{j \neq i}^{N} A_{ij}\, x_j^{(k)} + b_i \Big\} ; \qquad k \geqslant 0;\ i=1,\dots,N$$

or again :

$$(D-\omega L)\, x^{(k+1)} = \Big\{ \omega U + (1-\omega)\, D \Big\}\, x^{(k)} + \omega b; \qquad k \geqslant 0$$

with :

$$D = \begin{bmatrix} A_{11} & & \mathbb{O} \\ & \ddots & \\ \mathbb{O} & & A_{NN} \end{bmatrix} ; \quad L = - \begin{bmatrix} \mathbb{O} & & & \mathbb{O} \\ A_{21} & \ddots & & \\ \vdots & \ddots & \ddots & \\ A_{N1} & \cdots & A_{N\ N-1} & \mathbb{O} \end{bmatrix} ; \quad U = - \begin{bmatrix} \mathbb{O} & A_{12} & \cdots & A_{1N} \\ & \ddots & \ddots & \vdots \\ & & \ddots & A_{N-1.N} \\ \mathbb{O} & & & \mathbb{O} \end{bmatrix}$$

Let us note that the theorem given in section 2.3.1.1.c is valid for this partitioning which means that if A is real, symmetric and positive semi-definite, (D positive definite), then the points or blocks method of successive relaxations is convergent for $\omega \in]0,2[$ such that the matrix $(D - \omega L)$ is regular.
In general terms, the problem posed is naturally the determination of N and it is a delicate problem which calls into play both the methods of calculation (capacity of the calculators, number of units of calculation) and the performance of these methods, (dimensions and structure of the sub-systems, convergence of the process of iterative co-ordination).
This last point illustrates the usefulness of comparing several partitionings.
In fact, to take larger blocks does not always give rise to an improvement of the asymptotic rate of convergence. However, under certain conditions this may be the case. From among these conditions let us quote, for example, the case in which the matrix A^{-1} has non-negative elements, which corresponds in particular to the M - matrices.

<u>Definition : M - matrix</u>

A real matrix Anxn with $a_{ij} \leqslant 0\ \forall_{i \neq j}$ is a M - matrix if A is non singular and A^{-1} has non-negative elements.
Before giving some theorems of convergence and of comparison, we are going to present one of the basic principles used in the analysis and synthesis of iterative methods for linear equations, namely the principle of "splitting" attributed to Varga [VAR - 62].

Definition : Splitting of a matrix

For the resolution of the linear system Ax = b where A is a non singular nxn matrix, the decomposition A = M - N where M is a non singular nxn matrix represents a splitting of the matrix A, to which is associated the iterative method :

$$Mx^{(k+1)} = Nx^{(k)} + b; \qquad k \geq 0$$

Let us draw attention to the extension of this principle to pseudo-inverse matrices, in the case where A is non singular [GOL - 81],[LAU - 78].

Let us finally note that this principle of decomposition enables us to return to the block relaxation methods previously described :

- Bloc method of the Jacobi type or of parallel calculation :

$$M = D; \qquad N = L + U$$

- Block method of the Gauss-Seidel type or of sequential calculation

$$M = D - L \qquad N = U$$

or

$$M = D - U \qquad N = L$$

- Block method of successive over-relaxations :

$$M = \frac{1}{\omega}(D - \omega L); \qquad N = \frac{1}{\omega}[\omega U + (1 - \omega) D] ; \qquad \omega \neq 0$$

Definition : Regular splitting

A = M - N is a regular splitting of A, if the coefficients of the matrices M^{-1} and N are negative.

2) Convergence study

Let us recall some theorems (the first from [GOL - 81] the others from [VAR - 62]) expressing sufficient conditions for the convergence of those methods.

Theorem 1

Let A be a Hermitian matrix and M (A = M - N) any non singular matrix for which the matrix $Q = M + \bar{M}' - A$ is positive definite.
If A is positive definite, then $\rho(M^{-1}N) < 1$ and the associated iterative method converges $\forall x^{(o)}$.

Theorem 2

If A = M - N is a regular splitting of A and if the matrix A^{-1} has non-negative elements, then $\rho(M^{-1}N) < 1$ and the associated relative method converges $\forall x^{(o)}$.

Theorem 3

Let A be an nxn M - matrix. If M is an nxn matrix obtained by cancelling certain off diagonal inputs of A, then A = M - N is a regular splitting of A and $\rho(M^{-1}N) < 1$.

Similarly, let us quote the theorem of comparison between different splitting

Theorem of comparison 4 :

Let $A = M_1 - N_1 = M_2 - N_2$, two regular splittings of A where the matrix A^{-1} has positive elements. If the elements of N_2 are greater than or equal to those of N_1, all of them being non-negative, (however $N_1 \neq 0$, $N_2 \neq N_1$), then $\rho(M_1^{-1} N_1) < \rho(M_2^{-1} N_2) < 1$, ie. the first decomposition has a greater asymptotic rate of convergence.

3) Structural analysis for partitioning

We have already been able to verify in the preceding sections the importance of the structural properties of the matrices considered in the studies of convergence (matrices with non-negative elements, diagonally dominant,...).
We should like to emphasize particularly on two aspects of structural analysis which we feel must play a determining role in the synthesis of methods of decomposition-co-ordination.
One of these concerns natural decomposition based on the weak interaction of interconnected systems, the practical importance of which is, we feel, very great. We shall try to justify this more specifically in sections 2.4.2. and 2.4.4. on problems of control.
The other concerns structural decomposition properly so called, exploiting properties of structure and scheduling, ie. applying to systems described by sparse matrices whose zero elements are arranged in a certain pattern.
We shall end this section by recalling the interest of representation by graph for these points of structural analysis.

a) Natural decomposition :

This methodology of decomposition has the advantage of applying to a wide class of interconnected systems found in practice [MAY - 76][LOO - 80] , (sections 2.4.2. and 2.4.4.) and of being able to base itself on the physical and experimental knowledge one may have of the process.
It is based on a hypothesis, on the whole a fairly realistic hypothesis when one is considering undertaking an act of decomposition, namely the weak coupling between sub-systems.
We shall therefore compare the search for a procedure of natural decomposition with the search for the ε-couplings in a connection of systems and we propose to derive, from the definition of decomposition of a matrix (section 2.3.2.1.), the following definition :

Definition : Natural splitting of A

This corresponds to the decomposition $A = M - \varepsilon\tilde{N}$ where ε is a sufficiently small scalar parameter to which is associated the iterative method :

$$Mx^{(k+1)} = \varepsilon\tilde{N}x^{(k)} + b \ ; \qquad k \geq 0$$

It therefore assumes the display of a multiplication factor ε in the expression of the matrix N, which therefore appears in the expression of the spectral radius of the matrix of iteration $M^{-1}N$.

$$\rho(M^{-1}N) = \varepsilon\rho(M^{-1}\tilde{N})$$

associated with the approximation :

$$x^{(k+1)} = M^{-1}Nx^{(k)} + M^{-1}b \qquad k > 0$$

and therefore guarantees its convergence for ε sufficiently small .

This display may be made by techniques of ε perturbations which consist of finding a perturbed solution $x(\varepsilon)$ of the system $Ax = b$.

In sections 2.4.2. and 2.4.4. we shall return to the subject with examples of application.

It is also interesting to note that for this type of method, the kth iterative of the algorithm of decomposition-co-ordination, ie. the kth approximation of the solution x, corresponds to the approximation in series to the degree k of this perturbed solution $x(\varepsilon)$, namely :

$$x^{(k)} = x^{(k)}(\varepsilon) = \sum_{n=0}^{k} \frac{1}{n!}\varepsilon^n \left[\frac{\partial^n x^{(k)}}{\partial\varepsilon^n}(\varepsilon)\right]_{\varepsilon = 0}$$

Finally, let us point out how all physical and experimental knowledge of the process can be integrated with a view to a partition into weakly coupled sub-systems from structural analysis by the datas[RIC - 75] , and particularly in the determination of static or dynamic coupling matrices based on the Information theory.

Let us emphasize, however, the restriction linked with application to continuous systems, which must be made discrete. This means a subsequent loss of information.

b) Structural decomposition :

Let us first recall the concept of reducibility :

Definition 1 : Reducible and irreducible matrices

For $n \geq 2$, an nxn matrix A is reducible if there exists an nxn matrix of permutation P such that :

$$P\,A\,P^T = \begin{bmatrix} \tilde{A}_{11} & 0 \\ \tilde{A}_{21} & \tilde{A}_{22} \end{bmatrix}$$

where the matrices $\tilde{A}_{ii}$ are square.

Otherwise, A is said to be irreducible.

Let us note that if we pursue this decomposition until the matrices $\tilde{A}_{ii}$ are either irreducible or zero scalar, we then obtain a lower block-triangular matrix said to be the normal form of A.

Definition 2 : Permutation matrix

This is an nxn matrix P with only one non-zero element exactly equal to 1 in each line and in each column. Let us note that

$P^{-1} = P^T$.

We should now like to draw attention to another scheme of decomposition based on the sparsity and the pattern of certain classes of matrix to which can be applied the concepts of "Property A" and of "Consistent orderings" attributed to Young [YOU - 71] , or again the different, though closely linked, concept of consistent ordering of p - cyclic matrices attributable to Varga [VAR - 62] .
These concepts are well-known to have facilitated the theoretical study (analysis of eigen values, graphs,...) of iterative methods of relaxation.

Definition 3 : Property A

An nxn matrix A has "property A", if there exists 2 unconnected sub-sets S1 and S2 of the set S of the n positive integers, such that S1 + S2 = S and if $i \neq j$ and $a_{ij} \neq 0$, or else $a_{ji} \neq 0$, then $i \in S1$ and $j \in S2$, or else $i \in S2$ and $j \in S1$.
Let us note that if S1 or S2 is empty, then A is diagonal.

Definition 4 : Consistent ordering

An nxn matrix A is consistently ordered if for any integer p, there exists unconnected sub-sets S1, S2,..., Sp of $S = \{1, 2, ..., n\}$ such that : $\sum_{k=1}^{p} Sk = S$ and such that if the integers i and j are "associated" in relation to A (ie. $a_{ij} \neq 0$ or $a_{ji} \neq 0$), then $j \in Sk+1$ if $j > i$ and $j \in Sk-1$ if $j < i$, where Sk is the sub-set containing i.
Let us recall two theorems on this subject from [YOU - 71] :

Theorem 1 :

An nxn matrix A has "property A", if, and only if, there exists a permutation matrix P such that $\tilde{A} = PAP^T$ is consistently ordered.

Theorem 2 :

Let A be a symmetric and consistently ordered matrix with positive diagonal elements. Then $\rho(J) < 1$ if and only if A is positive definite where J is the point Jacobi matrix.

$$J = - (\text{diagonal } A)^{-1} A + 𝟙 .$$

We return to these same references for a generalisation of this to the block methods.

Definition 5 : Weakly cyclic matrix of index p

An nxn matrix A is weakly cyclic with index $p \geq 2$, if there exists an nxn matrix with permutation P such that PAP^T has the following form in which the zero diagonal sub-matrices are square :

$$PAP^T = \begin{bmatrix} \mathbb{O} & \cdots & \cdots & \mathbb{O} & \tilde{A}_{1P} \\ \tilde{A}_{21} & \ddots & & & \mathbb{O} \\ & \ddots & \ddots & & \vdots \\ \mathbb{O} & & \ddots & \ddots & \vdots \\ \mathbb{O} & \mathbb{O} & & \tilde{A}_{p\,p-1} & \mathbb{O} \end{bmatrix}$$

<u>Definition 6 : p-cyclic matrix</u>

If the block Jacobi matrix $J = - D^{1} A + \mathbb{1}$ for a given partitioning of A is weakly cyclic with index $p \geqslant 2$, then A is p-cyclic relatively to this partitioning.

<u>Definition 7 : p-cyclic matrix consistently ordered</u>

If a matrix A is p-cyclic it is then consistently ordered if all the eigen values of the matrix :

$$J(\alpha) = \alpha L_J + \alpha^{-(p-1)} U_J \quad ; \qquad \alpha \neq 0$$

are independent of α, (where L_J and U_J represent the strictly lower and upper triangular matrix respectively of the Jacobi matrix J).
In [VAR - 62] it is pointed out that the property for A to be 2-cyclic is implied by the "property A". It is also demonstrated that the consistent orderings, when they exist, are optimal $\forall \omega$ in a relaxation procedure.
Linked to this, let us also mention the following theorem from [VAR - 62] which enables a comparison to be made between the Jacobi and Gauss-Seidel methods.

<u>Theorem 3</u> :

Let us consider a partitioned matrix A which is p-cyclic consistently ordered and in which the block diagonals A_{ii} are regular, then the block - Jacobi iteration method converges if and only if, the block Gauss-Seidel iteration method converges. If they both converge, then :

$$\rho(GS) = (\rho(J))^p < 1$$

and therefore :

$$R_\infty(GS) = p\, R_\infty(J),$$

which means that the Gauss-Seidel method is p times asymptotically faster than Jacobi method.

Two particular matrices A_1 and A_2 are of great practical interest.

$$A_1 = \begin{bmatrix} A_{11} & \mathbb{O} & \cdots & \mathbb{O} & A_{1P} \\ A_{21} & \ddots & \ddots & & \mathbb{O} \\ & \ddots & \ddots & \ddots & \vdots \\ \mathbb{O} & & \ddots & \ddots & \mathbb{O} \\ \mathbb{O} & \cdots & \mathbb{O} & A_{pp-1} & A_{pp} \end{bmatrix}, \quad A_2 = \begin{bmatrix} A_{11} & A_{12} & & & \mathbb{O} \\ A_{21} & A_{22} & A_{23} & & \\ & \ddots & \ddots & \ddots & A_{N-1.N} \\ \mathbb{O} & & & A_{N.N-1} & A_{NN} \end{bmatrix}$$

<u>Note 1</u> :

On the hypotheses of non-singularity of the block diagonals, the matrix A_1 is p-cyclic consistently ordered.

Note 2 :

On these same hypotheses, any block tridiagonal matrix of type A_2 is 2-cyclic consistently ordered.

Note 3 :

Any partitioning of A with N = 2, for which the block diagonals are regular, is such that the matrix A is 2-cyclic consistently ordered.

Theorem 4 :

Let A be a p-cyclic consistently ordered matrix with regular blocks A_{ii}. If all the eigen values of the pth power of the associated block Jacobi matrix J are real and non-negative, and if $0 \leq \rho(J) < 1$, then :

(i) $\rho(GS)_{\omega=\omega opt} = (\omega opt - 1)(p-1)$ with $\omega opt = 2/1 + \sqrt{1-\rho^2(J)}$

(ii) $\rho(GS)_{\omega} > \rho(GS)_{\omega opt} \quad \forall \omega \neq \omega opt$

Furthermore, the method of successive over-relaxation by blocks is convergent for any ω such that $0 < \omega < p / p - 1$.

As we have tried to show briefly, the presentation of consistent ordering is theoretically very important at the level of the effect this may have on the asymptotic rates of convergence, and also for the determination of the optimal relaxation factor. As Varga stressed, these properties are frequently found in practise. Let us draw particular attention in his book to the applications of this to the resolution of a broad class of discrete approximations of problems to equations with partial derivatives. See also the consistently ordered matrices arising from equations with differences and the associated sequential methods of decomposition in [WIS - 71] .
In section 2.4.2.2. we shall return to the possible exploitation of this type of structure.
Naturally the presentation of such types is certainly not trivial (nor unique). Let us draw attention in [YOU - 71] to the presentation of a systematic procedure and a calculation programme to test "property A" and consistent orderings.
In the next section we concentrate on the usefulness of graph representations for this type of structural analysis.

c) Usefulness of graph representation :

Structural analysis of models using graphs is the subject of much literature and contributes particularly effectively to the determination of partitions which facilitate the resolution of algebraic or differential equations.
To illustrate this we shall refer to the graph interpretation of the concept of reducibility, of consistent orderings of index 2, and, referring to [RIC - 75] , we shall show how the possibilities for partition of a system can be linked to the connexity properties of the associated graph.
For this, let us give a few definitions :

Definition 1 : Directed graph (Digraph)

Let there be an nxn matrix $A = (a_{ij})$ and n distincts points $P_1, \ldots, P_n$ in the plane called vertices or nodes. For each input $a_{ij} \neq 0$ of A, let us connect the node P_i to P_j by means of the directed path $\overrightarrow{P_iP_j}$ leading from P_i towards P_j thus defining a finite directed graph $\mathcal{G}(A)$:

ex. : $A = \begin{bmatrix} 1 & 2 \\ 0 & 3 \end{bmatrix}$ $\quad \mathcal{G}(A) =$

Definition 2 : Simple connexity

A digraph is simply connected if for some pairs of nodes P_i and P_j there exists a path (not necessarily directed) which links P_i to P_j.

Definition 3 : Strong connexity

A digraph is strongly connected if for P_i and P_j there exists a directed path linking P_i to P_j :

ex. : $\mathcal{G}(A)$ is simply connected.

The following theorems make it possible to interpret graphically the respective concepts of irreducibility and of consistent ordering of order 2.

Theorem 1 :

An nxn matrix A is irreducible if, and only if, its digraph $\mathcal{G}(A)$ is strongly connected.

Definition 4 : p-cyclic graph

A strongly connected digraph $\mathcal{G}$ of a matrix A is a cyclic graph with index $p \geq 2$ if the highest common divisor of all the lengths of its closed paths is $p \geq 2$.

Definition 5 : Digraph of type 2

The digraph of type 2 : $\mathcal{G}2$ of a matrix $A = (a_{ij})$ is constructed in such a way that if $a_{ij} \neq 0$, then the path of the node P_i to node P_j is represented by a double arrow line only if $j > i$, otherwise a single arrow is used. The first paths are called major paths, the others minor paths.

Theorem 2 :

The matrix A is consistently ordered only if each closed path of $\mathcal{G}2$ (A) has an equal number of major and minor paths.

ex. :

$$A = \begin{bmatrix} 0 & 0 & 1 & 1 \\ 0 & 0 & 1 & 1 \\ 1 & 1 & 0 & 0 \\ 1 & 1 & 0 & 0 \end{bmatrix} \qquad \mathcal{G}2(A) =$$

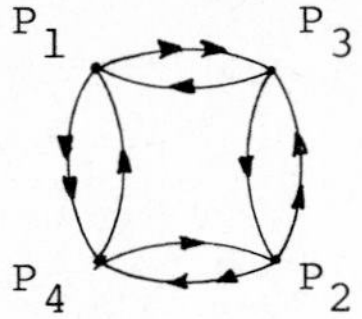

$\mathcal{G}(A)$ is strongly connected, 2-cyclic.

→ A is a 2-cyclic matrix consistently ordered.

We are now going to recall some definitions and properties of graphs inducing particular schemes of structural decomposition.

Definition 6 : Separating set

In a simply connected graph with nodes P : $\mathcal{G}_P$, any sub-set C of nodes such that the sub-graph $\mathcal{G}_{P-C}$ is not simply connected is called separating set.

Property 1 :

For each separating set C of a simply connected graph, there exists a permutation matrix P such that $\tilde{A} = PAP^T$ is a matrix with the following form, said to be arrow-shaped, or having a block-angular structure.

$$\tilde{A} = \begin{bmatrix} A_{11} & \ddots & \mathbb{O} & A_{1N+1} \\ \mathbb{O} & & A_{NN} & \vdots \\ A_{N+1\ N} & \cdots & \cdots & A_{N+1\ N+1} \end{bmatrix}$$

Definition 7 : Associated tree

A separating set C of a simply connected graph defines a partition of nodes $P = \{P_1, \ldots, P_N, C\}$ such that the sub-graphs $\mathcal{G}_{P_i}$ are simply connected. From separating sets we thus define increasingly fine partitions of P and it is agreed to represent this hierarchical structure induced by simple connexity by a tree-like diagram (see figure 2.3).

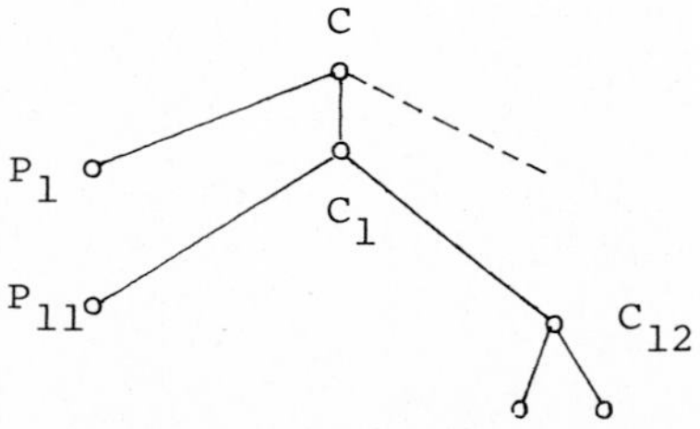

Figure 2.3 - Tree-like diagram

Let us note that the structural decompositions thus proposed are often called "nested decomposition", [WIN - 74]. See also section 2.4.2.3.

Property 2 :

A simply connected component of $\mathcal{G}$ has a hierarchical structure if, and only if, it is not strongly connected.
Furthermore, if the graph $\mathcal{G}(A)$ is not simply connected, there is a possibility of block-diagonalisation.
On the other hand, if $\mathcal{G}$ is simply connected but not strongly connected there is a possibility of block-triangularisation.
It is evident from this that the identification of strongly connected sub-graphs of $\mathcal{G}$ determines the irreducible structures of the system and that the relationships of simple connexity reveal general structures of hierarchical interconnections.
This graphic analysis therefore makes it possible to show partitions into irreducible sub-systems connected in series-parallel form in a hierarchical manner, suggesting parallelo-sequential phases of calculation in an overal procedure which could be qualified as distributed co-ordination.
Determination of the sub-systems with a view to structural decomposition is thus brought back to the search for the strongly and simply connected components of a graph.
We refer to [RIC - 75] for the presentation of systematic methods of identification of such components and separating sets
Naturally several partitionings are possible and so it is a difficult problem of choice based on the number of components which are strongly or simply connected, the number of separating sets, the number of elements in these sets... which express graphically the number of sub-systems, their size, the number of coupling variables...

2.3.2.2. Generalised semi-iterative methods

1) Generalised Richardson methods

These can be formulated as follows :

$$x^{(k+1)} = x^{(k)} + D_{k+1} \; r^{(k)} \quad ; \quad k \geq 0 \quad ; \text{ arbitrary } x^{(o)}$$

where D_{k+1} are non singular diagonal matrices and where $r^{(k)}$ expresses the residual vector at step k :

$$r^{(k)} \triangleq b - A\, x^{(k)}$$

namely :

$$x^{(k+1)} = (\mathbb{1} - D_{k+1} A)\, x^{(k)} + D_{k+1}\, b \quad ; \; k \geq 0$$

Note 1 - It is also possible to conceive a generalised Richardson method of order 2 such as :

$$x^{(k+1)} = G_{k+1}\, x^{(k)} + H_{k+1}\, x^{(k-1)} + \tilde{b} \qquad k \geq 0$$

Let us note that the conjugate gradient method and the cyclic iterative methods which we shall shortly present fit into this formalism.

2) Conjugate gradient method

For a detailed an complete description of the method refer to [GOL - 81] . It is a semi-iterative method, which consists in generating a sequence $\{y(k)\}$, $k \geq 0$, converging in a finite number of steps to the true solution of $Ax = b$ ($A = M - N$; A and M positive definite, possible generalisation given in the reference). The sequence is such that if setting

$$M\, z^{(k)} = r^{(k)} \triangleq b - A\, y^{(k)} \; ; \qquad k > 0$$

then

$$z^{(p)T}\, M\, z^{(q)} = 0 \qquad p \neq q \text{ and } z^{(k)} = 0 \text{ for some } k \leq n$$

This leads to the conjugate gradient method (standard case with $M = \mathbb{1}$);

$$Mz^{(k)} = b - Ay^{(k)} \; ; \quad k > 0 \quad : y^{(o)} \quad \text{arbitrary}$$

$$y^{(k+1)} = y^{(k-1)} + \omega_{k+1} [\alpha_k\, z^{(k)} + y^{(k)} - y^{(k-1)}];$$

$$k \geq 1;\; y^{(1)} = \alpha_o\, z^{(o)} + y^{(o)}$$

$$\alpha_k = \frac{z^{(k)T}\, M\, z^{(k)}}{z^{(k)T}\, A\, z^{(k)}} \quad ; \quad k \geq 0$$

$$\omega_{k+1} = \left[1 - \frac{1}{\omega_k}\, \frac{\alpha_k}{\alpha_{k-1}}\, \frac{z^{(k)T}\, M\, z^{(k)}}{z^{(k-1)T}\, M\, z^{(k-1)}} \right]^{-1} \quad ; \; k \geq 1; \; \omega_1 = 1$$

Note 1 - This method minimizes, for a given initial condition $\varepsilon^{(o)}$ the quadratic norm : $\frac{1}{2}\; \varepsilon^{(k+1)}\, A\, \varepsilon^{(k+1)}$; ($\mathcal{E} \equiv y - x^*$)

Note 2 - If the ($\mathbb{1} - M^{-1} N$) matrix has d distinct eigen values it can be shown that the convergence can be achieved in at most d iterations.

3) Alternating - directions methods

a) Peaceman-Rachford method

From a decomposition of $A = A_1 + A_2$, this can be put in the following form, where the acceleration parameters ω may vary with the iterations :

$$\begin{cases} (A_1 + \omega\, \mathbb{1}\,)\; x^{(k+1/2)} = (\,\omega \mathbb{1} - A_2) x^{(k)} + b \\ (A_2 + \omega \mathbb{1})\; x^{(k+1)} = (\,\omega\, \mathbb{1} - A_1)\; x^{(k+1/2)} + b \end{cases} \qquad k \geq 0$$

It is assumed that the linear systems $(A_1 + \omega \mathbb{1}\,)\; x = b_1$ and $(A_2 + \omega\, \mathbb{1}\,)\; x = b_2$) are easy to resolve, A_1 and A_2 being in a triadiagonal or equivalent form in many cases of interest. See for example in [VAR - 62] an application to the resolution of partial differential equations of elliptical type approximated by finite differences.

Let us observe that in this type of method the vector $x^{(k+1/2)}$,

appears as an auxiliary vector which is not retained during the complete iteration and therefore does not increase the volume of calculation.
Finally let us draw attention to the generalisation of this procedure attributed to HABETLER and WACHSPRESS [LIO - 74] and the corresponding theorem expressing a sufficient condition of convergence :

$$\begin{cases} (A_1 + D)\, x^{(k+1/2)} = (D - A_2)\, x^{(k)} + b \\ (A_2 + D)\, x^{(k+1)} = (D - A_1)\, x^{(k+1/2)} + b \end{cases} \qquad k \geq 0$$

where D is a symmetric and positive definite matrix of acceleration.

Theorem :

If A_1 and A_2 are positive definite (not necessarily symmetric), then the method converges.

b) Symmetric successive over-relaxations

The efficiency of symmetrisable methods is well known to numerical analysts, this being linked to the particular properties of real positive matrices.
We give below a symmetrisation of a process of over-relaxation :

$$\begin{cases} (D - \omega L)\, x^{(k+1/2)} = [\omega U + (1 - \omega)\, D]\, x^{(k)} + \omega b \\ (D - \omega U)\, x^{(k+1)} = [\omega L + (1 - \omega)\, D]\, x^{(k+1/2)} + \omega b \end{cases} \qquad k \geq 0$$

which could also be written :

$$x^{(k+1)} = T\, x^{(k)} + t.$$

In the case where A is a real symmetric matrix with positive diagonal elements, we can show that the method converges only if A is positive definite and if $0 < \omega < 2$. Thus T has real and positive eigen values and the convergence can be activated by semi-iterative methods. Even if the symmetric method of over-relaxation with ω optimal is generally slower than the method of optimal over-relaxation, it does not appear possible on the other hand, to accelerate the convergence of the latter by semi-iterative methods.

4) Cyclic iterative methods

These methods make use of the structural properties of cyclicity which we mentioned in section 2.3.2.1..
In fact, if the choice of the parameters of acceleration of the semi-iterative methods is in general made in a heuristic and experimental way, it can also be dictated by particular structural properties and especially in the case of p-cyclic matrices consistently ordered.

a) Semi-iterative cyclic Chebyshev method

This method applies to the case where the iteration matrix T is hermitian and weakly cyclic with index 2, of the form :

$$T = \begin{bmatrix} \mathbb{O} & T_1 \\ \overline{T_1}' & \mathbb{O} \end{bmatrix}$$

where the zero sub-matrices are square.
This partitioning corresponds to :

$$x_1 = T_1 x_2 + t_1$$

$$x_2 = \overline{T_1}' x_1 + t_2$$

As was shown in section 2.3.1.2.-b, Chebyshev iterative method is characterised by the use of a sequence of parameters of acceleration ωk defined from Chebyshev polynomials, contrary to the method of successive over-relaxation where ω is fixed. The cyclic semi-iterative method of Chebyshev applies the following scheme :

$$x_1^{(2k+1)} = \omega_{2k+1} \left[T_1 x_2^{(2k)} + t_1 - x_1^{(2k-1)}\right] + x_1^{(2k-1)} ; \quad k \geqslant 1$$

$$x_1^{(1)} = T_1 x_2^{(o)} + t_1$$

$$x_2^{(2k+2)} = \omega_{2k+2} \left[\overline{T_1}' x_1^{(2k+1)} + t_2 - x_2^{(2k)}\right] + x_2^{(2k)} ; \quad k \geqslant 0$$

$$x_2^{(2)} = \overline{T_1}' x_1^{(o)} + t_2$$

Figure (2.4) shows this iterative process.

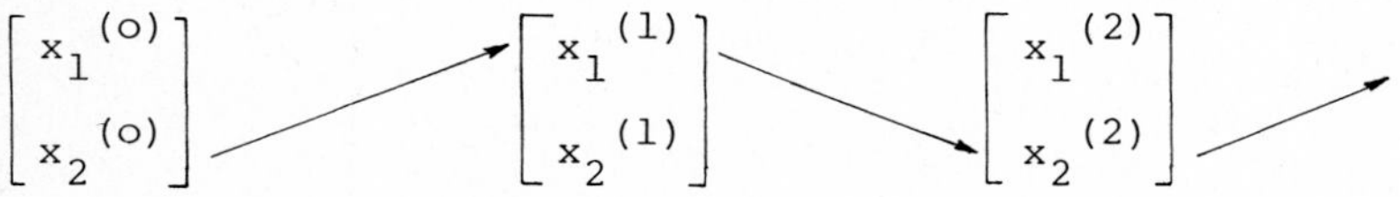

Figure 2.4

This method has an asymptotic rate of convergence equivalent to that of the successive over-relaxation method [VAR - 62].
Let us note that by applying theorem 4, (section 2.3.2.1. - 3b), we know that the optimal acceleration parameter of the process of over-relaxation applied to the matrix T is given by the relation $\omega\text{opt} = 2 / 1 + \sqrt{1 - \rho^2(T)}$, which corresponds to $R_\infty = - \text{Log} (\omega\text{opt} - 1)$.
However, whereas this method requires an initial vector approximation of x_1 and x_2, Chebyshev's cyclic method requires only the approximation of the 2nd half of x, namely x_2.

b) Cyclic iterative method of an expanded system

When the matrix of iteration T does not have the preceding characteristics, but is a convergent nxn matrix with real eigen values, we can consider the following expansion of the initial scheme.

$$\begin{cases} x = Ty + t \\ y = Tx + t \end{cases}$$

namely : $z = \tilde{T}z + \tilde{t}$

with : $z = \begin{bmatrix} x \\ y \end{bmatrix}$; $\tilde{T} = \begin{bmatrix} 0 & T \\ T & 0 \end{bmatrix}$; $\tilde{t} = \begin{bmatrix} t \\ t \end{bmatrix}$

We note that $\tilde{T}$ is weakly cyclic with index 2, that $\rho(\tilde{T}) = \rho(T)$, which causes the convergence of the process towards the unique solution x = y of the initial process x = Tx + t.
In virtue of this, if we apply a scheme of over-relaxation, namely :

$$\begin{cases} x^{(k+1)} = \omega[Ty^{(k)} + t - x^{(k)}] + x^{(k)} \\ y^{(k+1)} = \omega[Tx^{(k+1)} + t - y^{(k)}] + y^{(k)} \end{cases} \qquad k \geq 0$$

we then have the theorem [VAR - 62] :

Theorem :

Let T be a convergent nxn matrix with real eigen values. The iterative process above derived from the expanded matrix T is such that the optimal asymptotic rate of convergence is given by :

$$R_\infty\left(\omega_{opt} = \frac{2}{1 + \sqrt{1 - \rho^2(T)}}\right) = -\operatorname{Log}(\omega_{opt} - 1)$$

The interest of this procedure is therefore to give a means of calculating ω_{opt}, the disadvantage is to double the size of the system to be resolved (as does Chebyshev method).
Let us observe that by putting :

$$\begin{cases} s^{2k} = x^k \\ s^{2k+1} = y^k \end{cases} \qquad k \geq 0$$

we get :

$$s^{k+1} = \omega[Ts^k + t - s^{k-1}] + s^{k-1}; \qquad k \geq 1$$

which is no more than a Chebyshev type method where ω would be fixed all along the iterative process (see section 2.3.1.2.-b). In fact, one can show that the ω_k calculated from the Chebyshev polynomials decrease in a monotone fashion towards ω_{opt} [Var - 62].

2.4 Application to control problems

The importance of the results obtained in the iterative methods

of decomposition of algebraic equations systems naturally leads to an interest in an extension of their applications to systems of continuous differential equations.
The most immediate extension, which we have already shown in previous sections, consists of going back to a system of algebraic equations by a process of discretisation. Let us point out, however, the limitations of this procedure which reverts to a considerable increase in the number of variables and to the artificial creation of a large scale problem.
Another possible extension, just as natural and which we shall enlarge on in section 2.4.4., concerns the resolution of the system of algebraic equations corresponding to the stationary case of a differential system.
Let us return to the extension properly so called, to systems of differential equations, ie. to the extension of methods of decomposition in R^n to decomposition in function space. This will be the subject of section 2.4.2. which will deal with linear differential systems.
First we shall approach the main numerical difficulty linked with problems of optimal control, namely the integration of a differential system with boundary conditions at several points defined by boundary conditions and conditions of transversality (if necessary, of continuity) [GEL - 63].
Section 2.4.1. presents a problem with two point boundary conditions.

2.4.1. Two-point-boundary-value problem

It can, in fact, be considered that there are no major theoretical problems, apart from dimensionality, in the resolution of a continuous differential system with one single point boundary conditions. This finds expression in the resolution of a Volterra-type integral which is a Cauchy problem and of which one knows, by application of the generalised principle of contraction (cf. section 2.2.3.) that the solution can always be obtained after a sufficient number of iterations [KOL - 57].
This is no longer the case of the problems of optimisation which we shall illustrate by the following problem (Pb of Bolza).

$$\left[\begin{array}{l} \min\limits_{u} \int_{t_o}^{t_f} p\,(x, u, t)\, dt \\ \text{s.t.} \quad \dot{x} = f\,(x,u,t)\ ;\ x\,(t_o) = x_o;\ t \in [t_o, t_f] \\ x \in \mathbb{R}^n, \quad u \in \mathbb{R}^m \end{array}\right. \tag{2.3}$$

where :
- f is a vector function whose components are defined in the class of functions which are continuously differentiable in relation to all its arguments which guarantees in particular the uniqueness of the solution of the differential equations for $x\,(t_o)$ and $u\,(t)$ given ;
- p is a continuously differentiable scalar function.

It is well known that the necessary conditions of stationarity of this problem give rise to a differential system with boundary conditions at two points [ATH - 66].

$$\begin{cases} \dot{x} = f(x, u, t) \; ; \; x(t_o) = x_o \\ \dot{\psi} = -\mathcal{H}_x (x, u, \psi, t) \; ; \; \psi(t_f) = 0 \\ \mathcal{H}_u (x, u, \psi, t) = 0 \end{cases} \tag{2.4}$$

where $\mathcal{H}[x, u, \psi, t] = p(x, u, t) + \psi(t)^T f(x, u, t)$ is the Hamiltonian function.
We can consider that what is involved is a system of differential equations with two-boundary conditions linked by an algebraic type system of equations corresponding to the conditions of stationarity relating to u. Generally one is induced to use procedures of iterative solution and one can, for example, bring this system back to a differential system with conditions at only one single point by a process of quasi-linearisation, at the cost however of a notable increase in the number of variables. [CAL - 79] . We shall not cover here the different techniques of solution of two-point-boundary-value problems [KIR - 70],[BER - 76],...,... and we are simply accentualing the possibility of applying a process of relaxation such as a Richardson type method of first or second order to the resolution of the conditions of stationarity $\mathcal{H}_u = 0$.
For example, a method of first order would apply the following iterative scheme :

$$\begin{cases} \dot{x}^{(k)} = f(x^{(k)}, u^{(k)}, t) \; ; \; x^{(k)}(t_o) = x_o \\ \dot{\psi}^{(k)} = -\mathcal{H}_x (x^{(k)}, u^{(k)}, \psi^{(k)}, t); \; \psi^{(k)}(t_f) = 0 \\ u^{(k+1)} = u^{(k)} + \alpha k \mathcal{H}_u (x^{(k)}, u^{(k)}, \psi^{(k)}, t) \end{cases} \tag{2.5}$$

This decomposition implies a sequential solution of the system of differential equations, the state equations being integrated in the forward sense, the adjoint equations in the retrograde sense. Let us note that this has the advantage of integrating these equations in the sense of their asymptotic stability.
The most currently used of the possibilities for determining the αk parameters is to use a gradient type method, aiming to minimise the Hamiltonian in relation to u; the αk being chosen negative and the solution in the sense steepest descent of a unidirectional problem of optimisation.
It is interesting to note that this type of procedure guarantees a monotone decrease of the criterion to be optimised and makes possible an "on-line" implementation of the algorithm. Naturally in the general case where the strict convexity of the problem is not guaranteed, this leads to a local solution.
Let us also mention, when one is sufficiently close to the solution, the possibility of accelerating the convergence and determining the αk from a method of Newton-Raphson :

$$\alpha k = -c[\mathcal{H}_{uu} (x^{(k)}, u^{(k)}, \psi^{(k)}, t)]^{-1} \qquad 0 < c < 2$$

as well other more sophisticated predictive methods such as that proposed in [SEW - 75].
Let us finally mention the approach by G. COHEN [COH - 78] di-

rectly related to the Prediction Principle of Interactions [MES - 70] which will be discussed in the section 2.4.2.1.

The possibilities which we have just raised ("on line" application, local convergence, acceleration of convergence) suggest a particularly interesting area of application of this process of relaxation to the iterative amelioration of a sub-optimal control, the sub-optimality coming, for example, from a linearisation of the system [CAL - 76], from a choice of structure of decentralised control [GER - 79], from the constant approximation of a feedback gain, from a combination of all this...
We refer to [CAL - 76][HUR - 77] for numerical and (or) analog simulations of such structures of hierarchical controls, causing to appear a "splitting" of the control function u (t) into a linear (decentralised) regulation and an "on line" improved open loop complement :

$$u(t) = \hat{G}\, x(t) + u_o(t)$$

2.4.2. Partitioned methods in the case of Linear Quadratic Systems (LQS)

As with the algebraic systems, we shall restrict ourselves from a methodological viewpoint to the description of partitioned methods for the solution of linear differential systems. We refer in particular to [SAN - 79] for extensions to non-linear systems.
As we have already mentioned, since the main difficulty appearing in the numerical resolution of different systems is linked with the boundary conditions at several points, we are therefore concerned with optimisation problems of linear dynamic systems. The choice of a quadratic criterion will determine linear conditions of stationarity.
Let us therefore take up again the preceding optimisation problem where :

$$f(x, u, t) = A(t)\, x(t) + B(t)\, u(t)$$

$$p(x,u,t) = \frac{1}{2}\left(x(t)^T\, Q\, x(t) + u(t)^T\, R\, u(t)\right); \quad \begin{matrix} Q \geq 0 \\ R > 0 \end{matrix} \text{ symmetric}$$

The necessary conditions of stationarity (2.4) become :

$$\left[\begin{array}{ll} \dot{x}(t) = A(t)\, x(t) + B(t)\, u(t); & x(t_o) = x_o \\ \dot{\psi}(t) = -\, Q\, x(t) - A^T(t)\, \psi(t); & \psi(t_f) = 0 \\ R\, u(t) + B^T(t)\, \psi(t) = 0 & \end{array}\right. \qquad (2.6)$$

As opposed to the preceding section, we are not going to separate the state system and the adjoint system, but we shall predict the variable u(t) from the last condition of stationarity.

We thus define the system of linear differential equations with two-boundary-conditions :

$$\begin{cases} \dot{x}(t) = A(t)\,x(t) - B(t)\,R^{-1}\,B^T(t)\,\psi(t); & x(t_o) = x_o \\ \dot{\psi}(t) = -\,Q\,x(t) - A^T(t)\,\psi(t)\;; & \psi(t_f) = 0 \end{cases}$$

which can also be written as :

$$Dz - Hz = 0;\ +\ \text{boundary conditions (B.C.)} \qquad (2.7)$$

- $z(t)$ is the composite vector $\begin{bmatrix} x \\ \psi \end{bmatrix}$
- $D(.)$ is the diagonal operator of differentiation in relation to the time t.
- H is the hamiltonian matrix of the system :

$$H(t) = \begin{bmatrix} A(t) & -\,B(t)\,R^{-1}\,B^T(t) \\ -\,Q & -\,A^T(t) \end{bmatrix}$$

The operator D being diagonal, the partitioned methods which we shall present will therefore be based on a partitioning and a "splitting" of the hamiltonian matrix H.
Let us draw attention to a slightly more general presentation in [LHO - 79] which conserves the equation of stationarity related to the definite variable u(t) in a certain convex area of admissibility U.
Let us finally observe that the decomposition method presented in the previous section (equation (2.5) where $\alpha k = -\mathcal{H}_{uu}^{-1} = -\,R^{-1}$ can also be defined as a Gauss-Seidel type decomposition scheme of the system (2.7) where H (t) is decomposed in the following way :

$$H = H_o + H_1$$

$$H_o(t) = \begin{bmatrix} A\,(t) & 0 \\ -\,Q & -A^T(t) \end{bmatrix}$$

$$H_1(t) = \begin{bmatrix} 0 & -\,B(t)\,R^{-1}\,B^T(t) \\ 0 & 0 \end{bmatrix}$$

The iterative process being expressed as follows :

$$Dz^{(k+1)} - H_o z^{(k+1)} = H_1 z^{(k)} \quad ; \qquad k \geqslant 0$$

2.4.2.1. Parallel decomposition

1) <u>Method of prediction of interactions</u>

Let us now consider the following partitioning of the matrices A, B, Q and R where the last three matrices are chosen block-diagonal in a non-restrictive way in order to simplify the exposition.

$$A = \begin{bmatrix} A_{11} & \cdots & A_{1N} \\ \vdots & \ddots & \vdots \\ A_{N1} & \cdots & A_{NN} \end{bmatrix}; \quad B = \begin{bmatrix} B_1 & & \mathbb{O} \\ & \ddots & \\ \mathbb{O} & & B_N \end{bmatrix}; \quad Q = \begin{bmatrix} Q_1 & & \mathbb{O} \\ & \ddots & \\ \mathbb{O} & & Q_N \end{bmatrix}; \quad R = \begin{bmatrix} R_1 & & \mathbb{O} \\ & \ddots & \\ \mathbb{O} & & R_N \end{bmatrix} \tag{2.8}$$

defining the sub-matrices A_{ii} (nixni); Q_i (nixni) ; R_i (mixmi); B_i (nixmi)

with $\sum_{i=1}^{N} n_i = n$; and $\sum_{i=1}^{N} m_i = m$.

From the possible methods of parallel decomposition based on this partitioning, we present here a relaxation scheme leading to a decomposition of the system (2.7) into N two-point-bondary-value-sub-problems, ie. we are re-writing this system as follows :

$$Dz - Hz = 0 + \text{boundary conditions}$$

where :

$$* \quad z = \begin{bmatrix} x_1 \\ \psi_1 \\ \vdots \\ x_N \\ \psi_N \end{bmatrix}; \quad x_i, \psi_i \in \mathbb{R}^{ni}$$

$$* \quad H = \begin{bmatrix} \begin{bmatrix} A_{11} & -B_1 R_1^{-1} B_1^T \\ -Q_1 & -A_{11}^T \end{bmatrix} & \cdots & \begin{bmatrix} A_{1N} & \mathbb{O} \\ \mathbb{O} & -A_{N1}^T \end{bmatrix} \\ \vdots & \ddots & \vdots \\ \begin{bmatrix} A_{N1} & \mathbb{O} \\ \mathbb{O} & -A_{1N}^T \end{bmatrix} & \cdots & \begin{bmatrix} A_{NN} & -B_N R_N^{-1} B_N^T \\ -Q_N & -A_{NN}^T \end{bmatrix} \end{bmatrix} \tag{2.9}$$

If we then consider a splitting of the matrix $H = H_o + H_1$ where H_o is the block-diagonal part of H, we then obtain the following iterative scheme of parallel decomposition :

$$D_i \, (z_i)^{(k+1)} - H_{oi} \, z_i^{(k+1)} = H_{1i} \cdot z^{(k)} \ (+\text{B.C.}); \tag{2.10}$$

$$k \geqslant 0; \quad i = 1, \ldots, N$$

or again

$$\begin{bmatrix} \dot{x}_i \\ \dot{\psi}_i \end{bmatrix}^{(k+1)} - \begin{bmatrix} A_{ii} & - B_i R_i^{-1} B_i^T \\ -Q_i & - A_{ii}^T \end{bmatrix} \begin{bmatrix} x_i \\ \psi_i \end{bmatrix}^{(k+1)} \tag{2.11}$$

$$= \begin{bmatrix} \sum_{j \neq i\, 1}^{N} A_{ij} x_j \\ - \sum_{j \neq i\, 1}^{N} A_{ji}^T \psi_j \end{bmatrix}^{(k)} ; \quad \begin{matrix} x_i (t_o) = x_{io} \\ \psi_i (t_f) = 0 \end{matrix} \qquad i=1,\ldots,N$$

Let us first of all observe that the hypothesis of block-diagonality of the matrices B, Q and R has only one incidence which is to make zero sub-matrices appear in H, but is at no point necessary with respect to the principle of application.
Let us also note that this method of decomposition is also known by the name "method of prediction of Interactions" by Takahara [TAK - 65]. Indeed, if one adopts the more conventional formalism of the hierarchical control, this decomposition is made to appear as a decomposition into N sub-problems of optimisation corresponding to the N systems with conditions at both ends (2.11), made independent by the fixing of artificial parameters of co-ordination at a higher level.
For this, let us consider the following partitioned expression of the initial problem of optimisation :

$$\begin{cases} \min_u J \text{ where } J = \sum_{i=1}^{N} \frac{1}{2} \int_{t_o}^{t_f} (x_i^T Q_i X_i + u_i^T R_i u_i) dt \\ \text{s. } \dot{x}_i = A_{ii} x_i + B_i u_i + v_i ; \; x_i (t_o) = x_{io} \\ (*) \; v_i = \sum_{\substack{j=1 \\ j \neq i}}^{N} A_{ij} x_j \end{cases}$$

to which is associated the Lagrangian $\mathcal{L}$.

$$\mathcal{L} = \sum_i \Big[\frac{1}{2} (x_i^T Q_i x_i + u_i^T R_i u_i) + \psi_i^T (A_{ii} x_i + B_i u_i + v_i - \dot{x}_i) + \beta_i^T (v_i - \sum_j A_{ij} x_j) \Big]$$

where β_i is the vector of the Lagrange multipliers associated with the coupling constraint (*).
Bearing in mind that

$$\sum_{i=1}^{N} \Big[\beta_i^T (v_i - \sum_{j=1}^{N} A_{ij} x_j) \Big] = \sum_{i=1}^{N} \Big[\beta_i^T v_i - \sum_{j=1}^{N} \beta_j^T A_{ji} x_i \Big]$$

we find that the fixing of β as a parameter of co-ordination decouples the Lagrangian. This observation is at the basis of decomposition methods termed dual [LAS - 70] , nofeasible or of co-ordination by the criterion [TIT - 75],of decomposition by decoupling or of co-ordination by the prices [BER - 76]... The determination of β at a higher level corresponds to the resolution of an algebraic system of interactions : φ_β (x, v) which can be carried out, for example, under the hypothesis of strict convexity by application of theorem of strong duality which leads to the resolution of the dual problem of optimisation [CAL - 76] :

$$\underset{\beta}{\text{Max}} \left[\underset{u,\, v}{\text{Min}}\ \varphi \right]$$

β can then be calculated by techniques of extremisation of functions (alg. of gradient, of Fletcher-Reeves conjugate gradient ...).
The "Prediction of Interactions" method consists of fixing, in addition to the parameters β , the coupling variables v at a higher level, ie. of considering the algebraic system of interconnection redundant :

$$\varphi_{\beta i} = v_i - \sum_{j=1;j\neq i}^{N} A_{ij}\, x_j = 0 \qquad\qquad i = 1,\dots,N$$

$$\varphi_{v_i} = \beta_i + \psi_i = 0$$

the consequence of which is to make possible a direct procedure of determination of the co-ordination parameters in accordance with a Newton Raphson type iterative scheme, namely :

$$\begin{cases} v_i^{(k+1)} = \sum\limits_{\substack{j=1 \\ j\neq i}}^{N} A_{ij}\, x_j^{(k)} & \\ & i = 1,\dots, N; \quad k \geqslant 0 \\ \beta_i^{(k+1)} = -\ \psi_i^{(k)} & \end{cases} \tag{2.12}$$

Having defined $v_i^{*} \triangleq v_i^{(k+1)}$ and $\gamma_i^{*} \triangleq -\sum\limits_{\substack{j=1 \\ j\neq i}}^{N} A_{ji}^{T}\, \beta_j^{(k+1)}$

the independent sub-problems of optimisation handled at a lower level at step (k+1) correspond to :

$$\begin{cases} \underset{u_i^{(k+1)}}{\text{Min}} \displaystyle\int_{t_o}^{t_f} \left[\frac{1}{2} \left(x_i^{T(k+1)} Q_i x_i^{(k+1)} + u_i^{T(k+1)} R_i u_i^{(k+1)}\right) \right. \\ \left. \qquad + \gamma_i^{T*}\, x_i^{(k+1)} \right] dt \\ \text{s.}\quad \dot{x}_i^{(k+1)} = A_{ii} x_i^{(k+1)} + B_i u_i^{(k+1)} + v_i^{*};\ x_i^{(k+1)}(t_o) = x_{io} \end{cases} \tag{2.13}$$

which tends to justify the name "prediction of interactions"

method, also called : mixed co-ordination method by modification of the criterion (γ_i^*) and of the local model (v_i^*) [TIT - 75] .
One can easily verify that the resolution of these sub-problems and of the interconnection system (2.12) corresponds to the decomposition scheme (2.11)
Figure 2.5 represents the decentralisation of data and the exchanges of information in the parallel decomposition method.

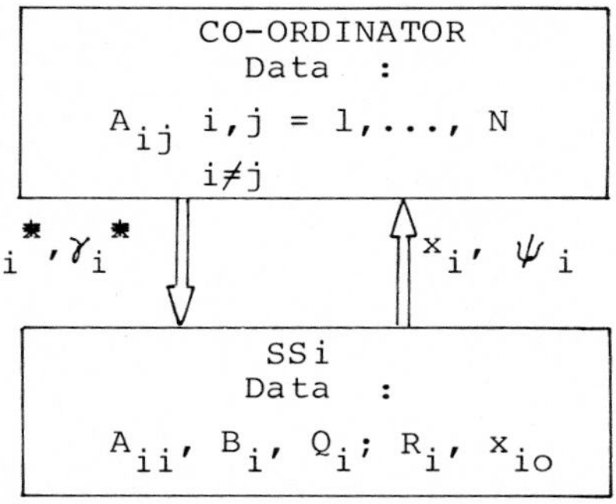

Figure 2.5

Let us note finally that if the dual method mentioned above does not appear explicity in the formalism (2.10), this is due to the fact that it can only be applied to L.Q. Systems after additional modification of the initial formulation in the introduction of the artificial variables v and β because of problems of singularity [GÅL - 73].

2) Study of convergence

Study of the convergence of this type of method, approached in different ways [TAK - 65] , [PEA - 71] , [COH - 80] ,[LHO - 79], [SAN - 79] , ... has given more or less exploitable results.
From a methodological viewpoint we are concerned with the relationship linking the convergence to the couplings which appear in such decompositions and which are of two types :
- the coupling between sub-systems linked to the partitioning and, under the circumstances, to the matrices A_{ij};
- the coupling between state and adjoint equations inherent to the multi-point-boundary-conditions linked, in the circumstances to the matrices $B_i R_i^{-1} B_i^T$ and Q_i as to the optimisation horizon time $(t_f - t_o)$.
In order to see this, let us consider the following iterative scheme of decomposition of a differential linear system with two-point-boundary conditions :

$$\begin{cases} \dot{z}^{(k+1)}(t) = + H_o(t)\, z^{(k+1)}(t) + H_1(t)\, z^{(k)}(t)\,; \quad z = \begin{bmatrix} x \\ \psi \end{bmatrix}; \quad k \geq 0 \\ t \in [t_o, t_f]\;;\; x^{(k)}(t_o) = x_o\;;\; \psi^{(k)}(t_f) = 0 \end{cases}$$

By defining $\varepsilon^{(k)} \triangleq z^{(k)} - z^{*}$ where $z^{*}(t)$ is the solution to the two-point-boundary value problem as the error of convergence at the step k, we obtain :

$$\varepsilon^{(k+1)}(t) = \int_{t_o}^{t} \Phi(t,\tau)\, H_1(\tau)\, \varepsilon^{(k)}(\tau)\, d\tau - \int_{t_o}^{t_f} \theta(t,\tau,t_o,t_f)\, H_1(\tau)\, \varepsilon^{(k)}(\tau)\, d\tau$$

with :

$$\dot{\Phi}(t,t_o) = H_o(t)\, \Phi(t,t_o) \ : \ \Phi(t_o,t_o) = \mathbb{1}_{2n}$$

$$\theta(t,\tau,t_o,t_f) = \begin{bmatrix} \phi_{12}(t,t_o) \\ \phi_{22}(t,t_o) \end{bmatrix} \phi_{22}^{-1}(t_f,t_o) \left[\phi_{21}(t_f,\tau)\ \ \phi_{22}(t_f,\tau) \right]$$

in which the partitioning of the matrix Φ is induced by that of the vector z in x and ψ satisfying the initial and final conditions.
Namely :

$$\varepsilon^{(k+1)}(t) \triangleq T\left[\varepsilon^{(k)}(t)\right] = T_1\left[\varepsilon^{(k)}\right] + T_2\left[\varepsilon^{(k)}\right] ; \ k \geqslant 0$$

with :

$$T_1[\varepsilon] = \int_{t_o}^{t} \Phi(t,\tau)\, H_1(\tau)\, \varepsilon(\tau)\, d\tau$$

$$T_2[\varepsilon] = -\int_{t_o}^{t_f} \theta(t,\tau,t_o,t_f)\, H_1(\tau)\, \varepsilon(\tau)\, d\tau$$

The operator $T_1[\varepsilon]$ is a Volterra type operator and $T_2[\varepsilon]$is a Fredholm type operator. If as we shall see in section 2.4.3., the first type of operator is always convergent after a sufficient number of applications, difficulties of convergence are more particularly inherent to the second type of operator[KOL - 57] . Let us note that $T_2[\varepsilon]$is directly linked by $\theta(t,\tau,t_o,t_f)$ to the coupling between state and adjoint equations subject to the two point boundary conditions.
By choosing, for example, a type of vector norm $|||x||| = (\|x_1\|,\ldots,\|x_n\|)^T$ where $\|x_i\| = \max_{t\in[t_o,t_f]} |x_i(t)|$, one can easily show that :

$$|||T(\varepsilon_1) - T(\varepsilon_2)||| \leqslant (t_f-t_o)\,(M_1+M_2)\,|||\varepsilon_1 - \varepsilon_2|||$$

where the matrices M_1 and M_2 are such that their elements $m_{ij} = \max_{t,\tau \in [t_o,t_f]} |k_{ij}(t,\tau)|$

where $k_{ij}(t,\tau)$ are the elements of the respective kernels $K_1(t,\tau)$ and $K_2(t,\tau)$ of the operators T_1 and T_2 :

$$K_1(t,\tau) = \Phi(t,\tau)\, H_1(\tau)$$

$$K_2(t,\tau) = \theta(t,\tau, t_o, t_f)\, H_1(\tau).$$

A sufficient condition of convergence can be provided by application of the contraction principle (see section 2.2.3.), namely :

$$(t_f - t_o)\ \rho(M_1 + M_2) < 1$$

So there always exists a theoretical possibility of reducing the optimisation horizon to ensure convergence of the algorithm.

ε- decoupling :

In order to help the convergence of this type of algorithm, this study is emphasizing the interest of the decoupling and offers a few methodological indications.

Firstly, concerning physical coupling between sub-systems, a first step consists of searching among the possible partitions for those which may be able to cause the appearance of an H_1 matrix whose spectral radius is very small. The direct effect of this is the diminution of $\rho(M_1 + M_2)$ and may be obtained, for example, by the introduction of a perturbation factor ε which is sufficiently small, in the expression of $H_1(t)$, namely :

$$H_1(t) = \varepsilon\, \tilde{H}_1(t)$$

$$\Longrightarrow \quad \rho(M_1 + M_2) = \varepsilon\rho(\tilde{M}_1 + \tilde{M}_2)$$

Let us note that for the given example this technique, known by the name of "regular perturbations", corresponds to the search for partitions such that $A_{ij} = \varepsilon\tilde{A}_{ij}$, and that in a rather more general case this type of coupling would also be linked to the matrices B_{ij}, Q_{ij} and R_{ij}^{-1}.

To this end we feel it is important to stress the usefulness of the application of precompensation techniques for dynamic systems in order to reduce interactions between sub-systems [MAY - 76] .

Finally, concerning coupling between state equations and adjoint equations at the level of each sub-problem (2.13), coupling linked to the optimisation horizon $(t_f - t_o)$ and to the matrix $H_o(t)$, this analysis proposes, for example, so far as the final solution is physically acceptable, a reduction of this horizon, a diminution of the weighting coefficients on the state (matrix Q_i) or again an increase of the weighting coefficients on the control (matrix R_i in $B_i R_i^{-1} B_i^T$).

In a rather more general presentation already mentioned in section 2.4.2., Lhote and Miellou give another sufficient condition of convergence based on the following definition of uniform vector minoration in relation to the vector norm :

$$|||x||| = (\|x_1\|, \ldots, \|x_N\|)^T$$

where $\|x_i\|$ is the Euclidean norm :

$$\|x_i\| \triangleq \sqrt{<x_i, x_i>} = [\int_{t_o}^{t_f} x_i^T(\tau)\, x_i(\tau)\, d\tau\,]^{1/2}$$

Definition : Uniform vector minoration

The partitioned matrix L (t) allows a uniform vector minoring K if, and only if :

$$k_{ij}\,\|x_i\|\,\|x_j\| \leq <L_{ij}(t)\, x_j(t),\, x_i(t)> \qquad \forall t \in [t_o, t_f] \quad \forall_{i,j} = 1, \ldots, N$$

In the case of the partitioned system (2.9), by putting L = - H, we show that a uniform vector minoring K of L (t) is also minoring of D + L where the terminal conditions are brought down to zero, and that a sufficient condition of convergence is that K is an M-matrix.
In this condition we can show that the Jacobi application is a P-contraction, ie :

$$|||T(\varepsilon_1) - T(\varepsilon_2)||| \leq P\,|||\varepsilon_1 - \varepsilon_2|||;\ \forall \varepsilon_1, \forall \varepsilon_2 \in D_o$$

$$\text{where } P = [\text{Diag. } (K)]^{-1}\ [\text{Diag. } (K) - K]$$

We can easily show in the context of the ultimate decomposition into scalar sub-systems and when H (t) is a constant matrix H, that the greatest uniform vector minoring K of - H is such that:

$$k_{ii} = - h_{ii} \qquad i = 1, \ldots, N$$

$$k_{ij} = - |h_{ij}| \qquad i,j = 1, \ldots, N;\ j \neq i$$

2.4.2.2. Sequential decomposition

If as we have seen in section 2.3, the methods of sequential (generally over-relaxed) decomposition are much used for the resolution of algebraic systems, it is curious to note that little literature concerning the optimisation of continuous dynamic systems accords them a particular usefulness [COH - 80]. In saying this we exclude of course the discrete resolutions of certain optimisation problems which we have previously mentioned [WIS - 71].
The aim of this section is to present and discuss the details of certain sequential schemes of decomposition applied to systems in a continuous form. We are particularly emphasizing the structural exploitations which can be envisaged in cases where systems have a cascade type interconnection structure. Finally, we shall extend the methodologies presented to discrete systems.

1) Splitting methods

We shall not return here to the sequential decomposition scheme presented in section 2.4.1., whose major interest was, to separate the resolution of the state system from that of the adjoint system. We are then concerned, as in the case of parallel decompositions in partitioned methods of L.Q.S. problems leading to the resolution of N two-point-boundary-value-subproblems.
Resuming the partitioning defined in section 2.4.2.1. - 1, the decomposition methods of the system (2.9) which we shall present are therefore based on different "splittings" of the hamiltonian matrix $H = H_o + H_1$.

a) "Direct" method

The method which we describe in this way corresponds to the case where H_o represents the lower block-triangular part of H, ie. where :

$$H_o = H_D + H_L$$

which $$H_D = \text{Block Diag.}\begin{bmatrix} A_{ii} - B_i R_i^{-1} B_i^T \\ -Q_i & -A_{ii}^T \end{bmatrix} \quad i = 1,\ldots,N$$

H_L : strictly lower block-triangular part of H.

$H_1 = H_U$: strictly upper block-triangular part of H.

whence the sequential iterative process :

$$D_i\ (z_i^{(k+1)}) - H_{oi}\ z^{(k+1)} = H_{1i} z^{(k)} \quad (+\text{B.C.});\ k \geq 0; i=1 \rightarrow N$$

or again for the example chosen :

$$\begin{bmatrix} \dot{x}_i \\ \dot{\psi}_i \end{bmatrix}^{(k+1)} - \begin{bmatrix} A_{ii} - B_i R_i^{-1} B_i^T \\ -Q_i & -A_{ii}^T \end{bmatrix}\begin{bmatrix} x_i \\ \psi_i \end{bmatrix}^{(k+1)} - \begin{bmatrix} v_i \\ \theta_i \end{bmatrix}^{(k+1)} = \begin{bmatrix} \omega_i \\ \gamma_i \end{bmatrix}^{(k)}$$

whith

$$v_i^{(k)} = \sum_{j=1}^{i-1} A_{ij}\ x_j^{(k)};\ \omega_i^{(k)} = \sum_{j=i+1}^{N} A_{ij}\ x_j^{(k)}$$

$$\gamma_i^{(k)} = -\sum_{j=i+1}^{N} A_{ji}^T\ \psi_j^{(k)};\ \theta_i^{(k)} = -\sum_{j=1}^{i-1} A_{ji}^T\ \psi_j^{(k)}$$

b) "Dual" method

This method, whose name we shall justify in the next section, corresponds to the splitting of $H = H_o + H_1$.

where : $H_o = H_D + H_U$

$$H_1 = H_L$$

ie. for the example given it develops the sequential iterative process in the reverse direction :

$$\begin{bmatrix} \dot{x}_i \\ \dot{\psi}_i \end{bmatrix}^{(k+1)} - \begin{bmatrix} A_{ii} - B_i R_i^{-1} B_i^T \\ -Q_i \quad -A_{ii}^T \end{bmatrix} \begin{bmatrix} x_i \\ \psi_i \end{bmatrix}^{(k+1)} - \begin{bmatrix} \omega_i \\ \gamma_i \end{bmatrix}^{(k+1)} = \begin{bmatrix} v_i \\ \theta_i \end{bmatrix}^{(k)} \quad (+B.C.)$$

$$k \geq 0 \; ; \; i = N \longrightarrow 1$$

c) Alternating-directions method

In a certain way this method constitutes a generalisation of the preceding methods alternating the directions of direct and retrograde searching, which can be written as :

$$\begin{cases} Dz^{(k+1/2)} - (H_D + H_L)\, z^{(k+1/2)} = H_U z^{(k)} \quad (+B.C.) \\ Dz^{(k+1)} - (H_D + H_U)\, z^{(k+1)} = H_L z^{(k+1/2)} \quad (+B.C.) \end{cases} \quad k \geq 0$$

In addition to the property particular to sequential methods of reducing by half the storage volume of co-ordination parameters, this symmetrised method involves half as much calculation per "iteration", which causes the sub-systems to be scanned twice as rapidly.
Figure (2.6) shows the particular structure (in relation to Figure (2.2) section 2.3.1.1. - b) of the information exchanged in this type of method where, properly speaking, no upper level appears, one part of the information being itself stored from one "iteration" to the other, at the level of each sub-system, the other part being exchanged sequentially between sub-systems.

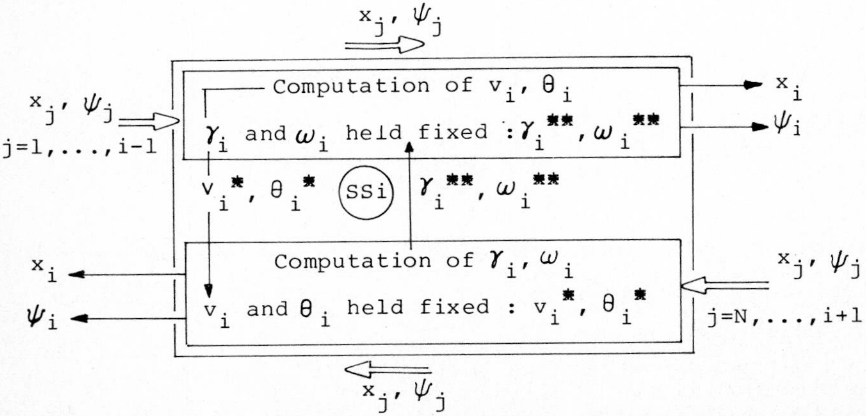

Figure 2.6 Communications in the alternating direction method

Finally, by analogy with section 2.3.2.2. (3-b), let us draw attention to the interest which a symmetric scheme of relaxation may have which, in the circumstances, would come back to application of the following algorithm :

$$(D - H_D - \omega H_L) z^{(k+1/2)} = [\omega H_U + (1-\omega)(D-H_D)] z^{(k)}$$
$$(D - H_D - \omega H_U) z^{(k+1)} = [\omega H_L + (1-\omega)(D-H_D)] z^{(k+1/2)} \qquad k \geqslant 0$$

For the three types of methods presented, the study of convergence can be made in the same way as in the previous section, with new definitions of the matrices H_o and H_1, to give analogous conclusions with respect to decoupling. In general the comparisons that can be made between these sequential approaches and a parallel approach enter into the general context of the comparisons which we were able to make in section 2.3.
However, we shall concern ourselves in the following sections with some aspects which are peculiar to problems of optimal control of systems which are interconnected in a particular manner. It is also interesting in this type of sequential method to return to the hierarchical character of the process of decomposition already mentioned in section 2.3.1.1.-b, to observe that, contrary to the parallel approach of section 2.4.2.1 -1, the different calculation levels (or sequences) involve here a double optimisation and co-ordination function.
Let us recall, finally, that this sequentialisation of the calculation scheme may be very advantageous in the case of a monoprocessor calculation scheme, more especially as the sub-problems here are of the same nature and can therefore be resolved by the same type of programme.

2) Application to the optimisation of multi-stage systems

Figure (2.7) represents the interconnection structure of these systems, also called parallel-serie structure.

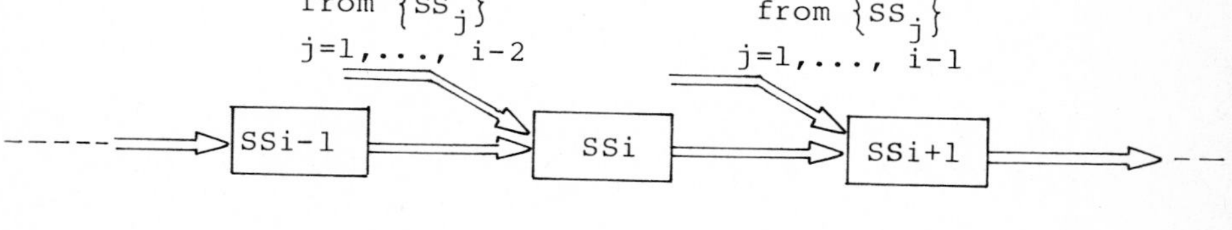

Figure 2.7 Multi-stage structure

Let us note that this kind of configuration appears frequently in practice where these systems are numerous (transport systems, hydraulic systems, chemical processes,...) [SAN - 76], [CHU - 76],...

a) Triangular structure

Still with reference to partitioning (2.8), we shall illustrate this type of structure by considering the A matrix as lower block-triangular type, i.e. :

$$A_{ij} = 0 \qquad j > i$$

Let us note that here again the example chosen is not restrictive and that we could have considered in particular matrix B,
itself lower block triangular type.
The effect of this type of structure is to eliminate the vectors
ω_i and θ_i in the expression of the algorithm, and the previous
"direct" method (respectively "dual") consists of fixing predic-
tively   either just the set of γ , the sequential resolution of
the N sub-problems being carried out for i : 1 → N, (or the set
of v (resolution for i : N → 1)).
These two methods appear to be definitely dual with regard to
the nature of the variables fixed at the upper level and to the
ordering of the calculation sequences. We note on the following
figures that they have, in relation to the TAKAHARA algorithm,
the interest of reducing by half the number of predicted variables at the upper level, the other half being calculated and
moved sequentially. The figures also show the types of sub-problems of optimisation resolved at step (k+1) at the level of
each of the sub-systems

a-1 "Direct" method

$$\gamma_i^* = - \sum_{j=i+1}^{N} A_{ji}^{T} \psi_j^{(k)}$$

$$i = 1, \ldots N$$

$$\gamma_i^* \qquad \psi_i (k+1)$$

$$\underset{u_i^{(k+1)}}{\text{Min}} \int_{t_0}^{t_f} \left[\frac{1}{2}(x_i^{(k+1)T} Q_i x_i^{(k+1)} + u_i^{(k+1)T} R_i u_i^{(k+1)} + \gamma_i^{*T} x_i^{(k+1)} \right] dt$$

$$\text{s. } \dot{x}_i^{(k+1)} = A_{ii} x_i^{(k+1)} + B_i u_i^{(k+1)} + v_i^*; \ x_i^{(k+1)}(t_0) = x_{io}$$

$$v_i^* = \sum_{j=1}^{i-1} A_{ij} x_j^{(k+1)} \qquad \underline{\text{S Pbi}}$$

$x_j^{(k+1)}$ j=1,...i-1 $x_i^{(k+1)}$

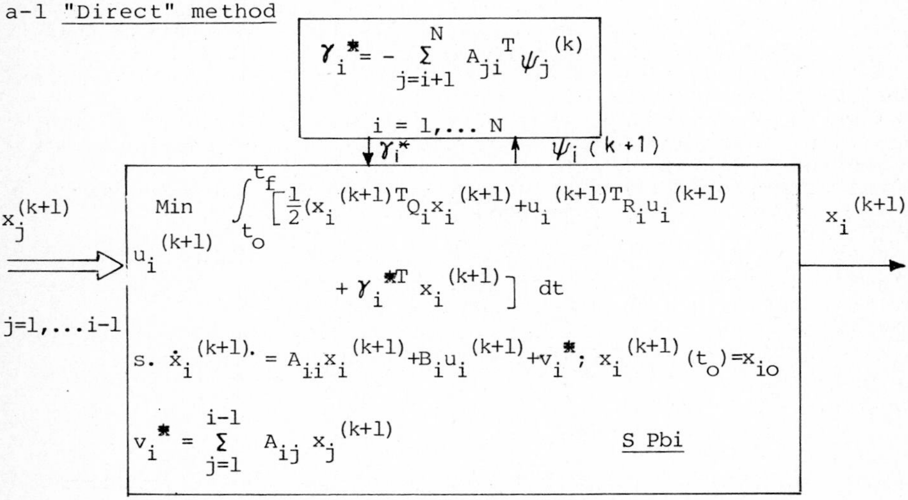

It is interesting to note tnat this algorithm follows a feasible path (the constraints of·physical coupling are satisfied
at each interaction) and therefore make it possible to envisage
an "on-line" implementation on concrete processes if the calculation time per iteration is sufficiently small in relation
to the optimisation horizon time.
Let us also observe for this particular interconnection structure, that this method may be viewed as a "Costate prediction
method".

a-2 "Dual" method

$$v_i^* = \sum_{j=1}^{i-1} A_{ij} x_j^{(k)}$$

$$i=1,\ldots,N$$

$$v_i^* \qquad x_i^{(k+1)}$$

$$\varphi_i^{(k+1)}$$

$$\underset{u_i^{(k+1)}}{\text{Min}} \int_{t_0}^{t_f} \left\{ \frac{1}{2}\left[x_i^{(k+1)T} Q_i x_i^{(k+1)} + u_i^{(k+1)T} R_i u_i^{(k+1)} \right] \right.$$

$$\left. + \gamma_i^{*T} x_i^{(k+1)} \right\} dt$$

$$\text{s. } \dot{x}_i^{(k+1)} = A_{ii} x_i^{(k+1)} + B_i u_i^{(k+1)} + v_i^* ; \ x_i^{(k+1)}(t_0) = x_{io}$$

$$\gamma_i^* = - \sum_{j=i+1}^{N} A_{ji}^{T} \psi_j^{(k+1)}$$

$$\psi_j^{(k+1)}$$
$$j=i+1,\ldots,N$$

S Pbi

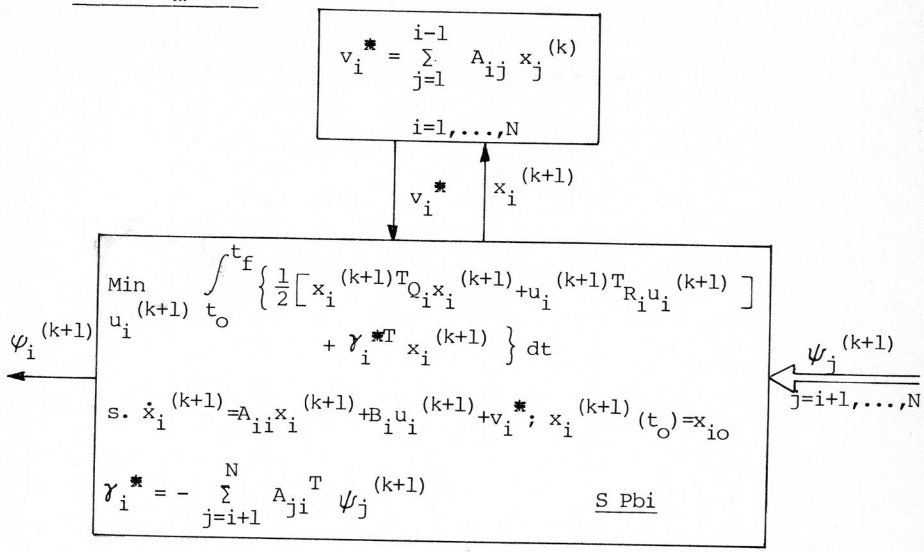

This second scheme is interesting to the extent where the co-ordination algorithm used requires to initialise (then to pre-dict) only the v parameters which are physical variables of the system and are therefore a priori easier to estimate than the dual variables γ.

Note :

Let us note that in the case where the control matrix B would itself have an inferior block-diagonal structure, the conside-red splitting of the hamiltonian matrix H no longer decouples the state (adjoint) equations of a subsystem SSi from the adjoint (state) variables of the other sub-systems.

We then get back to this situation by conserving the initial more general formulation of the conditions of stationarity (2.6) and by applying the following decomposition co-ordination sche-me.

$$\begin{bmatrix} \dot{x}_i \\ \dot{\psi}_i \\ 0 \end{bmatrix}^{(k+1)} - \begin{bmatrix} A_{ii} & \mathbb{O} & B_{ii} \\ -Q_i & -A_{ii}^{T} & \mathbb{O} \\ \mathbb{O} & B_{ii}^{T} & R_i \end{bmatrix} \begin{bmatrix} x_i \\ \psi_i \\ u_i \end{bmatrix}^{(k+1)} - \begin{bmatrix} v_i \\ \gamma_{1i} \\ \gamma_{2i} \end{bmatrix}^* = 0$$

$$i = 1,\ldots,N$$

where $v_i^{(k)} = \sum_{j=1}^{i-1} A_{ij} x_j^{(k)} + B_{ij} u_j^{(k)}$

$$\gamma_{1i}^{(k)} = - \sum_{j=i+1}^{N} A_{ji}^{T} \psi_j^{(k)}$$

$$\gamma_{2i}^{(k)} = + \sum_{j=i+1}^{N} B_{ji}^{T} \psi_j^{(k)}$$

the "direct" method then corresponds at the step (k+1) to :

$$v_i^{*} = v_i^{(k+1)}$$

$$\gamma_{1i}^{*} = \gamma_{1i}^{(k)}$$

$$\gamma_{2i}^{*} = \gamma_{2i}^{(k)}$$

the "dual" method :

$$v_i^{*} = v_i^{(k)}$$

$$\gamma_{1i}^{*} = \gamma_{1i}^{(k+1)}$$

$$\gamma_{2i}^{*} : \gamma_{2i}^{(k+1)}$$

b) Staircase structure

This is a particular case of block-triangular structure where the sub-systems are connected sequentially, which corresponds, for the example under consideration, to the case where :

$$A_{ij} = \mathbb{O} \quad j \neq i; \quad j \neq i-1 \qquad i = 1,\ldots,N.$$

In addition to the aforementioned properties, the application of sequential methods to this type of structure has the advantage of necessitating the initialisation of only n_1 $(x_1^{(o)}(t))$ or n_N $(\psi_N^{(o)}(t))$ variables according to the ordering of the method.

Furthermore, it is interesting to note that this particular configuration gives rise to a process of distributed co-ordination which operates at the level of the sub-system SSi through a communication with only the adjacent sub-systems SSi - 1 and SSi + 1. This is due to the block-tridiagonal structure of the hamiltonian matrix H which concerns a case of optimal ordering for a procedure of successive relaxation. From this also comes the usefulness in a similar case of the application of the alternating directions method of successive over-relaxations. Finally, it could easily be verified that the properties mentioned in these last three sentences also apply to the case where the matrix A would itself have a block-tridiagonal structure.

3) Application to the optimisation of discrete systems

Apart from the possibilities of discretised resolution of the conditions of stationarity of problems of dynamic optimisation,

methods of the sequential type are also currently used for the optimisation of discrete systems.
As with the previous methods, we shall confine ourselves here, to L.Q.S. type problems for the sake of simplicity including however the possibility of distributed delays, namely :

$$\left|\begin{array}{l} \text{Min} \sum_{n=0}^{N-1} \frac{1}{2} (x_{n+1}^T Q_{n+1} x_{n+1} + U_n^T R_n U_n); \qquad Q_N = S \\ U_n : n = 0,\ldots, N-1 \\ \text{s. } x_{n+1} = \Phi(n,n)\, x_n + \ldots + \Phi(n,0)\, x_o + \ldots + \Phi(n,-m) x_{-m} \\ \qquad + \Gamma(n,n)\, U_n + \ldots + \Gamma(n,o)\, U_o \\ x_j = x_{jo} \text{ for } j = -m, -m+1, \ldots, 0 \\ Q_n \geq 0 \qquad n = 1,\ldots, N \\ R_n > 0 \qquad n = 0,\ldots, N-1 \end{array}\right. \qquad (2.14)$$

Methods applying the Principle of Optimality seem to be particularly well adapted to this type of system [BEL - 65] . Based on an implicit enumeration of solutions, dynamic programming can be applied in a similar case after quantification of the admissible values of the state variables x_n, but the first disadvantage is a considerable increase in the volume of calculation which is already large generally, when discrete systems are involved. However, as is well known, this volume may be found to be considerably reduced in cases where a large number of constraints appear on the states and the controls. Furthermore, the interest of this type of method is to define an accurate procedure of sequential decomposition, ie. "convergent" from the first interation towards the quantified optimum, even if the multi-stage process brought into play involves at each stage the resolution of a composite sub-problem whose dimensionality, although below that of the overall problem, may still seem prohibitive.

If, in this standard case of dynamic programming, the sub-system SSk is defined by the overall state of the system at the discrete instant k : x(k), which is described as temporal decomposition, another version proposed in [LAR - 79] , which should prove very interesting, shows a spatial decomposition of the process where the SSi is then defined by the discrete trajectory of the local state vector x_i : $x_i(k)$: k = 1,..., K. This last decomposition, where the sub-systems are arranged in an arbitrary sequence, is based on a hypothesis of decomposability which is satisfied, for example, when the interactions are additively separable.

A scheme of double spatio-temporal decomposition which naturally comes to mind, would consist of combining the previous two methods. However, this again involves a large volume of calculation and the authors propose dealing with local sub-problems of optimisation by a method of successive approximations put forward in [LAR - 70] . This method consists of decomposing a problem of optimisation of discrete systems into sub-problems containing one single state variable where, from an initial admissible approximation of state, the sub-problem i is defined as a problem of dynamic programming resolved sequentially in relation to the other sub-problems (but could also be resolved in a parallel manner), the other state variables being held fixed. Because of the quantification this procedure converges in monotone fashion after a finite number of iterations, but not necessarily towards the true optimal solution. Convergence towards this solution is guaranteed for certain classes of problems which appear to be quite restrictive [LAR - 70].

For lack of experimentations and adequate comparison it is difficult in the present situation to give preference to this type of approach, but to get around some of these difficulties, it may be convenient to define a spatio-temporal method of decomposition linking the same programme of dynamic spatial programming on the one part and the following principle of successive relaxations for the resolution of each of the sub-problems. This last principle corresponds to application in the discrete case of the decomposition schemes presented in the preceding paragraphs.
In order to develop this let us call P_n: n = 1,..., N the discrete adjoint vector associated with the optimisation problem (2.14) and let us write the necessary conditions of stationarity as :

$$\left|\begin{array}{l}
x_n = \sum_{j=0}^{n-1} [\Phi(n-1, j)\, x_j + \Gamma(n-1,j)\, U_j] + x_{on}^*; \quad n=1,\dots,N \\
P_n = Q_n\, x_n + \sum_{j=n+1}^{N} \Phi(j-1, n)^T P_j \quad ; \; n=1,\dots N-1 \\
P_N = S\, x_N \\
R_n\, U_n + \sum_{j=n+1}^{N} \Gamma(j-1,n)^T P_j = 0; \quad n = 0,\dots,N-1 \\
\text{with } x_{on}^* = \sum_{j=-m}^{-1} \Phi(n-1, j)\, x_{jo}
\end{array}\right.$$

This set of equations characterised by an inferior triangular structure of the system in relation to state and control, may be decomposed to advantage in a similar way to that of the continuous case (Section 2.4.2.2. - 2) by putting :

$$\left|\begin{array}{ll}
v_n = \sum_{j=0}^{n-1} [\Phi(n-1,j)\, x_j + \Gamma(n-1,j)\, U_j] & n=1,\dots, N \\
\gamma_{1n} = \sum_{j=n+1}^{N} \Phi(j-1,n)^T P_j & n=1,\dots, N-1
\end{array}\right.$$

$$\gamma_{2n} = - \sum_{j=n+1}^{N} \Gamma(j-1,n)^T P_j \qquad n=0,\ldots, N-1$$

At the step (k+1), the "direct" method can be defined as follows :

$$\begin{cases} x_n^{(k+1)} = v_n^{(k+1)} + x_{on}^{*} \; ; \; n = 1,\ldots, N \\ P_n^{(k+1)} = Q_n x_n^{(k+1)} + \gamma_{1n}^{(k)} \; ; \; n=1,\ldots,N-1; \; P_N^{(k+1)} = S\, x_N^{(k+1)} \\ U_n^{(k+1)} = R_n^{-1} \; \gamma_{2n}^{(k)} \; ; \; n = 0,\ldots,N-1 \end{cases}$$

The "dual" method would be defined in a similar way with $v_n^{(k)}$, $\gamma_{1n}^{(k+1)}$ and $\gamma_{2n}^{(k+1)}$ An alternated use of the "direct" and the "dual" methods corresponds for that case to the algorithm proposed in [PAP - 80].
Figure (2.8) represents the direct procedure in the particular case (staircase structure) of linear systems :

$$x_{n+1} = \Phi_n \, x_n + \Gamma_n U_n; \; n = 0,\ldots, N-1 \qquad (2.15)$$

x_o given

Let us note that one could have chosen a parallel decomposition method by fixing the parameters v, γ_1 and γ_2 at the upper level, which would constitute an application in the discrete case of the Takahara "prediction of interactions method" (section 2.4.2.1.-1).
Similarly, still in a parallel procedure, on the hypothesis of inversibility of the matrices Q_n, one could have determined at the upper level the parameters γ_1 and γ_2, or which comes down to the same thing, the adjoint vector p by maximisation of the dual function, which corresponds to the Tamura algorithm [TAM - 75].
Figure (2.9) gives an example of application of this latter "unfeasible" method, in the particular case of linear systems (2.15), where the parameters p are calculated by a gradient technique.
These various methods of "temporal" decomposition have the advantage of requiring a smaller volume of calculation and of applying to a broader class of problems in relation to the method of successive approximations previously mentioned [LAR - 70].

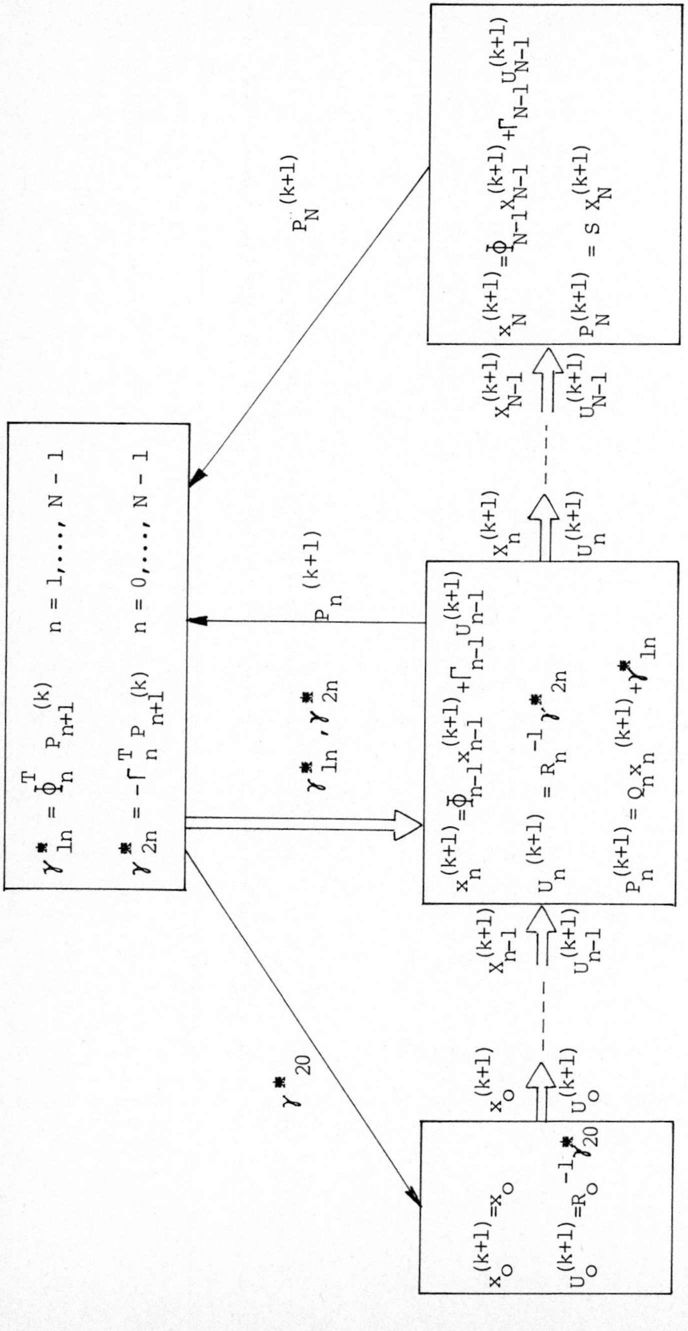

Figure 2.8 - "Direct" method

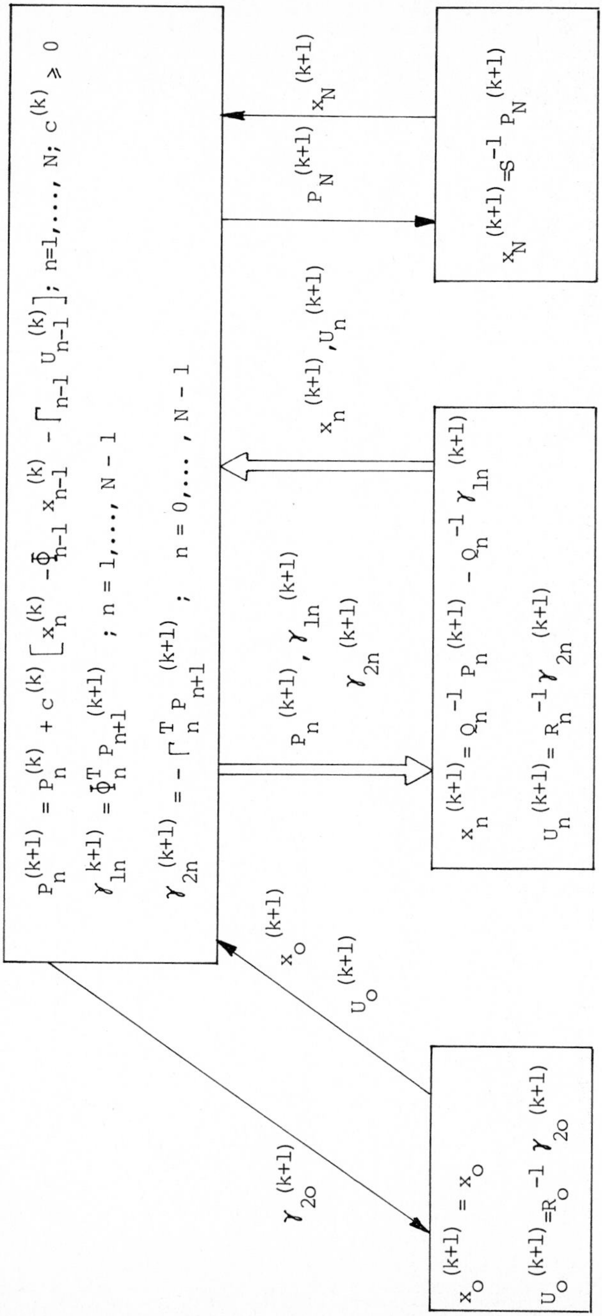

Figure 2.9 - Tamura algorithm

These various methods, as the Tamura method , which is particularly adapted to problems of optimisation with constraints on states and controls, can be combined after quantification of the results obtained (a procedure which it is advisable to study in more detail) to a method of dynamic spatial programming to constitute an efficient method of decomposition at two levels (spatio-temporal).
Figure 2.10 shows this structure of decomposition co-ordination :

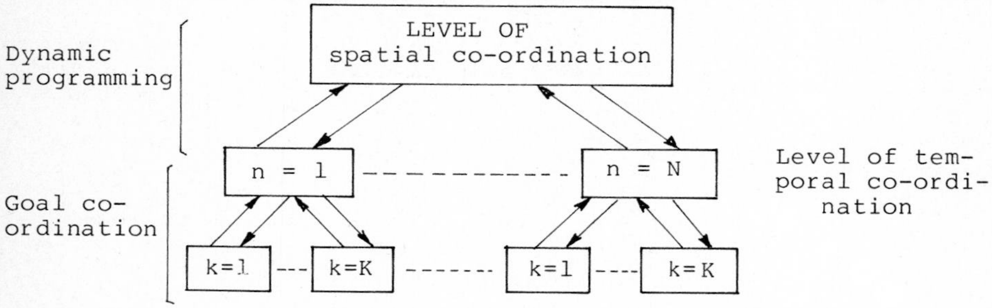

Figure 2.10 - Spatio-temporal decomposition co-ordination.

It would, then, be interesting to compare this mixed approach with that proposed by Tamura [SIN - 78] which combines at the first level the same method of space decomposition co-ordination of the K points of discretisation of each of the local state vectors $x_i(k)$ with a second level of decomposition co-ordination on the space of the variables $X = X_1 \times ... \times X_n$, $U = U_1 \times ... \times U_n$, this also applying a process of successive approximations of the "goal-co-ordination approach" type.
Let us finally note that out of many numerical simulations we were able to find great improvements, in relation to Tamura's temporal decomposition, linked to sequential methods ("direct", "dual", alternating direction, over-relaxed,...) [CAL - 82]. However, their application to problems of constraints is not straightforward.

2.4.2.3. Structural decomposition

As we pointed out in section 2.3.2.1-3 and as the previous section also proves, structural analysis constitutes an efficient way to exploit the structure of models and to suggest partitions facilitating the resolution of algebraic or differential equations. Let us mention in particular the works of Harary [RIC - 75], Varga [VAR - 62], Young [YOU - 71]... on the decomposition of linear algebraic systems, those of Ortega and Rheinbold [ORT - 70] on non-linear algebraic systems, those of Dantzig-Wolfe, Rosen [LAS - 70], Winkler [WIN - 74]..., on the decomposition of linear programmes, of Rosen, Benders [LAS - 70], Findeisen [FIN - 68] , for non-linear programmes, those of Kevorkian

[KEV - 75], Benrejeb [BEN - 80], Sezer and Siljak [SEZ - 81]..., emphasizing facilities induced by the concept of partitioning for the numerical resolution, controlability, observability and stability analysis of the large dynamic systems, those of Takahara [TAK - 65], Calvet et al.[CAL - 80][CAL - 82],... on partitioned optimisation of large dynamic systems (LQS),...

Although incomplete, this list of works which approach very different types of problems nevertheless has a certain unity in the methodology of decomposition, namely the exploitation of some structural properties which appear in the problem. This is the basis of the methods of decomposition and partitioning which we shall illustrate here by application to the problem of optimal control with a quadratic criterion for linear interconnected dynamical systems via a dynamic interconnection system. This decomposition, based on the search for a separating set between sub-systems (cf. section 2.3.2.1-3) corresponds to the concept of partitioning into general interconnected sub-systems as opposed to partitioning into sub-systems connected in series-parallel [RIC - 75].
Let us therefore consider the L.Q.S. problem given in section 2.4.2. and the partitioning (2.8) where the matrix A has what is described as an arrow-shaped structure [BEN - 80], or again a block-angular structure [WIN - 74], namely :

$$A = \begin{bmatrix} A_1 & & \mathbb{O} & A_{1N} \\ & \ddots & & \vdots \\ \mathbb{O} & & & \\ & & A_{N-1} & A_{N-1.N} \\ A_{N1} & \cdots & A_{N.N-1} & A_N \end{bmatrix}$$

If, as in section 2.4.2.1. equation (2.9) we adopt the following written expression of the conditions of stationarity :

$$Dz - Hz = 0 \quad \text{with } z = \begin{bmatrix} x_1 \\ \psi_1 \\ \vdots \\ x_{N-1} \\ \psi_{N-1} \\ x_N \\ \psi_N \end{bmatrix} \triangleq \begin{bmatrix} \tilde{z} \\ \\ z_N \end{bmatrix}$$

$$+ \text{ boundary conditions}$$

we can easily verify that the matrix H also has an arrow-shaped structure corresponding to the following written expression of these conditions. Let us note that this would also be true if the structure of B was also of this type.

$$\left|\begin{array}{l}\begin{bmatrix}\dot{x}_i \\ \dot{\psi}_i\end{bmatrix} - \begin{bmatrix}A_i - B_iR_i^{-1}B_i^T & \\ -Q_i & -A_i^T\end{bmatrix}\begin{bmatrix}x_i \\ \psi_i\end{bmatrix} = \begin{bmatrix}A_{iN}x_N \\ -A_{Ni}^T\psi_N\end{bmatrix} ; \quad \begin{array}{l}x_i(t_o)=x_{io} \\ \psi_i(t_f)=0\end{array} ; i=1,\dots,N-1 \\ \\ \begin{bmatrix}\dot{x}_N \\ \dot{\psi}_N\end{bmatrix} - \begin{bmatrix}A_N-B_NR_N^{-1}B_N^T & \\ -Q_N & -A_N^T\end{bmatrix}\begin{bmatrix}x_N \\ \psi_N\end{bmatrix} = \begin{bmatrix}\sum_{i=1}^{N-1} A_{Ni}x_i \\ -\sum_{i=1}^{N-1} A_{iN}^T\psi_i\end{bmatrix} ; \quad \begin{array}{l}x_N(t_o)=x_{No} \\ \psi_N(t_f)=0\end{array}\end{array}\right. \tag{2.16}$$

This particular expression of the conditions of stationarity shows the articulation of sub-problems SPi i=1,..., N-1 around the sub-problem SPN, which suggests an interesting scheme of hierarchical co-ordination based on the following sequential algorithm :

$$\left|\begin{array}{ll}D_i\ (z_i)^{(k+1)} - H_{oi}\ z_i^{(k+1)} = \tilde{H}_{1i}\ z_N^{(k)} & i=1,\dots, N-1 \\ D_N\ (z_N)^{(k+1)} - H_{oN}\ z_N^{(k+1)} = \tilde{H}_{1N}\ \tilde{z}^{(k+1)} & \end{array}\right.$$

where the matrices H_{Oi}, $\tilde{H}_{1i}$, H_{oN}, $\tilde{H}_{1N}$ are directly deduced from (2.16).

It is evident that we could just as well have applied a parallel algorithm here, but we observe that the interest of such an interconnection structure is to make possible a sequential decomposition-co-ordination process which, with the advantages that this type of (sequential) algorithm may have, notably for the convergence behaviour, associates practically the advantages linked to parallelisation since N-1 sub-problems can be resolved independently of each other.

Thus, at each co-ordination step, this algorithm involves a parallel decomposition between N-1 sub-problems at a "lower" level, combined with a sequential decomposition between these same sub-problems and a problem of optimisation on the interconnection system at a "higher" level, as is depicted in figure (2.11).

We refer to [CAL - 80] for a more detailed development and applications of this (see also chap. 1).

Let us draw attention to a rather more general formulation of this type of algorithm in [SAN - 79] and let us also stress a certain similarity of this particular decomposition with that proposed in [FIN - 68] , givinq a relaxation algorithm on two levels, derived from a "primal method', in a static case.

Let us finally note the usefulness of the particular case dealt which corresponds to numerous concrete examples such as power systems [CAL - 80], mechanical articulated systems, hierarchical systems,...

Furthermore, the algorithm proposed can be usefully adapted in a modular way to a tree type of interconnection structure, thus defining the multi-level optimisation scheme represented in figure (2.12), where different levels of parallelisation are

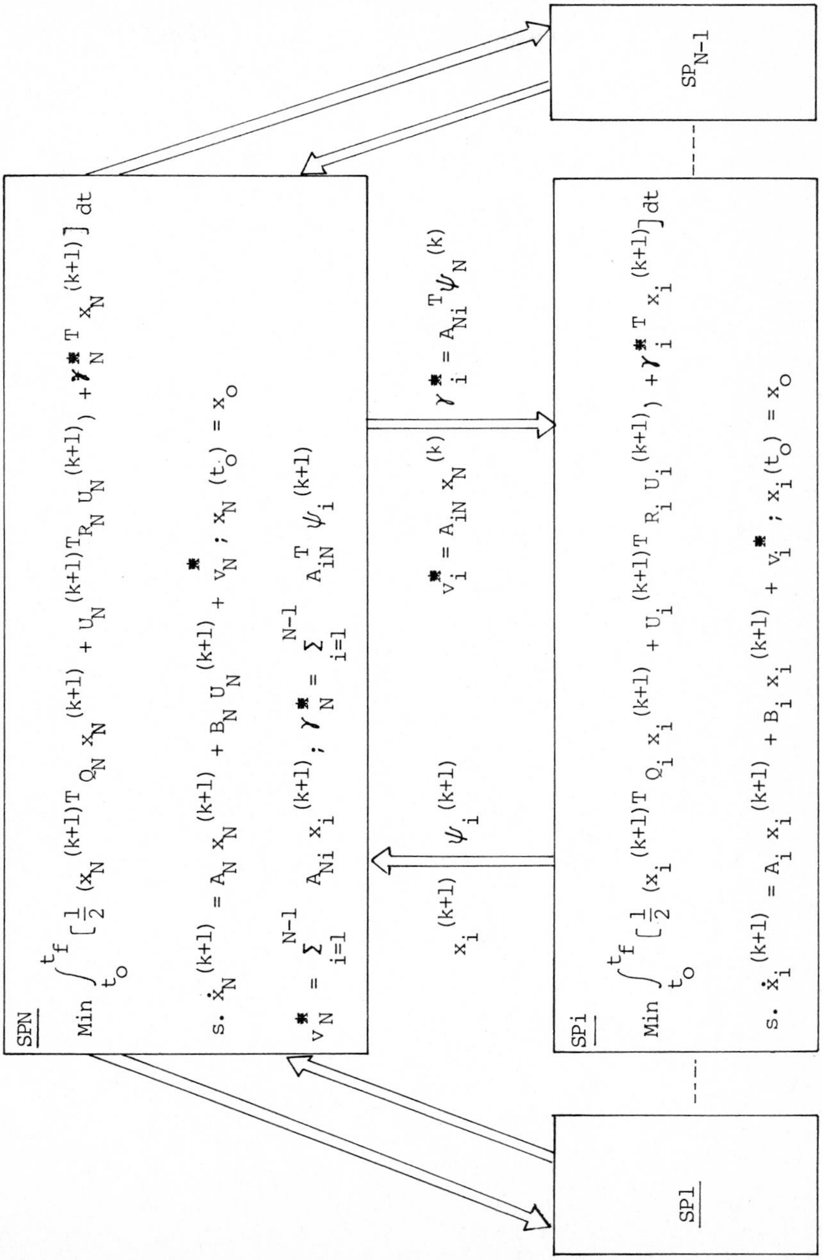

Figure 2.11 - Two-level optimisation scheme

handled sequentially upward the tree.

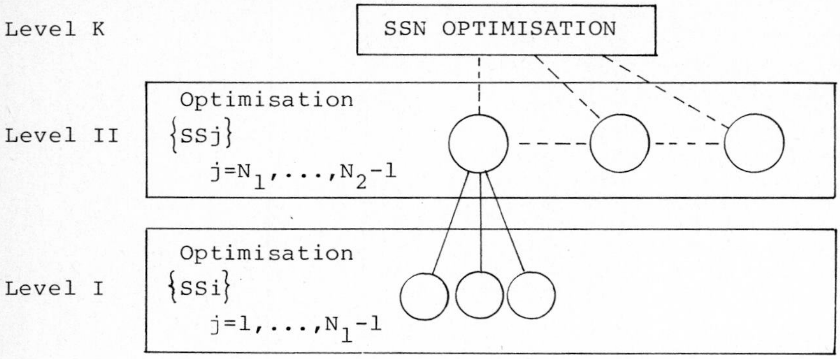

Figure 2.12 - Multi-level optimisation scheme

Naturally we could also envisage applying this type of algorithm (2.16) which consists of attributing a hierarchical position to a sub-system SSN, in the general case where the matrix A is full, which would give in figure (2.11) :

$$v_i^* = \sum_{\substack{j=1 \\ j \neq i}}^{N} A_{ij} \, x_j^{(k)} \quad ; \quad \gamma_i^* = \sum_{\substack{j=1 \\ j \neq i}}^{N} A_{ji}^T \, \psi_j^{(k)}$$

but would impose an exchange of information between SSi i=1,..., N-1.

Our conclusion is that the decomposition methods presented in this section and the previous one do not necessitate, from a fundamental viewpoint, an ad hoc structure for the process, but can be applied to advantage when this is the case.

Structural analysis and "natural decomposition" research must, we feel, be combined with a view to making the systems appear as approximately block-diagonal, block triangular, block-angular, cyclic,... in order to better adapt the structures of calculation, communication, control,... to the process structure.

2.4.2.4. Other decompositions

1- Aperiodic decomposition

Although the versatility of group iterative methods offers multiple choices of decomposition linked to the different possibilities of partitioning, parallelisation, sequentialisation, ordering, these choices are infinitely multiplied if we exploit the possibility of non-periodic scheduling schemes. This extension corresponds to the concept of chaotic iterations to which is even associated, by the introduction of delays in the information systems, the possibility of asynchronism in the handling of different sub-systems [MIE - 75] [LHO - 79].

We shall confine ourselves here to showing how structural analysis can assist the conception of an aperiodic decomposition scheme.

One can, in fact, imagine a scheduling which, rather than giving preference to a sub-system or set of sub-systems sequentially, as was done, for example, in the multi-level optimisation scheme in figure (2.12), would give preference to the resolution frequency of certain sub-systems, which would come down to relaxing the handling of the others on several iterations.
If, in terms of the convergence of the co-ordination algorithm, the interest of such an approach is very subjective, especially for linear systems, it is no less evident with respect to the reduction of communications involved between sub-systems.
If we consider, for example, a problem of optimal control of systems connected in series following a perturbation to the SSi, we can very easily imagine an iterative process (sequential or parallel) applied first to the SSi and its adjacents SSi-1 and SSi+1 (even SSi-2, SSi-1, SSi+1, SSi+2...), then scanning more or less progressively the set of sub-systems. This connection could itself act in an adaptive way at the end of the co-ordination process, or else with a certain periodicity.

This type of aperiodic decomposition, already validated by a few applications that we have been able to carry out, could, we find, be implemented to advantage in an approach which we shall describe as assisted decomposition co-ordination, according plenty of room to the operator capacity for analysis and imagination.

2) Overlapping decomposition

From the informational analysis of interconnected systems developed in [CHU - 76] it is clear in a first approximation that each local sub-system has a certain limited field of information (and reciprocally, of control), which can partially recover that of the other sub-systems, and that the information structure of the overall system is a network formed by the aggregation of such overlapped linkages.
This hypothesis has been considered in particular by Ikeda and Siljak [IKE - 80] in a spatial domain for the study of the decentralised stabilisation of dynamic systems, and in a temporal domain by Khalil and Kokotovic [KHA - 78] in an approach of the type order reduction of models.
We propose here a method of successive approximations based on this type of approximation defining a spatial overlapping decomposition.
This constitutes an extension of the concept of iterative methods by groups, in which an equation or set of equations could only be assigned to one single group.
The concept of overlapping decomposition introduces, for a set of equations, the possibility of belonging to several sub-systems, which leads to the notion of the expanded overall system defined in [IKE - 80] in which certain equations are redundant.
Although here again no specific structure is a priori necessary, we are going to illustrate this on the example dealt with in section 2.4.2.3. where the application of a method of overlapping decomposition can appear very natural, exploiting the existence of a dynamic interconnection system which will determine, in the circumstances, the set of equations which we shall associate with each of the sub-systems SSi i=1,..., N-1.

Contrary to the structural decomposition presented in the aforesaid section, where the sub-system of articulation SSN was treated independently of the others, this decomposition consists of dealing independently with the N-1 expanded sub-systems of dimension (n_i+n_N), namely, taking account of the block-angular form of A :

$$\begin{cases} \dot{x}_i = A_i\, x_i + A_{iN}\, x_N^i + B_i u_i \;;\; x_i(t_0) = x_{io} \\ \dot{x}_N^i = A_N\, x_N^i + \sum_{j=1}^{N-1} A_{Nj}\, x_j + B_N\, u_N ;\; x_N^i\,(t_0) = x_{No} \end{cases} \qquad i=1,\dots,N-1$$

If we rewrite the later equation :

$$\dot{x}_N^i = A_N\, x_N^i + A_{Ni}\, x_i + B_N\, u_N^i + \sum_{\substack{j=1\\ j\neq i}}^{N-1} \left[A_{Nj}\, x_j + B_N\, u_N^j \right]$$

and if we choose a parallel decomposition mode, we get the following iterative scheme :

$$\begin{bmatrix} \mathring{\tilde{x}}_i \\ \\ \mathring{\tilde{\psi}}_i \end{bmatrix}^{k+1} - \begin{bmatrix} A_i & A_{iN} & -B_i\, R_i^{-1}\, B_i^T & \mathbb{O} \\ A_{Ni} & A_N & \mathbb{O} & -B_N\, R_N^{-1}\, B_N^T \\ -Q_i & \mathbb{O} & -A_i^T & -A_{Ni}^T \\ \mathbb{O} & -\alpha_i Q_N & -A_{iN}^T & -A_N^T \end{bmatrix} \begin{bmatrix} \tilde{x}_i \\ \\ \tilde{\psi}_i \end{bmatrix}^{k+1} =$$

$$= \begin{bmatrix} \mathbb{O} \\ \sum_{\substack{j=1\\ j\neq i}}^{N-1} \left[A_{Nj}\, x_j - B_N\, R_N^{-1}\, B_N^T\, \psi_N^j \right] \\ -A_{Ni}^T \sum_{\substack{j=1\\ j\neq i}}^{N-1} \psi_N^j \\ \mathbb{O} \end{bmatrix}^k . \qquad i=1,\dots,N-1$$

with : $\tilde{x}_i \triangleq \begin{bmatrix} x_i \\ x_N^i \end{bmatrix}$; $\tilde{\psi}_i \triangleq \begin{bmatrix} \psi_i \\ \psi_N^i \end{bmatrix}$; $\sum_{i=1}^{N-1} \alpha_i = 1$

we can easily verify that at the convergence we have $x_N^i = x_N \;\forall i$

and $\psi_N = \sum_{i=1}^{N-1} \psi_N^i$.

Similarly the N - 1 sub-problems of optimisation treated independently at step (k+1) correspond to :

$$\underset{\tilde{u}_i^{(k+1)}}{\text{Min}} \int_{t_o}^{t_f} \left\{ \frac{1}{2} \left[\tilde{x}_i^{T(k+1)} \begin{bmatrix} Q_i & 0 \\ 0 & \alpha_i Q_N \end{bmatrix} \tilde{x}_i^{(k+1)} + \tilde{u}_i^{T(k+1)} \begin{bmatrix} R_i & 0 \\ 0 & R_N \end{bmatrix} \tilde{u}_i^{(k+1)} \right] + \gamma_i^{*T} \tilde{x}_i^{(k+1)} \right\} dt$$

$$\text{s. } \dot{\tilde{x}}_i^{(k+1)} = \begin{bmatrix} A_i & A_{iN} \\ A_{Ni} & A_N \end{bmatrix} \tilde{x}_i^{(k+1)} + \begin{bmatrix} B_i \\ B_N \end{bmatrix} \tilde{u}_i^{(k+1)} + v_i^*;\ \tilde{x}_i^{(k+1)}(t_o) = \begin{bmatrix} x_{io} \\ x_{No} \end{bmatrix}$$

$$\text{with} \quad \tilde{u}_i = \begin{bmatrix} u_i \\ u_N^i \end{bmatrix};\ v_i^* = \begin{bmatrix} 0 \\ \sum_{\substack{j=1 \\ j \neq i}}^{N-1} \left[A_{Nj} x_j^{(k)} - B_N R_N^{-1} B_N^T \psi_N^{j(k)} \right] \end{bmatrix}$$

$$\text{and} \quad \gamma_i^* = \begin{bmatrix} -A_{Ni}^T \sum_{\substack{j=1 \\ j \neq i}}^{N-1} \psi_N^{j(k)} \\ 0 \end{bmatrix}$$

2.4.3. Initial-valued problems

2.4.3.1. Block integration of a linear differential system

1) Volterra operator and convergence study

Contrary to the preceding section, we are concerned here with the partitioned resolution, after decomposition of a dynamic matrix $E = E_o + E_1$ of a differential linear system with conditions at only one point of the type :

$$\dot{\varepsilon}(t) = E(t)\ \varepsilon(t);\ \varepsilon(t_o) = \varepsilon_o\ ;\ t \in [t_o, t_f]$$

$$\longrightarrow \dot{\varepsilon}^{(k+1)}(t) = E_o(t)\ \varepsilon^{(k+1)}(t) + E_1(t)\varepsilon^{(k)}(t); \qquad (2.17)$$

$$\varepsilon^{(k+1)}(t_o) = \varepsilon_o\ ;\ k \geq 0$$

the solution of which can be expressed as follows :

$$\varepsilon^{(k+1)}(t) = T\,[\varepsilon_t^{(k)}(t)] + v(t)$$

$$\text{with : } T[\varepsilon] = \int_{t_o}^{t} \Phi(t,\tau)\, E_1(\tau)\, \varepsilon(\tau)\, d\tau \qquad (2.18)$$

$$v(t) = \Phi(t, t_o)\, \varepsilon_o$$

$$\dot{\Phi}(t,t_o) = E_o(t)\, \Phi(t,t_o) \;;\quad \Phi(t_o,t_o) = \mathbb{1}$$

where T is a Volterra type mapping of the complete space $C[t_o,t_f]$ into itself, of which the kernel of the integral: $K(t,\tau) \equiv \Phi(t,\tau)\, E_1(\tau)$ is continuous for t and $\tau \in [t_o,t_f]$, and where v(t) is also continuous for $t \in [t_o,t_f]$.

As in the preceding section, one sufficient condition of convergence is :

$$(t_f - t_o)\ \rho(M) < 1$$

where the elements m_{ij} of the matrix are expressed as follows from the elements k_{ij} of the matrix K :

$$m_{ij} \equiv \max_{t,\tau \in [t_o,t_f]} |\, k_{ij}(t,\tau)\,|$$

Let us note the damping character of the matrix Φ in the kernel of the integral if E_O is a stable matrix.

This condition of convergence expressing T as a P-contraction is also known by the name of the PICARD Theorem [KOL - 57]:

$$|||\, T(\varepsilon_1) - T(\varepsilon_2)\,||| \leqslant M\,(t_f - t_o)\, |||\,\varepsilon_1 - \varepsilon_2\,|||$$

Although this result may appear to be conservative in relation to the one given inthe case "with two-boundary-conditions" however it can be demonstrated that T^k is now a contraction mapping $\forall t_f$ for a sufficiently large number of iterations k.

In fact, we can easily show that :

$$|||\, T^k(\varepsilon_1) - T^k(\varepsilon_2)\,||| \leqslant M^k\, \frac{(t_f - t_o)^k}{k!}\, |||\,\varepsilon_1 - \varepsilon_2\,|||$$

and that :

$$\rho(M^k)\, \frac{(t_f - t_o)^k}{k!} < 1 \text{ for a sufficiently large number } k;$$

k! growing more quickly than any geometrical progression.

So we find again here an application of the generalised contraction principle which therefore indicates that the system (2.18) converges after a certain number of iterations towards the unique solution of $\varepsilon = T\varepsilon + v$; any fixed point for the mapping T is fixed for the contraction mapping T^k which can therefore have only one fixed point.

Note 1 :

Although convergence of this type of process is guaranteed from a theoretical point of view, from a practical point of view the choice of the latter must be such that one does not at first move too far away from the solution, which could cause "overflow" type numerical incidents.

Note 2 :

$$\text{If } E_0 = \mathbb{O} \,,\; E_1 = \mathbb{C}^{te} = E$$

then :

$$|||\, T^k(\varepsilon_1) - T^k(\varepsilon_2) \,||| \leq \frac{E^k (t_f - t_o)^k}{k!} \, |||\, \varepsilon_1 - \varepsilon_2 \,|||$$

2) Block resolution of a L.Q.S. type problem of optimal control

Let us reconsider the problem of L.Q.S. optimisation presented in section 2.4.2.. We recall that its solution may be brought down to the solution of a linear differential system of dimension 2n with conditions at the two-points t_o and t_f (2.7)

As is well known, one can imbed this two-point-boundary valued system in a one-point-valued differential non-linear system of dimension n^2 ($\frac{(n^2 + n)}{2}$ if the Q and R matrices are symmetric) leading to a Riccati type equation :

$$\begin{aligned} \dot{K}(t) &= K(t)\, B(t)\, R^{-1}(t)\, B^T(t)\, K(t) - A^T(t)\, K(t) - K(t) A(t) \\ &\quad - Q(t) \\ K(t_f) &= \mathbb{O}_n \quad ; \quad t \in [t_o, t_f] \end{aligned} \qquad (2.19)$$

and to a closed loop control law :

$$u(t) = - R^{-1}(t)\, B^T(t)\, K\,(t)\, x(t)$$

In the following section we shall return to some possible partitioned solutions of this differential non-linear equation. Let us emphasize finally that one can still, at the cost of a new increase in the number of variables and a matrix inversion, return to the solution of a linear differential system of dimension $4n^2$ with one-point-boundary condition [CAL - 79].

Let us note that for the type of problem under consideration this linear differential system integrates more equations than necessary and that it can be reduced to the following differential system of dimension $2n^2$ [BER - 78] :

$$\left[\begin{aligned} \dot{X}(t) &= A(t)\, X(t) - B(t)\, R^{-1}(t)\, B^T(t)\, \Lambda(t), \quad X(t_f) = \mathbb{1}_n \\ \dot{\Lambda}(t) &= -\, Q(t)\, X(t) - A^T(t)\, \Lambda\,(t) ; \quad \Lambda(t_f) = \mathbb{O}_n \end{aligned}\right.$$

which also leads to a closed loop control law :

$$u(t) = - R^{-1}(t)\; B^T(t)\; K(t)\; x(t)$$

with $$K(t) = \Lambda(t)\; X^{-1}(t)$$

By putting $Z_{2nxn} \triangleq \begin{bmatrix} X \\ \Lambda \end{bmatrix}$, the resolution of the optimisation problem is therefore equivalent to the matrix integration of the following system :

$$\dot{Z}(t) = H(t)\; Z(t);\; Z(t_f) = \begin{bmatrix} \mathbb{1}_n \\ \mathbb{0}_n \end{bmatrix};\; t \in [t_o, t_f]$$

where H(t) is the Hamiltonian matrix.
Passage to a vector formulation can be carried out as follows via KRONECKER matrix product :

$$\begin{cases} \dot{z}(t) = [H(t) \otimes \mathbb{1}_n]\; z(t) & ;\; t \in [t_o, t_f] \\ z(t_f) = (\underbrace{1\;\overbrace{0\; \ldots\; 0}^{n}\; 1\; 0\; \ldots\; 0\; \ldots\; 0\; 1}_{n^2}\; \underbrace{0\; \ldots\; 0}_{n^2})^T \end{cases} \quad (2.20)$$

where

$$z(t) = (x_{11}, \ldots, x_{1n}, \ldots, x_{n1}, \ldots, x_{nn}, \lambda_{11}, \ldots, \lambda_{1n}, \ldots, \lambda_{n1}, \ldots, \lambda_{nn})^T$$

the x_{ij} and λ_{ij} being the respective elements of the matrices X and Λ.

Definition : Kronecker matrix product

Kronecker matrix product of two matrices A and B or the direct product is defined as follows :

$$A_{mxn} \otimes B_{lxp} = (A \otimes B)_{mlxnp} \triangleq \begin{bmatrix} a_{11}B & \ldots & a_{1n}B \\ \vdots & & \vdots \\ a_{m1}B & \ldots & a_{mn}B \end{bmatrix}$$

The system (2.20) can then be resolved in a partitioned manner after decomposition of the matrix $H \otimes \mathbb{1}n = H_o(t) + H_1(t)$ leading to an iterative scheme of the type :

$$\dot{z}^{(k+1)}(t) = H_o(t)\; z^{(k+1)}(t) + H_1(t)\; z^{(k)}(t) \; : \; k \geq 0$$

3) Integration of a LYAPUNOV type equation :

Let us consider the differential equation associated with the autonomous linear system of the L.Q.S. example above, namely :

$$\dot{P}(t) = -A^T(t)\; P(t) - P(t)\; A(t) - Q(t)$$

$$P(t_f) = \mathbb{0}_n; \qquad t \in [t_o, t_f].$$

It is shown [BAR - 71] that this one-point-valued matricial system can be expressed in the following vector form of dimension n^2 :

$$\left[\begin{array}{l} \dot{p}(t) = -(A^T \otimes \mathbb{1}n + \mathbb{1}n \otimes A^T)\ p(t) - q(t) \equiv -Up-q \\ \\ p\ (t_f) = (0 \ldots 0)^T \\ \\ \text{where} \\ \\ p = (p_{11} \cdots p_{1n}\ p_{21} \cdots p_{2n} \cdots p_{n1} \cdots p_{nn})^T ;\ q = (q_{11} q_{12} \cdots q_{nn})^T \end{array}\right. \tag{2.21}$$

This system can then be resolved in a partitioned manner after decomposition of the matrix $U \rightarrow U_0 + U_1$.

Let us note that an order reduction of the system (2.21) is possible beforehand, related to the symmetry of the solution $P(t)$ in the case where the weighting matrix Q is symmetric $(Q = Q^T)$. For example, the system (2.21) can thus be reduced to the following system :

$$\left[\begin{array}{l} \dot{s}(t) = -S(t)\ s(t) - r(t) \qquad ;\ S_{\frac{n(n+1)}{2} \times \frac{n(n+1)}{2}} \\ \\ s\ (t_f) = (\underbrace{0 \ldots 0}_{n(n+1)/2})^T \\ \\ \text{where } s = (p_{11}\ p_{12}\ p_{22}\ p_{13}\ p_{23}\ p_{33} \cdots p_{nn})^T \end{array}\right.$$

2.4.3.2. Block integration of non-linear differential systems

1) <u>Study of convergence and decomposition method</u>

In the case of non-linear Volterra integrals :

$$T[\varepsilon] = \int_{t_o}^{t} K[t, \tau, \varepsilon(\tau)]\, d\tau$$

it is shown [KOL - 57],[LUE - 69], ... that the contraction principle is still applicable if the kernel $K[t, \tau; \varepsilon(\tau)]$ satisfies the LIPSCHITZ condition in relation to its functional variable $\varepsilon(\tau)$, ie. if :

$$| K[t, \tau; \varepsilon_1(\tau)] - K[t, \tau; \varepsilon_2(\tau)] | \leq M | \varepsilon_1(\tau) - \varepsilon_2(\tau) |$$

We can therefore apply partitioned integration methods to non-linear systems of the type :

$$\varepsilon(t) = E(t)\ \varepsilon(t) + e[\varepsilon(t), t] \ ;\ \varepsilon(t_o) = \varepsilon_o \ ;\ t \in [t_o, t_f]$$

which, after decomposition of the linear component ($E = E_o + E_1$) and relaxation of $E_1\ \varepsilon(t)$ and of the non-linear component, leads to the decomposition scheme :

$$\dot{\varepsilon}^{(k+1}(t) = E_o(t)\ \varepsilon^{(k+1)}(t) + E_1(t)\ \varepsilon^{(k)}(t) + e[\varepsilon^{(k)}(t), t]\ ;$$

$$\varepsilon^{(k+1)}(t_o) = \varepsilon_o \qquad k \geq 0$$

namely :

$$\varepsilon^{(k+1)} = T\left[\varepsilon^{(k)}\right] + v; \qquad \varepsilon^{(k+1)}(t_o) = \varepsilon_o; \qquad k \geq 0$$

where T is a Volterra type non-linear integral with the kernel :

$$K\ [t, \tau\ ;\ \varepsilon\ (\tau)] = \Phi(t, \tau)\ \ [E_1(\tau)\, \varepsilon\ (\tau)\ + e\ [\varepsilon(\tau), \tau]]$$

with

$$\dot{\Phi}\ (t,\ t_o) = E_o(t)\ \ \Phi\,(t,\ t_o)\ ;\ \Phi\,(t_o,\ t_o) = \mathbb{1}$$

and where :

$$v\ (t) = \Phi\,(t,\ t_o)\ \varepsilon_o$$

2) <u>Integration of a RICCATI type equation</u>

With respect to our example of a L.Q.S. optimisation problem, we saw that its resolution could be brought down to the resolution of a final-valued differential system (2.19) which we shall re-write as follows :

$$\dot{K}(t) = -\ A^T(t)\ K(t)\ -\ K(t)\ A(t)\ -\ C(t)\ ;\quad K(t_f) = \mathbb{0}_n$$

with

$$C\ (t) \triangleq\ -\ K(t)\ B(t)\ R^{-1}(t)\ B^T(t)\ K(t)\ +\ Q(t)$$

or again in vector form via the Kronecker product :

$$\dot{k}(t) = -\ (A^T \otimes\ \mathbb{1}\ n\ + \mathbb{1}\ n \otimes A^T)\ k(t)\ -\ c(t)$$

$$k\ (t_f) = \underbrace{(0\ldots0)}_{n2}{}^T$$

with

$$k = (k_{11}\ \cdots\ k_{1n}\ \cdots\ k_{n1}\ \cdots\ k_{nn})^T\ ;\quad \dot{k}\text{ idem};\quad c\text{ idem}$$

Let us note that the relaxation operated here is different from that operated in [KLE - 68] via an algorithm of quasi-linearisation.

As with the Lyapunov equations, it is interesting to exploit the possibility of reduction of this system taking account of the symmetry of $K(t)$ in the case $Q = Q^T$ and $R = R^T$, which leads to the resolution of the reduced system :

$$\dot{s}(t) = -\ S(t)\ s(t)\ -\ r(t)\ ;\quad S\ \frac{n\ (n+1)}{2}\ \times\ \frac{n\ n+1)}{2}$$

$$s\ (t_f) = \underbrace{(0\ldots.0)}_{n(n+1)/2}{}^T$$

Partitioning can also be done on the matrix S to define the following decomposition algorithm :

$$\dot{s}^{(k+1)} = -S_o\, s^{(k+1)} - S_1 s^{(k)} - r \quad ; \quad s^{(k+1)}(t_f) = (0\ldots0)^T$$

$$k \geq 0$$

2.4.4. Stationary case

2.4.4.1. Approximation by an evolutive partitioned system

Here we are taking a step which is quite frequently used for the resolution of algebraic systems when these appear as the stationary approximation of asymptotically stable systems of evolution.
The partitioned methods of decomposition of differential systems described in the preceding section can then, in view of their convergence properties, be used to advantage for the resolution of such algebraic systems.
We illustrate this by three typical examples of problems to be solved in linear systems control.
Naturally we shall consider here some invariant dynamic systems, referring for example to [BAR - 71] for the proof :

Example 1 :

Resolution of a linear algebraic equation.
This can be carried out via partitioned resolution of the following differential system, for any initial condition on x :

$$\dot{x}(t) = A\, x(t) - b; \quad \forall\, x\,(t_o) = x_o; \quad t > t_o \qquad (2.22)$$

Theorem 1 :

It A is a stable matrix, then the solution x(t) of the differential system (2.22) converges when $(t-t_o) \to \infty$ towards the solution x^* of the algebraic equation :

$$Ax = b$$

Example 2 : Lyapunov equation

This resolution can be obtained from the differential system :

$$\begin{cases} \dot{P}(t) = -A^T P(t) - P(t)\, A - Q\ ; \quad Q = Q^T > 0 \\ P(t_f) = \mathbb{O}_n \text{ for example} \end{cases} \qquad (2.23)$$

Theorem 2 :

If A is a stable matrix, then the backward integration of the differential system (2.23) converges for $(t' - t_f) \to \infty$ towards the unique positive definite solution P^* of the Lyapunov equation.

$$A^T P + P A + Q = \mathbb{O}_n$$

Example 3 : RICCATI equation

From the partitioned backward integration of the differential system :

$$\begin{cases} \dot{K}(t) = K(t)\,BR^{-1}\,B^T\,K(t) - A^TK(t) - K(t)\,A - Q; \\ K(t_f) = \mathbb{O} \end{cases} \qquad \begin{matrix} Q = Q^T > 0 \\ R = R^T > 0 \end{matrix}$$

Theorem 3 :

If the pair (A, B) is controllable, then the solution K(t) of this backward integration converges for $(t' - t_f) \to \infty$ towards the unique, symmetric positive definite solution K^* of the associated RICCATI algebraic equation :

$$K\,BR^{-1}\,B^T\,K - A^T\,K - KA - Q = \mathbb{O}_n$$

2.4.4.2. Block solution of a Lyapunov type equation

Let us consider the matrix equation :

$$A^TP + PA + Q = \mathbb{O}_n \tag{2.24}$$

As was done in the differential case (section 2.4.3.1.-3) this system can be reduced to an algebraic vector system :

$$(A^T \otimes \mathbb{1}n + \mathbb{1}n \otimes A^T)\ p = -\ q$$

ie. a system of the Ax = b type which we could envisage solving by the decomposition methods presented in section 2.3.
We are more concerned here with the decomposition methods directly applied to the system (2.24) in its matrix form, which we shall re-write again as :

$$L\ (P) + Q = \mathbb{O} \tag{2.25}$$

which will allow in particular a better exploitation of the structural properties of the process.
We shall return to this point by presenting two examples of ε-coupling, the first based on the theory of regular perturbations, the second onthe theory of singular perturbations.
After partitioning of the diverses matrices A, P and Q, many decomposition methods of the system (2.25) can then be considered after decomposition of $L\ (P) = L_o(P) + L_1(P)$ according to an iterative scheme of decomposition :

$$L_o\,[P^{(k+1)}] + L_1\,[P^{(k)}\ ] + Q = \mathbb{O}; \quad k > 0$$

Many possibilities of decomposition (parallel, sequential) are presented in [SAN - 79].
Therefore we shall confine ourselves here to presenting a parallel decomposition method induced by a partitioning of the matrices A, P and Q into N sub-matrices such that the diagonal sub-matrices (A_{ii}) are square :

$$A : \begin{bmatrix} A_{11} & \cdots & A_{1N} \\ \vdots & & \vdots \\ A_{N1} & \cdots & A_{NN} \end{bmatrix} ; \quad P = \ldots \quad ; \quad Q = \ldots$$

This gives the following written expression of system (2.24) :

$$A_{ii}^T P_{ij} + P_{ij} A_{jj} + \sum_{\substack{\ell=1 \\ \ell \neq i}}^{N} A_{\ell i}^T P_{\ell j} + \sum_{\substack{\ell=1 \\ \ell \neq j}}^{N} P_{i\ell} A_{\ell j} + Q_{ij} = \mathbb{O} \tag{2.26}$$

$$i = 1, \ldots, N \quad ; \quad j = 1, \ldots, N$$

Choosing $L_o(P)$ such that composed of the sub-matrices :

$$L_{o_{ij}} (P_{ij}) = A_{ii}^T P_{ij} + P_{ij} A_{jj} \qquad \forall_{i,j} = 1, \ldots, N \, .$$

we define a method of decomposition of the system (2.26) into N^2 ($N(N+1)/2$ if $Q = Q^T$) Lyapunov type sub-equations which can be handled independently as follows :

$$A_{ii}^T P_{ij}^{(k+1)} + P_{ij}^{(k+1)} A_{jj} + Q_{ij}^{(k+1)} = \mathbb{O}$$

with

$$Q_{ij}^{(k+1)} = Q_{ij} + \sum_{\substack{\ell=1 \\ \ell \neq i}}^{N} A_{\ell i}^T P_{\ell j}^{(k)} + \sum_{\substack{\ell=1 \\ \ell \neq j}}^{N} P_{i\ell}^{(k)} A_{\ell j}$$

A necessary and sufficient condition of convergence is therefore (cf. section 2.3) :

$$\rho (L_o^{-1} L_1) < 1$$

which could be the case, notably when the partitioning chosen causes natural decouplings or ε-couplings to appear.
We are presenting here two examples of natural decoupling, taking some results developed in [SAN - 79] particularly for the second.

1) Weak-coupling Decomposition algorithm

We are concerned here with the case of a situation of ε-coupling between sub-systems, considered as a "regular" perturbation of a block-diagonal system, ie. that the configuration of the matrices A and Q can be considered block-diagonal as a first approximation.
We shall note respectively the off block-diagonal sub-matrices :

$$A_{ij} = \varepsilon \tilde{A}_{ij} \; ; \quad Q_{ij} = \varepsilon \tilde{Q}_{ij} \quad \forall_{i \neq j} \quad i,j = 1, \ldots, N$$

where ε is a small positive parameter.
If we look for a solution P of the Lyapunov equation (2.26) such that :

$$P_{ij}(\varepsilon) = \varepsilon \tilde{P}_{ij} \quad \forall_{i \neq j} \; i,j=1, \ldots, N; \quad P_{ii}(\varepsilon) = P_{ii} \forall_i = 1, \ldots, N$$

we easily verify that the sequential decomposition method based on the decomposition $L(P) = L_o(P) + L_1(P)$ defined hereafter, can again be written as :

$$L(P) = L_o(P) + L_1(\varepsilon\tilde{P}) = L_o(P) + \varepsilon L_1(\tilde{P})$$

and therefore makes it possible a natural decoupling to occur. This decomposition method is such that :

$$L_{Oii}(P_{ii}) \triangleq A_{ii}^T P_{ii} + P_{ii} A_{ii} \qquad i = 1,\dots,N$$

$$L_{Oij}(P_{ij},P_{ii},P_{jj}) \triangleq A_{ii}^T P_{ij} + P_{ij} A_{jj} + A_{ji}^T P_{jj} + P_{ii} A_{ij}$$

$$\forall_{i\neq j} \quad i,j = 1,\dots, N$$

and that the N following independent equations are treated in a first phase (Lower Level) :

$$L_{Oii}[P_{ii}^{(k+1)}] = - L_{1ii}[\varepsilon\tilde{P}^{(k)}] - Q_{ii} \qquad i=1,\dots, N$$

and in the second phase of the sequence (upper co-ordination level) the $N^2 - N$ (or $N(N-1)/2$ if $Q = Q^T$) independent equations :

$$L_{Oij}[P_{ij}^{(k+1)}, P_{ii}^{(k+1)}, P_{jj}^{(k+1)}] = - L_{1ij}[\varepsilon\tilde{P}^{(k)}] - Q_{ij} \quad ; i\neq j$$

2) Two-time-scales decomposition algorithm

We return this time to the case where N = 2 and the dynamic matrix A characterises the singularly perturbed system :

$$\dot{x}_1 = A_{11} x_1 + A_{12} x_2$$

$$\varepsilon\dot{x}_2 = \tilde{A}_{21} x_1 + \tilde{A}_{22} x_2$$

This induces the following partitioning of matrix A :

$$A = \begin{bmatrix} A_{11} & A_{12} \\ \frac{\tilde{A}_{21}}{\varepsilon} & \frac{\tilde{A}_{22}}{\varepsilon} \end{bmatrix}$$

From an adequate partitioning of matrix Q assumed to be symmetric :

$$Q = \begin{bmatrix} Q_{11} & Q_{12} \\ Q_{12}^T & Q_{22} \end{bmatrix}$$

if we look for the solution P of the Lyapunov equation in the following form, as per [KOK - 72].

$$P = \begin{bmatrix} \tilde{P}_{11} & \varepsilon \tilde{P}_{12} \\ \varepsilon \tilde{P}_{21} & \varepsilon \tilde{P}_{22} \end{bmatrix}$$

whe show[LEH - 78] that the computation of the P_{ij} sub-matrices may be carried out through a process of sequential ε-coupled decomposition :

$$L_o \left[P^{(k+1)}\right] + \varepsilon L_1 \left[\tilde{P}^{(k)}\right] + Q = \mathbb{O}_n \qquad k > 0$$

namely :

$$L_o \left[P^{(k+1)}\right] + L_1 \left[P^{(k)}\right] + Q = \mathbb{O}_n$$

with a choice of :

$$L_o = \begin{bmatrix} A_{11}^T P_{11} + P_{11} A_{11} + A_{21}^T P_{21} + P_{12} A_{21} & A_{21}^T P_{22} + P_{11} A_{12} + P_{12} A_{22} \\ P_{22} A_{21} + A_{12}^T P_{11} + A_{22}^T P_{21} & A_{22}^T P_{22} + P_{22} A_{22} \end{bmatrix}$$

and of :

$$L_1 = \begin{bmatrix} \mathbb{O} & A_{11}^T P_{12} \\ P_{21} A_{11} & A_{12}^T P_{12} + P_{21} A_{12} \end{bmatrix}$$

It can also be verified that this process of decomposition corresponds to the two-time-scales decomposition of the initial Lyapunov equation defining two sub-problems (one " fast", the other "slow") treated in a parallel manner and co-ordinated sequentially by the relaxed resolution of a problem which we shall describe as a co-ordination problem.
Figure (2.13) shows this process for which it will be noted that the initialisation $P_{12}^{(o)} = P_{21}^{T(o)} = \mathbb{O}$ corresponding to the classical decentralisation ($\varepsilon \to 0$), induces a first approximation $P_{11}^{(1)}$, $P_{22}^{(1)}$ which is no other than te usual approximation in singular perturbations of P.
Just as in the preceding algorithm based on regular perturbation, it is shown that this algorithm is equivalent to a P series expansion [SAN - 79].

2.4.4.3. Block resolution of an algebraic Riccati equation

As above, we are considering here the decomposition methods directly applied to the matricial expression of the algebraic RICCATI equation, namely :

(2.27) $$A^T K + KA - K BR^{-1} B^T K + Q = \mathbb{O}_n \qquad (2.27)$$

which we write again as :

$$F(K) + Q = \mathbb{O} \tag{2.28}$$

After adequate partitions of the matrices A, Q, K and S defined as follows :

$$S \triangleq BR^{-1}B^T = \begin{bmatrix} S_{11} & \cdots & S_{1N} \\ \vdots & & \vdots \\ S_{N1} & \cdots & S_{NN} \end{bmatrix}$$

we can rewrite the Riccati equation (2.27) in the partitioned form :

$$A_{ii}^T K_{ij} + K_{ij} A_{jj} + \sum_{\substack{\ell=1 \\ \ell \neq i}}^{N} A_{\ell i}^T K_{\ell j} + \sum_{\substack{\ell=1 \\ \ell \neq j}}^{N} K_{i\ell} A_{\ell j} + Q_{ij}$$

$$- K_{ij} \sum_{\substack{\ell=1 \\ \ell \neq i}}^{N} S_{j\ell} K_{\ell j} - K_{ij} S_{ji} K_{ij} - \Big(\sum_{\substack{\ell=1 \\ \ell \neq j}}^{N} K_{i\ell} S_{\ell i} \Big) K_{ij} \tag{2.29}$$

$$- \sum_{\substack{m=1 \\ m \neq i}}^{N} \Big(\sum_{\substack{\ell=1 \\ \ell \neq j}}^{N} K_{i\ell} S_{\ell m} \Big) K_{mj} = \mathbb{O} \quad \forall i, j = 1, \ldots, N$$

The non-linearity of the Riccati equation does not allow a separability of F to appear and although numerous decompositions can nevertheless still be envisaged, they imply a supplementary relaxation on the non-linear part of the equation.
This means that we shall be led to consider a decomposition of $F(K) = F_o(K) + F_1(K)$ such that, in general :

$$F_{o_{ij}} = F_{o_{ij}} (K_{ij}, K) \quad ; \quad F_{1_{ij}} = F_{1_{ij}} (K_{ij}, K)$$

leading to the process of approximations :

$$F_{o_{ij}} (K_{ij}^{(k+1)}, K^{(k)}) + F_{1_{ij}} (K_{ij}^{(k)}, K^{(k)}) + Q_{ij} = \mathbb{O} \qquad k > 0$$

$$\forall i, j = 1, \ldots, N$$

To illustrate this, let us mention the parallel decomposition method which appears as the most intuitive from equations (2.29) and where :

$$\begin{cases} F_{o_{ij}} = F_{o_{ij}} (K_{ij}) = A_{ii}^T K_{ij} + K_{ij} A_{jj} + K_{ij} S_{ji} K_{ij} \\ F_{1_{ij}} = F_{1_{ij}} (K_{ij}, K) \text{ cf } (2.29) \qquad \forall i, j = 1, \ldots, N \end{cases}$$

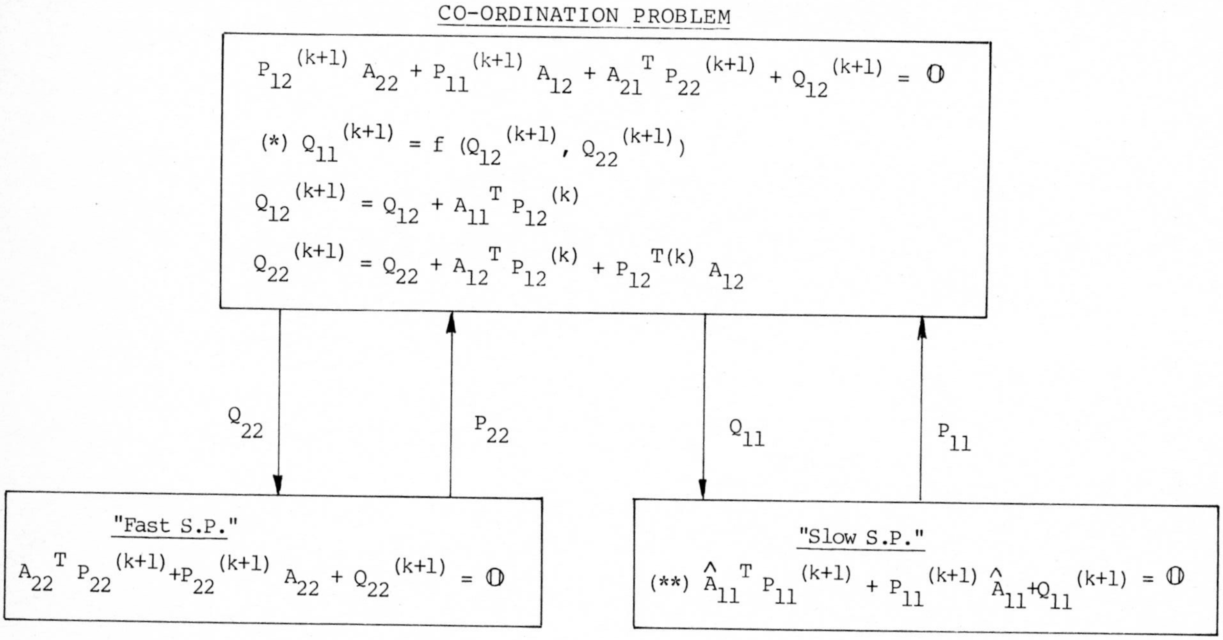

(*) $f(Q_{12}^{(k)}, Q_{22}^{(k)}) \triangleq Q_{11} - Q_{12}^{(k)} (A_{22}^{-1} A_{21}) - (A_{22}^{-1} A_{21})^T Q_{12}^{T(k)} + (A_{22}^{-1} A_{21})^T Q_{22}^{(k)} (A_{22}^{-1} A_{21})$

(**) $\hat{A}_{11} \triangleq A_{11} - A_{12} (A_{22}^{-1} A_{21})$

Figure 2.13

A consequence of this non-linear formulation is to lead only to the expression of a local condition of convergence.

1) Weak-coupling decomposition algorithm

In a similar way to the previous section, by application of a regular perturbation technique to the matrices A, Q, S :

$$A_{ij} = \varepsilon \tilde{A}_{ij} \; ; \; Q_{ij} = \varepsilon \tilde{Q}_{ij} ; \; S_{ij} = \varepsilon \tilde{S}_{ij} \quad \forall i \neq j \; i, j = 1, \ldots, N$$

we can show that the search for a perturbed solution $K(\varepsilon)$ of the Riccati equation, such that :

$$K_{ij}(\varepsilon) = \varepsilon \tilde{K}_{ij} \quad \forall i \neq j \; i, j = 1, \ldots, N$$

$$K_{ii}(\varepsilon) = K_{ii} \quad \forall i = 1, \ldots, N$$

can be carried out by a sequential decomposition method of the type :

$$F(K) = F_o(K) + F_1(\varepsilon \tilde{K}) = F_o(K) + \varepsilon F_1(\tilde{K})$$

This method is defined as follows :

1st phase : Lower level :

$$F_{o_{ii}}[K_{ii}^{(k+1)}] = - F_{1_{ii}}[\varepsilon \tilde{K}^{(k)}] - Q_{ii} \qquad i = 1, \ldots, N$$

with :

$$F_{o_{ii}}[K_{ii}] = A_{ii}^T K_{ii} + K_{ii} A_{ii} - K_{ii} S_{ii} K_{ii}$$

2nd phase : Upper co-ordination level :

$$F_{o_{ij}}[K_{ij}^{(k+1)}, K_{ii}^{(k+1)}, K_{jj}^{(k+1)}] = - F_{1_{ij}}[\varepsilon \tilde{K}^{(k)}] - Q_{ij} \qquad i \neq j$$

with :

$$F_{o_{ij}}[K_{ij}, K_{ii}, K_{jj}] = (A_{ii} - S_{ii} K_{ii})^T K_{ij} + K_{ij} (A_{jj} - S_{jj} K_{jj})$$

$$+ A_{ji}^T K_{jj} + K_{ii} A_{ij} - K_{ii} S_{ij} K_{jj} - K_{ij} S_{ji} K_{ij}$$

Finally let us point out, this being linked to the non-linearity of the applications, that here the kth approximation $K^{(k)}(\varepsilon)$ does not correspond to an asymptotic development of the perturbed solution.

2) Two-time scales decomposition algorithm

Let us now consider the Riccati equation associated with a singularly perturbed linear system and the partitioning induced in accordance with the matrices A, B, Q and S.

$$A = \begin{bmatrix} A_{11} & A_{12} \\ \frac{\tilde{A}_{21}}{\varepsilon} & \frac{\tilde{A}_{22}}{\varepsilon} \end{bmatrix} ; B = \begin{bmatrix} B_1 \\ \frac{\tilde{B}_2}{\varepsilon} \end{bmatrix} ; \; Q = \begin{bmatrix} Q_{11} & Q_{12} \\ Q_{21} & Q_{22} \end{bmatrix} ; S = \begin{bmatrix} S_{11} & S_{12} \\ S_{21} & S_{22} \end{bmatrix}$$

with $S_{11} = B_1 \; R^{-1} B_1{}^T$

$$S_{12} = B_1 \; R^{-1} \frac{\tilde{B}_2}{\varepsilon}^T = \frac{\tilde{S}_{12}}{\varepsilon}$$

$$S_{22} = \frac{\tilde{B}_2}{\varepsilon} R^{-1} \frac{\tilde{B}_2}{\varepsilon}^T = \frac{\tilde{S}_{22}}{\varepsilon^2} \longrightarrow S = \begin{bmatrix} S_{11} & \frac{\tilde{S}_{12}}{\varepsilon} \\ \frac{\tilde{S}_{21}}{\varepsilon} & \frac{\tilde{S}_{22}}{\varepsilon^2} \end{bmatrix}$$

On the hypothesis that the matrices Q and R are symmetric, it is shown in [LEH - 78] that a perturbed solution :

$$K(\varepsilon) = \begin{bmatrix} K_{11} & \varepsilon \tilde{K}_{12} \\ \varepsilon \tilde{K}_{12}{}^T & \varepsilon \tilde{K}_{22} \end{bmatrix}$$

can be obtained by a sequential decomposition method of the type :

$$F_o \; [K^{(k+1)}] + F_1 \; [\varepsilon \tilde{K}^{(k)} \;] + Q = 0$$

This method of relaxation implies at step (k+1) the resolution of :

- two independent Riccati equations at a lower level :

$$\hat{A}_{11}{}^T \; K_{11}{}^{(k+1)} + K_{11}{}^{(k+1)} \hat{A}_{11} + Q_{11}{}^{(k+1)} - [K_{11}{}^{(k+1)} \hat{B}1 + C^{(k+1)}]$$
$$R^{(k+1)^{-1}} [K_{11}{}^{(k+1)} \; \hat{B}1 + C^{(k+1)}]^T = 0$$

with $\hat{A}_{11} \triangleq A_{11} - A_{12} (A_{22}{}^{-1} \; A_{21})$; $\hat{B}_1 \triangleq B_1 - A_{12} \; (A_{22}{}^{-1} \; B_2)$

$$A_{22}{}^T \; K_{22}{}^{(k+1)} + K_{22}{}^{(k+1)} \; A_{22} + Q_{22}{}^{(k+1)} - K_{22}{}^{(k+1)}$$
$$S_{22} \; K_{22}{}^{(k+1)} = 0$$

- and of a coordination system at an upper level :

$$K_{12}{}^{(k+1)} [A_{22} - S_{22} K_{22}{}^{k+1)}] + K_{11}{}^{(k+1)} [A_{12} - S_{12} K_{22}{}^{(k+1)} \;]$$
$$+ A_{21}{}^T \; K_{22}{}^{(k+1)} + Q_{12}{}^{(k+1)} = 0$$

$$Q_{12}^{(k+1)} = Q_{12} + [A_{11}^T - K_{11}^{(k)} S_{11} - K_{12}^{(k)} S_{21}] K_{12}^{(k)}$$

$$Q_{22}^{(k+1)} = Q_{22} + A_{12}^T K_{12}^{(k)} + K_{12}^{(k)T} A_{12} - K_{22}^{(k)} S_{21} K_{12}^{(k)}$$
$$- K_{12}^{(k)T} S_{12} K_{22}^{(k)} - K_{12}^{(k)T} S_{11} K_{12}^{(k)}$$

$$Q_{11}^{(k+1)} = Q_{11} + (A_{22}^{-1} A_{21})^T Q_{22}^{(k+1)} A_{22}^{-1} A_{21} - Q_{12}^{(k+1)} (A_{22}^{-1} A_{21})$$
$$- (A_{22}^{-1} A_{21})^T Q_{12}^{(k+1)T}$$

$$R^{(k+1)} = R + (A_{22}^{-1} B_2)^T Q_{22}^{(k+1)} (A_{22}^{-1} B_2)$$

$$C^{(k+1)} = [(A_{22}^{-1} A_{21})^T Q_{22}^{(k+1)} - Q_{12}^{(k+1)}] (A_{22}^{-1} B_2)$$

We shall not cover here the conditions of convergence and applicability of such an algorithm, which are analysed in the aforementioned reference.

Let us emphasize, however, that this method is equivalent to solving in a parallel manner two optimisation sub-problems on a "slow" system and on a "fast" system, whose weighting coefficients in the criterion are iteratively improved.

Figure (2.14) represents this decentralised optimisation scheme for which it will be noted that the initialisation $K_{12}^{(o)} = \mathbb{O}$ corresponding to the complete decentralisation ($\varepsilon \to o$) induces a first approximation of $K_{11}^{(1)}$ and $K_{22}^{(1)}$ which is no other than the normal sub-optimal decentralisation of the singular perturbation approach in which :

$$u = u_1 + u_2$$

3) <u>Relaxation on two levels of the Riccati equation</u>

Another possibility which we propose here is to apply a generalised algorithm of quasi-linearisation to the Riccati equation (2.27) (which, in an overall approach, would correspond to the algorithm of [KLE - 68]) and then to solve in partitioned manner the successive linearisations which constitute Lyapunov type equations.

This superposition of a procedure of quasi-linearisation and partitioning leads in fact at step k to the partitioned resolution of :

$$[A - SK^{(k)}]^T K^{(k+1)} + K^{(k+1)} [A - SK^{(k)}] + K^{(k)} S K^{(k)} + Q = \mathbb{O}_n$$

namely, of the Lyapunov equation :

$$A^{(k+1)T} K^{(k+1)} + K^{(k+1)} A^{(k+1)} + Q^{(k+1)} = \mathbb{O}_n \qquad (2.30)$$

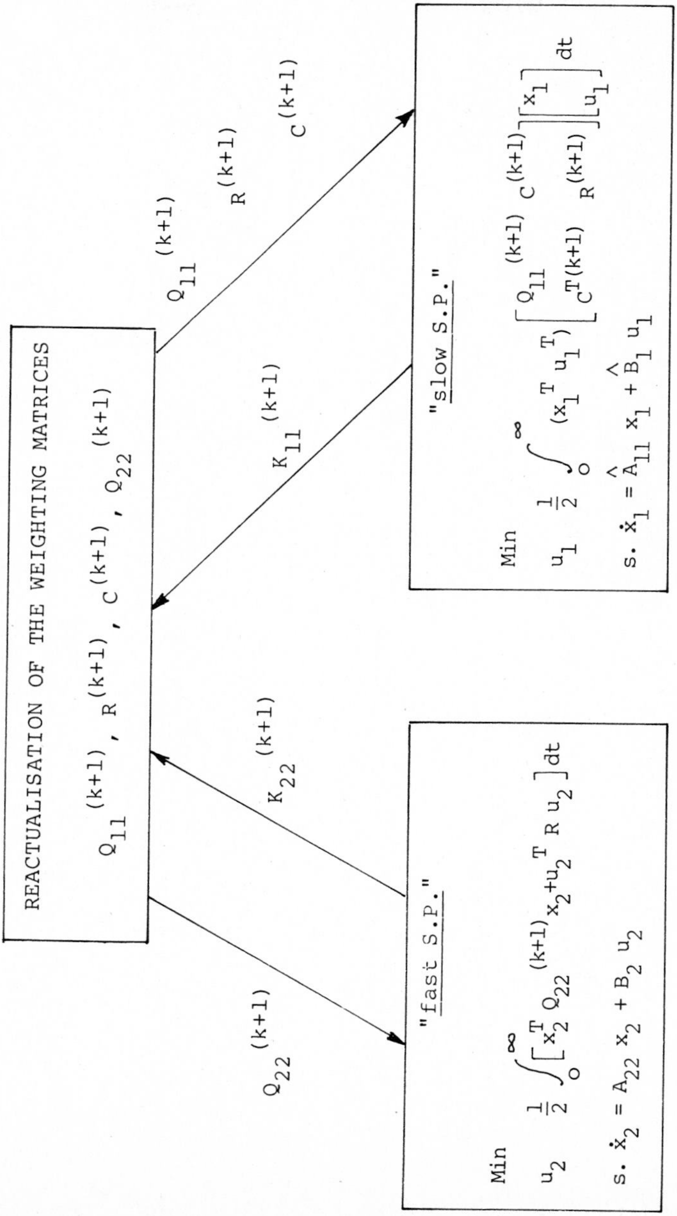
REACTUALISATION OF THE WEIGHTING MATRICES
Q11(k+1), R(k+1), C(k+1), Q22(k+1)
Q11(k+1) R(k+1) C(k+1)
K11(k+1)
K22(k+1)
Q22(k+1)
"slow S.P."
Min u1 1/2 ∫o∞ (x1T u1T) [Q11(k+1) C(k+1); CT(k+1) R(k+1)] [x1; u1] dt
s. ẋ1 = Â11 x1 + B̂1 u1
"fast S.P."
Min u2 1/2 ∫o∞ [x2T Q22(k+1) x2 + u2T R u2] dt
s. ẋ2 = A22 x2 + B2 u2

Figure 2.14

Let us remember that a property of the Kleinman algorithm is to guarantee a monotone and quadratic convergence, if one knows how to define an initial approximation $K^{(o)}$ such that the "closed loop" matrix : $A - SK^{(o)}$ is stable.
Whence the usefulness of applying one of the decomposition methods described in section 2.4.4.2. to each of the iterations of such an algorithm, all the more as each of the sub-systems now consists of resolving a Lyapunov type and no longer a Riccati type equation.
Figure (2.15) schematises the relaxed iterative scheme of decomposition co-ordination on two levels thus defined : the first level consisting of resolving equation (2.30) by a decomposition method the second level consisting of reactualising the matrices $A^{(k+1)}$ and $Q^{(k+1)}$ expressed from local solutions of the lower level as follows :

$$A_{ij}^{(k+1)} = A_{ij} - \sum_{\ell=1}^{N} S_{i\ell} K_{\ell j}^{(k)}$$

$$Q_{ij}^{(k+1)} = Q_{ij} + K_{ij}^{(k)} \sum_{\substack{\ell=1 \\ \ell \neq i}}^{N} S_{j\ell} K_{\ell j}^{(k)} + K_{ij}^{(k)} S_{ji} K_{ij}^{(k)} +$$

$$+ \left[\sum_{\substack{\ell=1 \\ \ell \neq j}}^{N} K_{i\ell}^{(k)} S_{\ell i} \right] K_{ij}^{(k)} + \sum_{\substack{m=1 \\ m \neq i}}^{N} \left[\sum_{\substack{\ell=1 \\ \ell \neq j}}^{N} K_{i\ell}^{(k)} S_{\ell m} \right] K_{mj}^{(k)}$$

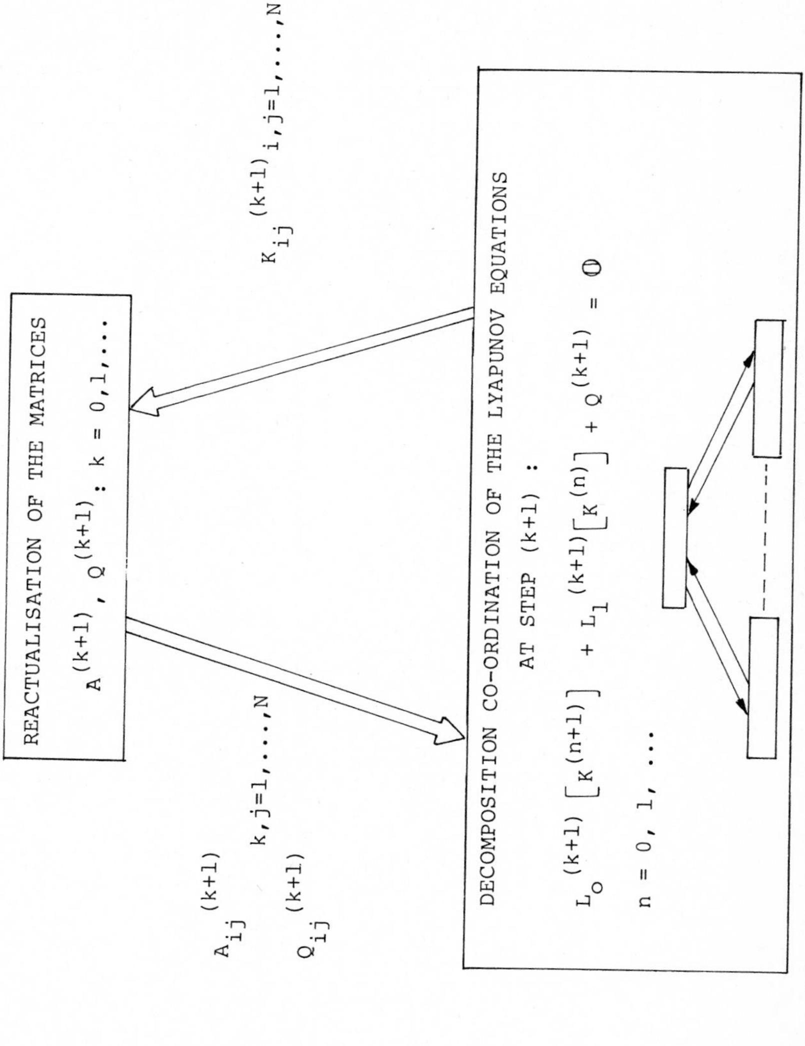
REACTUALISATION OF THE MATRICES
A(k+1), Q(k+1) ; k = 0,1,...
Aij(k+1) k,j=1,...,N
Qij(k+1)
Kij(k+1) i,j=1,...,N
DECOMPOSITION CO-ORDINATION OF THE LYAPUNOV EQUATIONS
AT STEP (k+1) :
Lo(k+1) [K(n+1)] + L1(k+1)[K(n)] + Q(k+1) = 0
n = 0, 1, ...

Figure 2.15

REFERENCES

ATH - 66 ATHANS M., P.L. FALB, "Optimal control" Mac Graw Hill, 1966

BAR - 71 BARNETT S., "Matrices in control theory", Van Nostrand Reinhold-Company, 1971

BEL - 65 BELLMAN R.E., S.E. DREYFUS, "La programmation dynamique et ses applications", Dunod 1965.

BEN - 80 BENREJEB M., "Sur l'analyse et la synthèse de processus complexes hiérarchisés. Application aux systèmes singulièrement perturbés", Doctoral thesis, Université des Sciences et Techniques, Lille, 1980.

BER - 76 BERNHARD P., "Commande optimale, décentralisation et jeux dynamiques", Dunod Automatique, 1976.

BER - 78 BERNHARD P., "La théorie de la seconde variation et le problème linéaire quadratique", Int. Report. n° 7803. Cahiers de mathématiques de la décision. Univ. Paris Dauphine, 1978.

CAL - 76 CALVET J.L., "Optimisation par calcul hiérarchisé et coordination en ligne des systèmes dynamiques de grande dimension". Doctoral thesis, Univ. Paul Sabatier, Toulouse, 1976.

CAL - 79 CALVET J.L., "Programmes numériques de résolution de problèmes d'optimisation dynamiques interconnectés, par quasi-linéarisation". Int. Report CDH n° 79.I.52. L.A.A.S. Toulouse, 1979.

CAL - 80 CALVET J.L., A. TITLI, "Hierarchical optimization and control of large scale systems with dynamical interconnection system", 2nd Symposium IFAC "Large Scale Systems : Theory and Applications", Toulouse, June 26-28, 1980.

CAL - 82 CALVET J.L., A. MANO, "Decomposition on the basis of structure of discrete systems. Application to hierarchical approach of a water quality control problem". 3rd IFAC/IFIP Symposium on "Software for computer control", Madrid, October 5-8, 1982.

CHU - 76 CHU K.C., "Comparisons of information structures in decentralized dynamic systems", in HO - 76 .

COH - 78 COHEN G. "Optimization by decomposition and coordination : a unified approach", IEEE Trans. on Autom. Control, vol. AC-23, n° 2, 222-232, 1978.

COH - 80 COHEN G., "Auxiliary problem principle and decomposition of optimization problems", J.O.T.A., vol. 32,1980.

COUR - 80 COURCOUBATIS C., P. VARAIYA, "A preliminary model for distributed algorithms". 2nd Symposium IFAC "LSSTA", Toulouse, June 26-28, 1980.

FIN - 68 FINDEISEN W., "Parametric optimization by primal method in multilevel systems". IEEE Trans. Syst. Sciences Cybern. Vol. SSC-4, n° 2, 155-164, 1968.

FIN - 80 FINDEISEN W. et al., "Control and coordination in hierarchical systems". IIASA - Wiley, 1980.

GAL - 73 GALY J., "Optimisation dynamique par quasi-linéarisation et commande hiérarchisée", Doctoral thesis 1973, Univ. Paul Sabatier Toulouse.

GEL - 63 GELFAND I.M., S. FOMIN, "Calculus of variations", Prentice Hall, 1963.

GER - 79 GEROMEL J.C., "Contribution à l'étude des systèmes dynamiques interconnectés. Aspects de Décentralisation", Doctoral thesis, Univ. Paul Sabatier Toulouse, 1979.

GOL - 81 GOLUB G.H., Lectures Notes : "Methodes itératives pour la résolution des systèmes linéaires" I.N.R.I.A. Seminar on "Analyse des Systèmes à Matrices Creuses" Nice, March 23 - 27, 1981.

HUR - 77 HURTEAU R., "Calcul et commande hiérarchisés des systèmes dynamiques : mise en oeuvre numérique et hybride", Doctoral thesis, Univ. Paul Sabatier, Toulouse, 1977.

HO - 76 HO Y.C., S.K. MITTER, "Directions in large scale systems : many-person optimization and decentralized control". Ed. Plenum Press - 1976.

IKE - 80 IKEDA M., D.D. SILJAK, "Overlapping decompositions, expansions, and contractions of dynamic systems", LSSTA, Vol. 1, N° 1, 1980.

KEV - 75 KEVORKIAN A.K., "Structural aspects of large dynamic systems", 6th triennal-world congress. IFAC - BOSTON, August 24 - 30, 1975.

KHA - 78 KHALIL H.K., P.V. KOKOTOVIC, "Control strategies for decision makers using different models of the same system", IEEE Trans. on Autom. Control, Vol. AC-23, N° 2, 1978.

KIR - 70 KIRK D.E., "Optimal control theory-an introduction", Prentice Hall, 1970.

KLE - 68 KLEINMAN D.L., "On an iterative technique for Riccati equation computations". IEEE, Trans. on Automatic. Control, vol. AC - 13, N° 1, 114-115, 1968.

KOK - 72 KOKOTOVIC P., R.A. YACKEL, "Singular perturbation of linear regulators : Basic theorems", IEEE Trans. on Automatic Control, Vol. AC - 17, N° 1, 1972.

KOL - 57 KOLMOGOROV A., S. FOMIN, "Elements of the theory of functions and functional analysis", Vol. 1, Graylock, 1957.

LAN - 77 LANG, B., "Calcul hybride décentralisé de la commande optimale d'un processus dynamique continu", Doctoral thesis. Faculté des Sciences et Techniques de l'Université de Franche- Comté, 1977.

LAR - 70 LARSON R.E., A.J. KORSAK, "A dynamic programming successive approximations technique with convergence proofs", Automatica, vol. 6, 245 - 260, 1970.

LAR - 79 LARSON R.E. et al., "Foundations of spatial dynamic programming". In "Systems Engineering for power : Vol. 1". Davos, Switzerland, September 30 - October 5, 1979.

LAS - 70 LASDON L.S., "Optimization theory for large systems" Mac Millan - 1970.

LAU - 78 LAUB A.J., F.N. BAILEY, "An iterative coordination approach to decentralized decision problems". IEEE, Trans. on Autom. Control, vol. AC-23, n° 6, 1031-1036, 1978.

LEH - 78 LEHTOMAKI N.A., "Iterative decomposition of the Lyapunov and Riccati equations". Report ESL - R - 818. Electronic Systems Laboratory - MIT Cambridge, 1978.

LHO - 79 LHOTE F., J.C. MIELLOU, "Algorithmes de décentralisation et de coordination par relaxation en commande optimale". Chapter VII in "Analyse et Commande des Systèmes Complexes". A. TITLI et al., Monographie AFCET, 1979, Cepadues Editions.

LIO - 74 LIONS J.L., G.I. MARCHOUK Editors, "Sur les méthodes numériques en sciences physiques et économiques" Dunod, 1974.

LIU - 65 LIUSTERNIK L.A., V.J. SOBOLEV, "Elements of functional analysis", Ungar, 1965.

LOO - 80 LOOZE D.P., N.R. SANDELL Jr., "Hierarchical control of weakly coupled systems". 2nd Symposium IFAC : Large Scale Systems : Theory and applications. June 24-26, 1980, Toulouse.

LUE - 69 LUENBERGER D.G., "Optimization by vector-space methods". Wiley - 1969.

MAY - 76 MAYNE D.Q., "Decentralized control and large scale systems", in [HO - 76].

MES - 70 MESAROVIC M.D., D. MACKO, Y. TAKAHARA, "Theory of hierarchical, multilevel systems", Academic Press, 1970.

MIE - 75 MIELLOU J.C., "Algorithmes de relaxation chaotique à retards". RAIRO, 9ème année, R.1, 55-82, 1975.

ORT - 70 ORTEGA J.M., W.C. RHEINBOLDT, "Iterative solutions of non linear equations in several variables", Academic Press - 1970.

PAP - 80 PAPAGEORGIOU M., G. SCHMIDT, "On the hierarchical solution of non linear optimal control problems" LSSTA, Vol. 1, N° 4, 265-271, 1980.

PEA - 71 PEARSON J.D., "Dynamic decomposition techniques" in "Optimization methods for large scale systems ... with applications". Wismer Editor, 1971.

RIC - 75 RICHETIN M., "Analyse structurale des systèmes complexes en vue d'une commande hiérarchisée", Doctoral thesis, Univ. Paul Sabatier, Toulouse, 1975.

SAN - 76 SANDELL N.R., Jr., "Information flow in decentralized systems", in [HO - 76] .

SAN - 79 SANDELL N.R. Jr., D.A. CASTANON, A.J. LAUB, "The development of decentralized organizational forms for the control of large scale interconnected power systems" in "Systems Engineering for Power : Vol. 1". Davos, Switzerland, Sept. 30, October 5, 1979.

SEW - 75 SEW HOY W., J.A. GIBSON, "An extension of the predictive Min-H method to multivariable control", Int. Jal. Control, Vol. 21, N° 5, 795 - 799, 1975.

SEZ - 81 SEZER M.E., D.D. SILJAK, "On structural decomposition and stabilization of large scale control systems", IEEE, Trans. on Autom. Control., vol AC-26, n° 2, 439-444, 1981.

SIN - 78 SINGH M.G., A. TITLI, "Systems : Decomposition, optimisation and control", Pergamon Press, 1978.

SUR - 78 SURI R., Y.C. HO, "Resources management in an automated warehouse". Colloque International sur l'analyse et l'optimisation des systèmes, Versailles - December 11-13, 1978.

TAK - 65 TAKAHARA Y., "A multi-level structure for a class of dynamical optimization problems", M.S. Thesis, Case Western Reserve University, Cleveland, 1965.

TAM - 75 TAMURA H., "Decentralized optimisation for distributed-lag models of discrete systems", Automatica, Vol. 11, 593-602, 1975.

TIT - 75 TITLI A., "Commande hiérarchisée et optimisation des systèmes complexes". Dunod Automatique, Paris, 1975.

VAR - 62 VARGA R.S., "Matrix iterative analysis", Prentice Hall, 1962.

WIN - 74 WINKLER C., "Basis factorization for block-angular linear programs : unified theory of partitioning and decomposition using the simplex method" T.R. SOL. 74-19 - Systems optim. Laboratory. Department of operation research. Standford Univ. California.

WIS - 71 WISMER D.A., "Optimization methods for large-scale systems ... with applications". Mac Graw-Hill, 1971.

YOU - 71 YOUNG D.M., "Iterative solution of large linear systems", Academic Press, 1971.

CHAPTER 3

SUB-OPTIMAL CONTROL AND PARAMETRIC OPTIMISATION SPECIAL CASE OF DECENTRALISED CONTROL

3.1. Introduction

Within the field of automatic control the L.Q.G. (linear quadratic gaussian) approach constitutes a highly efficient tool for the case of linear systems excited by gaussian white noise. The determination of control goes through the resolution of a dynamic auxiliary optimisation problem consisting of minimising the value of an integral quadratic criterion evaluated along the trajectories of the controlled system. According to the standard observability and controllability assumptions of the system, the efficiency of this approach is due to the fact that it allows the analytical determination of a feedback type controller resulting from the concatenation (principle of separation) of a "filtering" part and a control part. Each of these parts is obtained by taking into account separately problems of reconstruction of state and of control respectively, the optimal gains resulting from the solution of two matrix Riccati equations [KWA - 72],[ATH - 71] .

To carry out this type of control implies a certain degree of centralisation and, in general, one single decision maker should have all the information on the system to be controlled : a priori information on the mathematical model and up to date information on the state of the system (measured or reconstructed). This hypothesis of centralisation cannot realistically be accepted since we are interested in the field of large scale complex systems. For such processes it is more reasonable to see the problem of control as a problem with several decision makers, each having a partial and different view of the global process (decentralisation aspect) and even sometimes different objective functions. In the literature this type of problem is generally referenced in the category of "non-classical information pattern problems" in accordance with terminology attributable to Witsenhausen who was the first to draw attention to some of the consequences linked to the abandonment of standard hypotheses in L.Q.G. problems [WIT - 68],[WIT - 71] .
It must be said that these type of problems still remain broadly within the area of research and at present there is no general method which makes possible the calculation of optimal constrained solutions [COH - 81],[HO - 78] .
One approach which makes it possible to reduce the complexity of these problems considerably while still providing solutions which are interesting in practical terms consists of searching

for the best solution belonging to a set or category of controls chosen a priori and which is parametrisable. Of course we are talking about an approach which must be described as sub-optimal since it imposes, for the control, the constraint of belonging to a special class. The problem of the choice of this category remains open and of course the quality of the result concerning the controlled system depends on it. We shall make a few comments about a specific example dealt with in this section which may suggest an iterative method for the search for a satisfactory category of admissible controls.

Many works carried out in this field have confined themselves to the category of constant, linear feedbacks [GUA - 79], [COH - 75],[GER - 79-a]. This is justified by the importance assumed in practice by this category for control and regulation problems. Nevertheless it can be observed that this approach is in no way restricted to this category and that any other type of control can be envisaged, the only restriction being that it is formally described in a paramètrised fashion[QUA - 79]. For this approach, control problems are changed into parametric (static) optimisation problems whose resolution can be undertaken using appropriate methods and algorithms in non-linear programming.

Presentation of the parametric optimisation approach will be carried out by means of a special case concerning the decentralisation constraint imposed on the regulation structure of an interconnected linear system. A gradient type algorithm will be described for the resolution of the problem and the following questions posed by its implementation will be approached successively :

- determination of matrix gradients for which a simple method is given
- projection of matrix gradients
- convergence - stability of the controlled system
- degree of sub-optimality

We shall restrict ourselves to the deterministic case (L.Q.)

3.2. Decentralised regulation

3.2.1. Parametric optimisation problem and numerical algorithm

Let there be the invariant interconnected linear system described by the following differential system :

$$\begin{cases} \overset{\circ}{x}_i = A_i x_i + B_i u_i + z_i \\ z_i = \sum_{j=1}^{N} A_{ij} x_j \\ x_i(o) = x_{io} \ ; \quad i = 1,2, \ldots, N \end{cases} \qquad (3.1)$$

where $x_i \in R^{n_i}$, $u_i \in R^{m_i}$, $z_i \in R^{n_i}$ are respectively, the state, control and interconnection vectors of the sub-system i.

The decentralisation constraint is expressed by the fact that the local regulators only decide their controls u_i from the only knowledge of local measurements :

$$u_i = - K_i C_i x_i \tag{3.2}$$

where $C_i x_i$ is the measurable output vector of the sub-system i. Alongside this structural constraint for the global control of the system and for the purpose of greater generality, it will be supposed that the local feedback matrices must besides satisfy linear constraints of the type

$$G_i K_i H_i = 0 \tag{3.3}$$

G_i, H_i being matrices of appropriate dimension and of full rank. This type of constraint may serve to express the fact that the control u_i must use linear combinations of the measurable outputs (for example : area control error encountered in the power frequency control problem of electrical power networks - [ELG - 73]).

In the category of feedback type control defined by the constraints (3.2) and (3.3) it is a question of determining the gains K_i; i = 1, 2, ..., N while trying to minimise the quadratic criterion (chosen initially in an additive separable form).

$$J = \sum_{i=1}^{N} \int_{o}^{T} (x_i' Q_i x_i + u_i' R_i u_i) \, dt + x_i'(T) S_i x_i(T) \tag{3.4}$$

with $Q_i \geq 0$, $S_i \geq 0$ (positive semi-definite); $R_i > 0$.

The problem (3.1 - 3.4) can be re-written in condensed form as :

$$\left| \begin{array}{l} \min\limits_{K \in \mathfrak{D}} \; J(K) = \int_{o}^{T} (x' Qx + u' Ru) \, dt + x'(T) \, S \, x(T) \\ \text{subject to} \\ \overset{\circ}{x} = Ax + Bu \\ u = - KCx \\ GKH = 0 \end{array} \right. \tag{3.5}$$

where x, u are the state and control vectors of the global system, $A = \{A_{ij};\ i,j = 1, 2, ..., N\}$ and Q, R, S, B, C, G, H are block-diagonal matrices whose block diagonals are the matrices Q_i, R_i, S_i, B_i, C_i, G_i, H_i related to each of the sub-systems. $\mathfrak{D}$ represents the category of the block-diagonal matrices

$$\mathfrak{D} = \left\{ K;\ K = \text{block diag } (K_i),\ i = 1, 2, \ldots N \right\} \tag{3.6}$$

For the purpose of simplification of the calculations to follow, and using the properties of the matrix trace functions (noted as $T_r(.)$), (3.5) can be written in the same way

$$\left|\begin{array}{l} \min\limits_{K\in\mathcal{D}} \quad J(K) = \int_o^T T_r\,[(Q + C'K'RKC)\;X]dt + T_r\,[S\;X(T)] \\ \text{subject to} \\ \overset{\circ}{X} = (A-BKC)\;X + X\;(A-BKC)' \quad ; \quad X(o) = X_o \\ GKH = 0 \end{array}\right. \tag{3.7}$$

where $X = x\,x'$

In this form the problem is presented as a non-linear parametric optimisation problem which can be resolved by using a primal method (gradient - admissible directions), the numerical realisation of which gives the following fundamental stages :

stage 1 - initialisation by an admissible control K_o, ie. $K_o \in \mathcal{D}$ and $G\,K_o\,H = 0$

stage 2 - calculation of the matrix gradient

$$\frac{dJ}{dK} \triangleq \left\{\frac{dJ}{dk_{ij}}\right\}$$

stage 3 - determination of an admissible direction D ($D \in \mathcal{D}$ and $GDH = 0$) which is a direction of decrease for the objective function, and of a step α of progression in this direction.

Let us first concern ourselves with stages 2 and 3; calculation of the matrix gradient and the problem of search for admissible directions.

3.2.2. Determination of matrix gradients

One of the first articles to appear concerning this approach, which consists of searching for the best solution in a category given a priori is certainly that of Levine and Athans [LEV - 70] and concerns the problem of optimal output feedback. In this article the necessary conditions of optimality (cancellation of the gradient) are obtained in a direct way, by substitution in the quadratic criterion J (K) (3.5) of x(t) by its expression in terms of the transition matrix of the system. This then leads, for determination of the matrix gradient $\frac{dJ}{dK}$ to relatively long and tiresome calculations, although they are not fundamentally difficult. Such calculations can be avoided by using the variational approach and by writing Lagrangians and Hamiltonians associated with the problem [MEN - 74];[BER - 81]. Let f(K) be a scalar function of matrix differentiable in relation to its arguments. By definition the matrix gradient $\frac{df}{dK}$ is the matrix whose elements a_{ij} (i line, j column) are $\partial f/\partial k_{ij}$.

3.2.2.1. The static case

<u>Theorem 3.1</u>. Let f $[X, G(X)]$ be a scalar function derivable in

relation to all its arguments; X and G are matrices.

$$\frac{df}{dX}[X, G(X)] = \frac{\partial L}{\partial X}(X, Y, \Lambda)$$

where Y, Λ are the matrices obtained by resolving the necessary conditions of stationarity of the Langrangian function :

$$L = f(X,Y) + T_r\left[\Lambda'(G(X) - Y)\right]$$

Proof : Let $X = \{x_{ij}\}$, $G = \{g_{ij}\}$ Then

$$\frac{df}{dx_{ij}} = \frac{\partial f(X,G)}{\partial x_{ij}} + \sum_k \sum_\ell \frac{\partial f(X,G)}{\partial g_{kl}} \cdot \frac{\partial g_{kl}}{\partial x_{ij}} = \frac{\partial f}{\partial x_{ij}} + T_r\left[\frac{\partial f'}{\partial G} \cdot \frac{\partial G}{\partial x_{ij}}\right]$$

where $\frac{\partial f}{\partial G}$ is the matrix derivative of the scalar function f in relation to the matrix G (see definition above) and $\frac{\partial G}{\partial x_{ij}}$ is the matrix derivative of matrix in relation to a scalar.

Let us put $\Lambda = \frac{\partial f}{\partial G}$, we have

$$\frac{\partial f}{\partial x_{ij}} = \frac{\partial}{\partial x_{ij}}\left\{f(X,G) + T_r\left[\Lambda' G\right]\right\}\Bigg|_{\Lambda = \frac{\partial f}{\partial G}}$$

which, considering the matrix $\frac{df}{dX}$, can be written as

$$\frac{df}{dX} = \frac{\partial}{\partial X}\left\{f(X,Y) + T_r\left[\Lambda'(G(X) - Y)\right]\right\}\Bigg|_{\substack{Y = G(X) \\ \Lambda = \frac{\partial f}{\partial G}}}$$

This proves the theorem.

3.2.2.2. The dynamic case

Let the objective function be

$$J(K) = \int_0^T f(X,K)\,dt + g(X(T))$$

where f and g are scalar functions (differentiable) in relation to their arguments; X, K are matrices linked across the differential system

$$\overset{\circ}{X} = F\left[X(t), K\right] \quad ; \quad X(0) = X_0$$

<u>Theorem 3.2.</u>

$$\frac{dJ}{dK} = \int_0^T \frac{\partial H}{\partial K}(X, \Lambda, K)\,dt$$

where $H(X, \Lambda, K)$ is the hamiltonian function

$$H = f(X,K) + T_r [\Lambda' F(X,K)]$$

and X, Λ satisfy the conditions of stationarity

$$\overset{\circ}{X} = \frac{\partial H}{\partial \Lambda} \qquad X(o) = X_o$$

$$\overset{\circ}{\Lambda} = - \frac{\partial H}{\partial X} \qquad \Lambda(T) = \frac{dg}{dX(T)}$$

The proof is obtained very quickly by increasing the size of the differential system $\overset{\circ}{X} = F$ by adjunction of the dynamic matrix equation

$$\overset{\circ}{\tilde{K}} = 0 \qquad \tilde{K}(o) = K$$

Calling Γ the matrix of the multipliers associated with this new equation and noting, taking into account the nullity of its second member, that the hamiltonian function remains unchanged, we have :

$$\overset{\circ}{\Gamma} = - \frac{\partial H}{\partial \tilde{K}} \quad ; \quad \Gamma(T) = 0$$

It is known that

$$\frac{dJ}{d\tilde{K}(o)} = \Gamma(o)$$

which gives

$$\frac{dJ}{dK} = - \int_o^T \overset{\circ}{\Gamma}(t)\, dt = \int_o^T \frac{\partial H}{\partial K}\, dt \qquad \text{q.e.d.}$$

Extension to the discrete case is relatively direct and we can give the

Corollary 3.1.

Let $J(K) = \sum_{t=o}^{T-1} f(X_t, K) + g(X_T)$

with $X_{t+1} = F(X_t, K), \quad X_o$

then $$\frac{dJ}{dK} = \sum_{t=o}^{T-1} \frac{\partial H}{\partial K}(X_t, \Lambda_t, K)$$

where $H(X_t, \Lambda_t, K) = f(X_t, \Lambda) + T_r [\Lambda' F(X_t, K)]$

and

$$\frac{\partial H}{\partial X_t} - \Lambda_{t-1} = 0 \qquad \Lambda_{T-1} = \frac{\partial g}{\partial X_T}$$

$$\frac{\partial H}{\partial \Lambda_t} - X_{t+1} = 0 \qquad X_o$$

3.2.2.3. Application to the problem (3.7)

The Hamiltonian function of theorem 3.2 is written as

$$H(X,\Lambda,K)=T_r[(Q+C'K'RKC)\ X] + T_r[\Lambda'\ \{(A-BKC)X + X\ (A-BKC)'\}]$$

Applying the normal rules of trace functions

$$T_r\ [AB] = T_r\ [BA] = T_r\ [B'A'] = T_r\ [A'B']$$

and knowing that

$$\frac{d}{dX}\ \left[T_r\ [AX]\right] = A'$$

application of the results of theorem 3.2 makes it possible to write :

$$\frac{dJ}{dK} = 2\int_0^T (RKC - B'\Lambda)X\ C'\ dt$$

with

$$\mathring{\Lambda} + (A-BKC)'\Lambda + \Lambda(A-BKC) + Q + C'K'RKC=0 \qquad \Lambda(T)=S \tag{3.8}$$

and

$$\mathring{X} = (A - BKC)\ X + X\ (A - BKC)' \qquad X(o)=X_o$$

Later we shall have to consider the case with infinite horizon. Let us make T go towards infinity, then $\Lambda \to \Lambda_o$, which is the solution of the Lyapunov algebraic equation (we are leaving aside for the moment the problem of the uniqueness of such a solution)

$$(A-BKC)'\Lambda_o + \Lambda_o\ (A-BKC) + Q + C'K'RKC = 0$$

we then have

$$\frac{dJ}{dK} = 2\ (RKC - B'\Lambda_o)\int_0^\infty X(t)\ dt.\ C'$$

If the gain matrix K is such that (A - BKC) is asymptotically stable

$$\int_0^{\mathcal{D}} X(t)\ dt = V$$

and V is the solution of the algebraic Lyapunov equation

$$(A - BKC)\ V + V\ (A-BKC)' + X_o = 0$$

In the case of the infinite horizon, this result could have been established differently by noting that for this case the problem (3.7) can be written in the following equivalent static form

$$\begin{cases} \min_K\ J(K) = T_r\ [P\ X_o] \\ \text{subject to} \\ (A-BKC)'\ P + P\ (A-BKC) + Q + C'K'RKC = 0 \end{cases} \tag{3.9}$$

Applying the results of the theorem 3.1 with the Lagrangian function

$$L = T_r\,[P\,X_o] + T_r\left\{V'\left[(A-BKC)'\,P+P(A-BKC) + Q + C'K'RKC\right]\right\}$$

we get :

$$\left|\begin{array}{l} \frac{dJ}{dK}\;\left(=\frac{\partial L}{\partial K}\right) = 2\,[RKC - B'P]\;VC' \\ \\ (A-BKC)'\;P + P\;(A-BKC) + Q + C'K'RKC = 0 \\ (A-BKC)\;V + V\;(A-BKC)' + X_o = 0 \end{array}\right. \qquad (3.10)$$

with

Note : the derivation of the matrix gradient could also have been undertaken through the vector formulation (3.5). The interest of the matrix formulation (3.7) comes from the linear dependence of the Hamiltonian function in relation to the state matrix X (= x x'). This finds expression in the direct obtaining of the two uncoupled Lyapunov equations (Λ, X in the dynamic case; P, V in the static case) which figure in the calculation of the matrix gradient.

3.2.3. Admissible direction

Being in possession of the matrix gradient calculated at an admissible point (K satisfying the constraints), we now have to determine a direction of progression for the gain K : it must be such that, locally, the criterion decreases and it must satisfy the constraints (whence the term admissible). For the case which concerns us here, the constraints being linear equality constraints, it is natural to think of methods of the projected gradient type. We shall define below the problem of projection, in the form of an optimisation problem, which, formulated in the matrix context, will provide a further example of application of theorem 3.1.

The gain matrix K is subject to two constraints : the constraint of decentralisation and the constraint GKH = 0. The matrices G and H being chosen block diagonal, these constraints are not coupled. Projection of the matrix gradient dJ/dK can therefore be made sequentially and separately on each of the two sub-spaces defined by these constraints. Furthermore, the operation of projection on the sub-space $\mathcal{D}$ is trivial and consists simply of cancellation in the matrix gradient $\frac{dJ}{dK}$ (which is a full matrix) of the off diagonal terms, namely :

$$\bar{D} = \text{block diagonal}\left[\frac{dJ}{dK}\right]$$

As regards projection on the set defined by the constraint GKH = 0, this can be carried out from the resolution of the following optimisation problem :

$$\left[\begin{array}{l} \min\limits_{D}\;\; \frac{1}{2}\;T_r\;\left[(D-\bar{D})'\;(D-\bar{D})\right] \\ \\ \text{subject to}\quad GDH = 0 \end{array}\right. \qquad (3.11)$$

Expanding the criterion explicitly, it is easy to see that this problem corresponds to that of the orthogonal projection of $\bar{D}$ on the linear set defined by the constraint.
The Lagrangian is written as :

$$L = T_r \left[(D - \bar{D})' \ (D - \bar{D}) \right] + T_r \left[\Lambda' \ GDH \right]$$

whence the necessary (and here, sufficient) conditions of optimality

$$\frac{\partial L}{\partial D} = D - \bar{D} + G' \Lambda H' = 0$$

$$\frac{\partial L}{\partial \Lambda} = GDH = 0$$

They allow for the unique solution (G, H of full rank) :

$$\Lambda = (G \ G')^{-1} \ G \ \bar{D} \ H \ (H' \ H)^{-1}$$

$$D = \bar{D} - G' \ (G \ G')^{-1} \ G \ \bar{D} \ H \ (H' \ H)^{-1} \ H' = \mathcal{P}(\bar{D})$$

where $\mathcal{P}$ is the projection operator defined by

$$\mathcal{P}(.) = (.) - G' \ (G \ G')^{-1} \ G \ (.) \ H \ (H'H)^{-1} \ H'$$

Here we are talking about extension to the matrix case of the Rosen projection operator [ROS - 60] defined for the vector case.
Another delicate point raised by every gradient algorithm lies in the choice of a parameter α which fixes the step of progression along the admissible direction D. This choice presents itself at each iteration of the given algorithm in its general form given in section 3.2.1.. For reasons of rapidity of convergence, it is desirable that the step chosen be sufficiently large and one may consider, a priori, fixing it by resolution of the following single variable optimisation problem along the search direction (and this at each iteration).

$$\min_{\alpha \geq 0} \ J \ (K - \alpha \ D)$$

In fact, this is a problem which is very much equivalent from the point of view of complexity of calculations to the initial problem and its solution cannot be obtained analytically. This is why, practically speaking, it is generally advantageous to adopt a heuristic method of adaptation of the step during the iterations.

Note : the problem of non-linear optimisation (3.5) is not generally convex in relation to K. This means that at the convergence of the algorithm described previously, satisfying the necessary conditions of optimality, one obtains either a local minimum, or even a saddle point in the space of the parameters k_{ij}. This point of convergence depends of course on the initial condition K_o (stage 1). Naturally this is one of the standard characteristics of every non-convex multivariable optimisation problems.
There is another dependence which, from a practical point of view, proves to be just as awkward, and it is that of K in rela-

tion to the initial conditions chosen for the state vector $X_o = x_o x'_o$. We shall return to this last problem later, after we have developed the case of the infinite horizon which, as we shall see, poses a further basic problem.

3.3. The infinite horizon : the problem of stability

Let us recall briefly the formulation of the problem in the case of the infinite horizon

$$\begin{aligned}
&\min_{K\in\mathcal{D}} \quad J(K) = \int_o^\infty T_r \left[(Q + C'K' RKC)\ X(t)\right] dt \\
&\text{subject to} \\
&\overset{\circ}{X} = (A-BKC)\ X + X\ (A - BKC)'; \ X(o) = X_o \\
&GKH = 0
\end{aligned} \tag{3.12}$$

It is well known that in the non-constrained case (C = I, full matrix K) and by using the sufficient hypotheses of controllability of the pair (A, B) and observability of the pair $(A, Q^{\frac{1}{2}})$, the optimal control $u^*(t)$, which is unique, can be written as :

$$u^*(t) = - K^*\ x(t)$$

where K^* is a matrix with constant elements and which is expressed in terms of the matrix solution of an algebraic Riccati equation. In addition, this control is such that the matrix $(A - BK^*)$ is asymptotically stable [AND - 71] .
For the specialisation in the case of the infinite horizon of the iterative numerical algorithm presented in the previous paragraph, one of the crucial points consists of guaranteeing the obtaining of a stabilising gain at each step of the algorithm. This condition is necessary so that the integral defining J (K) is defined and therefore so that problem (3.12) remains meaningful.
Let us indicate by $\mathcal{D}_s$ the set of decentralised stabilising feedback gains

$$\mathcal{D}_s = \left\{ K;\ K\in\mathcal{D} \text{ and } (A-BKC) \text{ asymptotically stable}\right\}$$

Replacing the constraint $K \in \mathcal{D}$ by that of $K \in \mathcal{D}_s$, problem (3.12) can be formulated in the following static equivalent form :

$$\begin{aligned}
&\min_{K \in \mathcal{D}_s} \quad J(K) = T_r\ \left[P\ X_o\right] \\
&\text{subject to} \\
&(A-BKC)'\ P + P\ (A-BKC) + Q + C'K'RKC = 0 \\
&GKH = 0
\end{aligned} \tag{3.13}$$

The numerical algorithm which can be used for the resolution of the problem presents practically the same characteristics as the one described for the finite horizon with, in addition, further conditions and tests required by the fact, that it is necessary to obtain at each iteration a gain which stabilises the system.

Stage 1 - Initialisation by an admissible and <u>stabilising</u> control

$$GK_oH = 0 \text{ and } \underline{K_o \in \mathcal{D}_s}$$

Stage 2 - Calculation of the gradient

$$\left[\begin{array}{l} \frac{dJ}{dK} = 2\left[RKC - B'P\right] VC' \\ (A-BKC)'\ P + P\ (A-BKC) + Q + C'K'RKC = 0 \\ (A-BKC)\ V + V\ (A-BKC)' + X_o = 0 \end{array}\right.$$

Stage 3 - Determination of an admissible direction D

$$D = \mathcal{P}\left[\text{block diagonal } \frac{dJ}{dK}\right]$$

and choice of a step α of progression such that

$$J\ (K - \alpha D) < J(K)$$

and

$$(K - \alpha D) \in \mathcal{D}_s$$

3.3.1. The initial K_o stabilising gain; the problem of its choice

For its application, the primal method (projected gradient) requires the knowledge of an admissible stabilising gain (which satisfies the various structural constraints).
Now for a given linear system, to know whether $\mathcal{D}_s$ is empty or not is a very complex problem which is still wide open to research. The question has been studied by many authors; the degree of its complexity can be measured by the number of conjectures made on sufficient conditions of stabilisability under constraint of decentralisation; conjectures which have been rejected one by one by means of counter examples. Let us mention chronologically [AOK - 73],[WAN - 78],[FES - 79],[IKE - 79], [SEZ - 81]. These are all works whose aim was to demonstrate the existence of stabilising decentralised gains. From among those which provide some procedure for their determination we can quote [DAV - 76],[COR - 76],[ARM - 81] , the exception of those which are based on a hypothesis of weak coupling, most of them being given in reference in [SIL - 78].

Before closing this section we feel it is interesting to mention one method of dealing with the necessary condition of having an initial stabilising gain, a method which remains well within the framework of the approach in non-linear programming of control problems which has been developed in this section. It consists of using penalisation functions and augmented Lagrangian methods. In [WEN - 80] a discussion of such an approach is given. The success of the approach depends on the skill of the designer to choose the sequences on the multipliers and the penalisation coefficients well so that in the case of non-convergence one cannot be sure that $\mathcal{D}_s$ is empty.

3.3.2. Stability tests [GER - 79.b]

From an initial stabilising gain K_o it is necessary to ensure that the numerical algorithm determines at each iteration an admissible stabilising control which plays the role of initial condition for the following iteration. What are the conditions necessary in order that, having initialised the algorithm inside $\mathcal{D}_s$, the sequence of gains obtained stays there?

The lemma given below provides sufficient conditions for this and, expressed in terms of the characteristics of the optimisation problem (matrices), they make it possible to ensure convergence of the algorithm.

Lemma 3.1.

a) if $K \in \mathcal{D}_s$, then there exists a step $\alpha > 0$ such that

$$J(K - \alpha D) < J(K) \qquad (\text{for } D \neq 0)$$

b) if the pair $(A, Q^{\frac{1}{2}})$ is completely observable and X_o is a strictly positive definitive matrix.
Then $\forall \alpha > 0$ such that $J(K - \alpha D) < J(K) < \infty \Rightarrow (K - \alpha D) \in \mathcal{D}_s$

Proof :

a) linearising $J(.)$ around $\alpha = 0$ we have

$$J(K - \alpha D) = J(K) - \alpha \left.\frac{dJ(K - \alpha D)}{d\alpha}\right|_{\alpha = 0} + \ldots$$

$$= J(K) - \alpha\, T_r\left[\frac{dJ'}{dK} \cdot D\right] + \ldots$$

D being obtained from $\frac{dJ}{dK}$ by orthogonal projection on a linear space we have to first order

$$J(K - \alpha D) - J(K) = -\alpha\, T_r\,[D'D]$$

$T_r\,[D'D] > 0$ if $D \neq 0$. There exists therefore a $\alpha > 0$ such that a) is verified.

b) Let us note first of all that J(K) can be expressed thus

$$J(K) = T_r\,[F(K) \cdot X_o]$$

where

$$F(K) = \int_0^{\infty} e^{(A-BKC)'t}\,(Q + C'K'RKC)\,e^{(A-BKC)t}\,dt$$

The matrix is strictly positive definite for by hypothesis the pair $(A, Q^{\frac{1}{2}})$ is completely observable. In addition F(K) is finite if and only if (A-BKC) is asymptotically stable. Now for a positive definite matrix the condition "matrix with finite elements" is equivalent to the condition "finite trace" : $T_r\,[F(K)] < \infty$. It remains therefore to be proved that there exists a $\alpha > 0$ such that $T_r\,[F(K - \alpha D)]$ is finite. Now if

$$T_r\,[F(K - \alpha D) \cdot X_o] = J(K - \alpha D) \leq J(K) < \infty$$

then

$$T_r\,[F(K - \alpha D)] \leq \lambda_{min}^{-1}(X_o)\, J(K - \alpha D) < \infty$$

where $\lambda_{min}(X_o)$ indicates the smallest eigenvalue of X_o which is non zero since $X_o > 0$.

Let us note that the lemma above provides sufficient conditions for the existence of a step α in such a way that $(K-\alpha D)$ is a stabilising gain without however giving a practical rule to use for this test. Indeed, the value of the criterion for a gain K is calculated from the expression

$$J(K) = T_r \left[P(K) \,.\, X_o\right]$$

and it is not sufficient to verify that

$$T_r \left[P(K-\alpha D)\ X_o\right] \leq T_r \left[P(K)\ X_o\right]$$

to be sure that $(K-\alpha D) \in \mathcal{D}_s$. Indeed, as we have already said at the beginning of section (3.3), $T_r\ [P(K)\ X_o]$, only represents J(K) on the express condition that $K \in \mathcal{D}_s$. On the other hand, P (K) can very easily be defined, even if $K \notin \mathcal{D}_s$.

The choice of the step α (stage 3), must be carried out in such a way that :

i $(K-\alpha D) \in \mathcal{D}_s$

ii $T_r \left[P(K-\alpha D)\ X_o\right] \leq T_r \left[P(K)\ X_o\right]$

Since the algorithm necessitates calculation of the matrices P the test of condition i can be carried out by means of the definite-positiveness test of these matrices P.

$$i \longrightarrow i;\quad P\left[K-\alpha D\right] > 0$$

3.3.3. Degree of sub-optimality

It can prove to be of practical interest to have a way of testing and controlling the degree of sub-optimality of the decentralised control obtained in relation to the optimal centralised control, ie. that obtained by releasing all constraints of structure. It is to this problem that we shall apply ourselves in this section, but retaining only the decentralisation constraints :

$$C = I$$

$$G = H = 0$$

Let the problem therefore be

$$\left[\begin{array}{l} \min\limits_{K \in \mathcal{D}_s} \quad J(K) = T_r \left[P(K)\ X_o\right] \\ \text{subject to} \\ (A-BK)'\ P + P\ (A-BK) + Q + K'RK = 0 \end{array}\right. \qquad (3.14)$$

In the absence of a decentralisation constraint the optimal control u* can be written $u^* = -K^*x$ where

$$K^* = R^{-1}B'\ P^* \text{ and} \qquad (3.15)$$
$$A'P^* + P^*A - P^*B\ R^{-1}\ B'\ P^* + Q = 0$$

In the presence of the decentralisation constraint, $K \in \mathcal{D}$ let us denote by (2Λ) the matrix of the dual variables associated with this constraint. The decentralised solution obtained by the approach described above satisfies the following necessary conditions of optimality :

$$\begin{cases} 2\,(RK - B'P)\,V + 2\Lambda = 0 \\ (A - BK)'\,P + P\,(A - BK) + Q + K'RK = 0 \\ (A - BK)\,V + V\,(A - BK)' + X_o = 0 \\ K \in \mathcal{D} \quad , \Lambda \in \overline{\mathcal{D}} \end{cases} \tag{3.16}$$

where $\overline{\mathcal{D}}$ denotes the complementary set of $\mathcal{D}$.

Let us suppose that V is positive definite (a sufficient condition for this is that X_o is strictly positive definite) then the decentralised gain K satisfies.

$$K = R^{-1}\,B'\,P - \frac{1}{2}\,R^{-1}\,\Lambda\,V^{-1} \tag{3.17}$$

Carrying this expression of K in the second equation of (3.17), we have

$$A'\,P + PA - PBR^{-1}B'P + Q + V^{-1}\,\Lambda'\,R^{-1}\,\Lambda V^{-1} = 0 \tag{3.18}$$

We get a Riccati equation which only differs from that which defines P* (3.15) by one independent term.
The comparison between the non-constrained centralised control and the constrained control will have bearing on the value of the criterion. Let us remember that this value, with or without constraint, is given by

$$J\,(K) = J = T_r\,\left[P\,X_o\right]$$

and

$$J\,(K^*) = J^* = T_r\left[P^*\,X_o\right]$$

The problem is therefore to calculate, or more exactly to estimate, the difference

$$\Delta J = J - J^*$$

without, of course, having to resolve the Riccati equations (3.15) or (3.18)
Let us define

$$\begin{cases} r\,(\beta) = T_r\,\left[P(\beta)\,X_o\right] \\ A'\,P(\beta) + P(\beta)\,A - P(\beta)\,BR^{-1}B'P(\beta) + Q + \beta\,Q_\Lambda = 0 \\ Q_\Lambda = V^{-1}\,\Lambda'\,R^{-1}\,\Lambda V^{-1} \end{cases} \tag{3.19}$$

On the evidence

$$\Delta(J) = r(1) - r\,(o)$$

$r\,(\beta)$ being a differentiable function in relation to β, it comes to mind to approach the value ΔJ by means of an expansion in a Taylor series. Remembering that the numerical algorithm provides P (=P(1)) it is of course necessary to effect this expansion around the point $\beta = 1$. We then have :

$$\Delta J = -\sum_{j=1}^{n} \frac{(-1)^j}{j!}\;T_r\left\{\left[\frac{d^jP(\beta)}{d\beta^j}\right]_{\beta=1} \cdot X_o\right\} \tag{3.20}$$

Evaluation of (3.20) passes through the calculation of the successive derivatives of the matrix $P(\beta)$ in relation to β . Denoting by ∂P, $\partial^2 P$... the first, second derivative of $P(\beta)$ at the point $\beta = 1$, we have

$$\begin{cases} (A - BR^{-1}B'P)'\,\partial P + \partial P\,(A - BR^{-1}B'P) + Q_\Lambda = 0 \\ (A - BR^{-1}B'P)'\,\partial^2 P + \partial^2 P\,(A-BR^{-1}B'P) - 2\,\partial P\,BR^{-1}B'\partial P = 0 \\ \vdots \\ (A - BR^{-1}B'P)'\,\partial^n P + \partial^n P\,(A - BR^{-1}B'P) + \mathcal{G}(\partial^{n-1}P, \partial^{n2}P, .. \partial P) = 0 \end{cases} \quad (3.21)$$

This shows that if $(A - BR^{-1}B'P)$ is asymptotically stable all the derivatives of $r(\beta)$ are bounded and in addition $\partial P \geq 0$ $\partial^2 P \leq 0$ therefore $d^2\, r(\beta) \,/\, d\beta^2 \leq 0$: $r(\beta)$ is a concave function of β.

The interesting point in this method of estimating the degree of sub-optimality lies in the fact that the obtaining of the successive derivatives of $r(\beta)$ is made by the intermediary of the solution of Lyapunov equations always of the same type, which greatly facilitates the calculation of the expansion (3.20)

3.3.4. An example

Let there be the interconnected system (N = 3) :

$$\overset{\circ}{x}_1 = \begin{bmatrix} 0 & 1 \\ -2 & -3 \end{bmatrix} x_1 + \begin{bmatrix} 1 \\ 1 \end{bmatrix} u_1 + \begin{bmatrix} 0,5 & 1 \\ 1 & 0 \end{bmatrix} x_2 + \begin{bmatrix} 0,6 & 0 \\ 0 & 1 \end{bmatrix} x_3$$

$$\overset{\circ}{x}_2 = \begin{bmatrix} 0 & 2 \\ 1 & 3 \end{bmatrix} x_2 + \begin{bmatrix} 3 & 0 \\ 0 & 4 \end{bmatrix} u_2 + \begin{bmatrix} 0,5 & 1 \\ 0 & -0,5 \end{bmatrix} x_1 + \begin{bmatrix} 1 & 0,5 \\ 0 & -0,5 \end{bmatrix} x_3$$

$$\overset{\circ}{x}_3 = \begin{bmatrix} 0 & 1 \\ -3 & -4 \end{bmatrix} x_3 + \begin{bmatrix} 2 \\ 2 \end{bmatrix} u_3 + \begin{bmatrix} 1 & 0 \\ 0 & 0,5 \end{bmatrix} x_1 + \begin{bmatrix} 1 & 0 \\ 0,5 & 0 \end{bmatrix} x_3$$

and

$$J(K) = \int_0^\infty (I + K'K)\, X(t) \quad dt$$

$$X_o = I$$

The stabilising decentralised gain K_o has been taken equal to

$$K_o = \begin{bmatrix} 11,51 & -0,55 & 0 & 0 & 0 & 0 \\ 0 & 0 & 2,30 & 0,62 & 0 & 0 \\ 0 & 0 & 0,82 & 3,41 & 0 & 0 \\ 0 & 0 & 0 & 0 & 4,96 & 0,06 \end{bmatrix}$$

Application of the numerical algorithm shown above has given

$$K = \begin{bmatrix} 1,21 & 0,35 & 0 & 0 & 0 & 0 \\ 0 & 0 & 1,43 & 0,45 & 0 & 0 \\ 0 & 0 & 0,87 & 2,46 & 0 & 0 \\ 0 & 0 & 0 & 0 & 1,58 & 0,24 \end{bmatrix}$$

and

$$\Lambda = 2 \begin{bmatrix} 0 & 0 & 0{,}27 & 0{,}06 & 0{,}44 & -0{,}18 \\ 0{,}59 & 0{,}34 & 0 & 0 & 0{,}39 & 0{,}11 \\ 2{,}21 & -1{,}36 & 0 & 0 & 0{,}94 & -1{,}01 \\ 1{,}67 & -0{,}45 & 0{,}45 & 0{,}05 & 0 & 0 \end{bmatrix} .10^{-1}$$

The figure below gives the estimation of the degree of sub-optimality obtained.

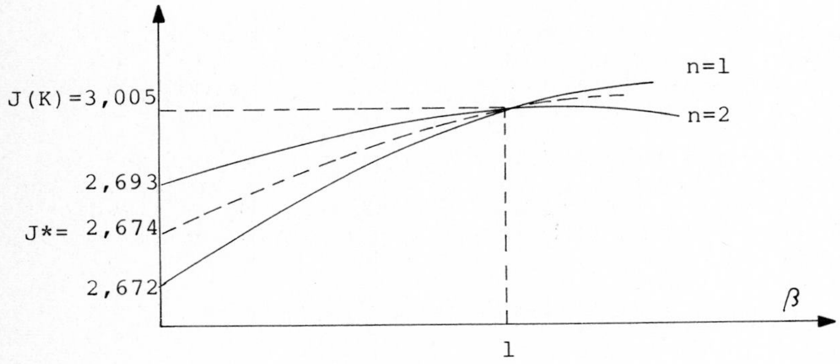

On this example the linear approximation provides J to 0,8% and the quadratic approximation to 0,07%. The good quality of the results obtained for n = 2 in this example is logical, taking into account the relative smallness of the matrix Λ.

Note : we have reached a point where we can take up again the problem raised in the general introduction of this section which concerned the link between the quality of the results obtained with the choice, made a priori, of the parametrisable category of controls. Without claiming to be able to give a definite response to such a problem, it is interesting to note that the proposed approach provides indications which can be used in order to attempt to guide the choice of this category.

Let us suppose that the problem is posed of the determination of a linear state feedback requiring as low a transfer of information as possible while still satisfying a maximum limit for the sub-optimality. The initial category of controls envisaged is normally that of decentralised controls. The approach presented here makes it possible to find in this category an admissible "optimal" solution (locally), then to test its sub-optimality in order to decide, in the event if the degree of sub-optimality being unsatisfactory, to augment the admissible category of controls.
This can be done after examination of the matrix Λ which is supplied at the convergence of the algorithm. In fact, it may then be decided to provide for a transfer of information between two sub-systems (i,j) in terms of the relative size of the corresponding sub-matrix Λ_{ij}. This size may in fact make it possible to provide for a considerable improvement of the value of the cri-

terion by the possibilities of using the sub-matrix of gain K_{ij}, taken as nil in the category previously chosen. This operation may be repeated as many times as necessary up to the obtaining of a satisfactory sub-optimality.

3.3.5. The problem of dependence vis à vis initial conditions

Examination of the necessary conditions of optimality which are resolved by means of the primal approach considered here makes it clear that the gain obtained at convergence depends on the choice of initial conditions made for the state vector. This constitutes a disadvantage from the practical viewpoint. The most commonly used method of getting round this problem consists of considering x_o as a random variable of given mean and covariance and then to minimise the mathematical expectation of the criterion J(K) (its mean taken on the space of the variables x_o)[LEV - 70].
Another way of proceeding is to use the now standard idea of the Worst case design.
Let us give a special application of this, and for this let us consider first of all a formulation in discrete time.
Let the problem be :

$$\left| \begin{array}{l} \min\limits_{K \in \mathcal{D}} \quad J(K) = \sum\limits_{t=0}^{T-1} \; T_r \left[(Q+K'RK)\, X_t \right] + T_r \left[S\, X_T \right] \\ \\ X_{t+1} = (A-BK)\, X_t\, (A-BK)' , \; X_o \end{array} \right. \qquad (3.22)$$

The idea of the worst case design consists, to use Kalman's terms, of defining a "game" between "man and nature". Man acts on the control K, nature on X_o. The objective of one is the minimisation of J(K), of the other, its maximisation.
Mathematically the problem can be defined thus :

$$\min_{K \in \mathcal{D}} \quad \omega(K) \qquad (3.23)$$

where

$$\omega(K) = \max_{X_o} \left\{ J(K, X_o) ; \; \| X_o \|^2 \leq 1 \right\} \qquad (3.24)$$

where $\| X_o \| = T_r \left[X_o' \, X_o \right]$. The constraint $\| X_o \|^2 \leq 1$ presents a certain degree of discretion, nevertheless its choice can be justified by the fact that the criterion J being linear in relation to X_o and the constraint strictly convex, the problem (3.24) allows for a unique solution. Let P_t ; t = 0,1,... T-1 be the matrices of the dual variables associated with the dynamic restriction and γ_o the dual variable associated with the constraint $\| X_o \|^2 \leq 1$, then the necessary (and here sufficient) conditions for the problem (3.24) are written as :

$$\left| \begin{array}{l} Q + K'RK + (A-BK)' \, P_o \, (A-BK) - 2\, \gamma_o X_o = 0 \\ \| X_o \|^2 \leq 1 \\ \gamma_o \geq 0 \\ \gamma_c \; (\| X_o \|^2 - 1) = 0 \end{array} \right. \qquad (3.25)$$

Writing

$$Q + K'RK + (A-BK)' P_o (A-BK) - P_{-1} = 0$$

the solution is written as

$$\left| \begin{array}{l} X_o = P_{-1} \left(\| P_{-1} \| \right)^{-1} \\ \gamma_o = \frac{1}{2} \| P_{-1} \| \end{array} \right. \tag{3.26}$$

Finally, by direct application of the corollary 3.1. and taking into account (3.26), we get

$$\left[\begin{array}{l} \frac{dw(K)}{dK} = 2 \Sigma \left\{ RK - B' P_t (A - BK) \right\} X_t \\ Q + K'RK + (A-BK)' P_t (A-BK) - P_{t-1} = 0 \\ \qquad t = 0, \dots T-1 \; ; \quad P_{T-1} = S \\ (A-BK) X_t (A-BK)' - X_{t+1} = 0 \quad X_o = P_{-1} (\| P_{-1} \|)^{-1} \end{array} \right. \tag{3.27}$$

Calculation of the matrix gradient is carried out here again by means of the resolution of two uncoupled matrix equations; the one relating to P_t must be resolved first.
By passing to the limit we deduce the case of the infinite horizon which gives :

$$\left[\begin{array}{l} \frac{dw}{dK} = 2 \left\{ RK - B'P (A - BK) \right\} V \\ (A-BK)' P (A-BK) - P + Q + K'RK = 0 \\ (A-BK) V (A-BK)' - V + P \| P \|^{-1} = 0 \end{array} \right. \tag{3.28}$$

Extension to the continuous case poses no special problems and we give below the corresponding results :
in the case of finite horizon :

$$\left[\begin{array}{l} \frac{dw}{dK} = 2 \int_o^T (RK - B'P(t)) X(t) dt \\ Q + K'RK + (A-BK)' P + P (A-BK) + \overset{\circ}{P} = 0 \qquad P(T)=S \\ (A-BK) X + X (A-BK)' - \overset{\circ}{X} = 0 \qquad X_o=P(o) \left[\| P(o) \| \right]^{-1} \end{array} \right.$$

in the case of infinite horizon

$$\left[\begin{array}{l} \frac{dw}{dK} = 2 (RK - B'P) V \\ (A-BK)' P + P (A-BK) + Q + K'RK = 0 \\ (A-BK) V + V (A-BK)' + P \left[\| P \| \right]^{-1} = 0 \end{array} \right.$$

3.4. Conclusion

The approach presented here (parametric optimisation - primal gradient method) is developed from a representation in the form of differential equations of state. The search for a control is done indirectly by definition of an auxiliary optimisation problem (quadratic criterion). This poses the standard problem of good choice of the weighting matrices which play a part in the criterion. Within the framework of multivariable linear systems the frequency approach [MAC - 72] does not involve this sort of problem and has the certain advantage for the designer of relying for the synthesis of feedback loops on classical notions such as the margin of phase, of gain. Nevertheless this approach has the disadvantage of a relative clumsiness in the case where a lot of loops have to be determined, ie. when the number of sub-systems is large. The approach presented here must also be seen within the framework of the methods termed C.A.D. (computer aided design) with critical intervention of the designer to adapt the problem of optimisation, after analysis of the results obtained. This analysis can certainly be conducted through frequential analysis which, as is mentioned above, makes it possible to use notions such as the gain margin, phase margin, which are well known to designers. In this sense the approaches "frequential" and "state space" can appear to be complementary.

One disadvantage of the approach presented which could become serious in the case of large scale systems, arises from the fact that it involves calculations affecting the global model. In fact, the various Lyapunov equations to be resolved are of the size of the global system. Here we enter into the very broad area of numerical analysis for the search for high performance algorithms. Some problems of this type have been discussed in chapter 2 of this book and among recent contributions in this area for decentralised regulation we can quote [PIE - 80]. In a number of physical systems one way of offsetting this disadvantage may be brought about by model reduction and temporal decomposition methods such as aggregation methods [AOK - 68] singular perturbation methods [KOK - 76] , etc.

REFERENCES

AND - 71 ANDERSON B.D.O. and J.B. MORRE, "Linear optimal control", Prentice Hall, Electrical Engineering series, 1971

AOK - 73 AOKI M. and M.T. LI, "Controllability and stabilizability of decentralized dynamic systems", Proc. 1973 jacc.

AOK - 68 AOKI M., "Control of large scale dynamic systems by aggregation", IEEE Trans. A.C., vol. AC-13, June 1968.

ARM - 81 ARMENTANO V.A. and M.G. SINGH ,"A new approach to the decentralized controller initialisation problem", 8th Triennal IFAC World Congress, Kyoto, 24-28 August 1981

ATH - 71 ATHANS M., "The role and use of the stochastic linear quadratic-gaussian problem in control system design", IEEE Trans. A.C, vol. A.C-16, n° 6, 71.

BER - 81 BERNUSSOU J. and J.C. GEROMEL, "An easy way to find gradient matrix of composite matricial functions", IEEE Trans. AC, vol. AC-26, n° 2, 1981.

COH - 75 COHEN G., "Commande par contre réaction sous contrainte de structure du gain d'un système stochastique linéaire. Résolution par une méthode de gradient", RAIRO, J.3, 1975.

COH - 81 COHEN G., "Partage de l'information entre systèmes stochastiques a priori indépendants", RAIRO, vol. 15, n° 1, 1981.

COR - 76 CORFMAT J.P. and A.S. MORSE, "Decentralized control of linear multivariable systems", Automatica, vol. 12, 1976.

DAV - 76 DAVISON E.J., "Decentralized stabilization and regulation in large multivariable systems, 'Directions in Decentralized control, many-person optimization and large scale systems'", HO Y.C. and MITTER S., Editors Plenum Press, 1976.

ELG - 73 ELGERD O.I., "Electric energy systems theory : an introduction", Mc Graw Hill Compagny, THM Edition, 1973.

FES - 79 FESSAS P., "A note on 'an example in decentralized control systems'", IEEE Trans. A.C., vol. A.C.-24, Aug. 1979.

GER - 79a GEROMEL J.C., "Contribution à l'étude des systèmes dynamiques interconnectés. Aspects de décentralisation", Thèse d'Etat, n° 903, Université Paul Sabatier, Toulouse, 1979.

GER - 79b GEROMEL J.C. and J. BERNUSSOU, "An algorithm for optimal decentralised regulation of linear quadratic interconnected systems", Automatica, vol. 15, 489-491, 1979.

GUA - 79 GUARDABASSI G. et al., "Parameter optimization in decentralized process control : a unified setting for multivariable feedback systems", IFAC Symp. 29-31 August 79, Zurich.

HO - 78 HO Y.C., M.P. KASTNER, E. WONG, "Teams, Signalling and Information Theory, IEEE Trans. A.C., vol. A.C.-23, n° 2, 1978.

IKE - 79 IKEDA M. and D.D. SILJAK, "Counterexamples to Fessas conjecture", IEEE Trans. A.C, vol. A.C 24, Aug. 79.

KOK - 76 KOKOTOVIC P.V., R.E. O'MALLEY, P. SANNUTI, "Singular perturbations and order reduction in control theory : an overview", Automatica, vol. 12, March 1976.

KWA - 72 KWAKERNAK H. and R. SIVAN, "Linear optimal control systems", J. WILEY, New-York, 72.

LEV - 70 LEVINE W.S. and M. ATHANS, "On the determination of the optimal output feedback gains for linear multivariable systems", Automatic Control, vol. A.C.-15, n° 1, 1970.

MAC - 72 MAC FARLANE A.G.T., "A survey of some recent results in linear multivariable theory", Automatica vol. 8, 1972.

MEN - 74 MENDEL J.M., "A concise derivation of optimal constant limited state feedback gains", IEEE Trans. A.C., vol. A.C.-19, 1974.

PIE - 80 PIETRAS J.V. and LOOZE D.P., "Decomposition of decentralized gain computation for interconnected systems", JACC Conference, San Francisco, August 1980.

QUA - 79 QUADRAT J.P., "Sur les feedback locaux optimaux en commande stochastique", Rapport interne INRIA, Le Chesnay, 1979.

ROS - 60 ROSEN J.B., "The gradient projection method for non linear programming : Part I - Linear Constraints" SIAM 8(1), 1960.

SEZ - 81 SEZER M.E and O. HUSEYIN, "Comments on decentralized state feedback stabilization", IEEE Trans. A.C., vol. A.C.-26, April 1981.

SIL - 78 SILJAK D.D., "Large scale dynamic systems : stability and structure", North Holland, New-York, 1978.

WAN - 78 WANG S.H., "An example in decentralized control systems", IEEE Trans. A.C., vol. A.C.- 23, October 1978.

WEN - 80 WENK C.J. and C.H. KNAPP, "Parameter optimization in linear systems with arbitrarily constrained controller structure", IEEE Trans. A.C., vol. A.C.-25, June 1980.

WIT - 68 WITSENHAUSEN H.S., "A counter-example in Stochastic Optimal Control", SIAM Journal of Control, vol. 6, n° 1, 1968.

WIT - 71 WITSENHAUSEN H.S., "Separation of estimation and control for discrete time systems", Proceedings of the IEEE vol. 59, n° 11, 1971.

CHAPTER 4

INTERCONNECTED SYSTEMS : STABILITY

Analysis of the stability properties of large scale systems has been the subject of a great many works over the past ten years. This chapter does not intend to provide an exhaustive summary of these, but rather to try to present the essential ideas arising from the consideration of a particular structure (sub-systems - network of interconnections) for the development of specific tools. We shall apply ourselves principally to the "Lyapunov functions" type approach.
The chapter is composed of five basic sections. In the first section a brief summary of the fundamental definitions and theorems concerning the second Lyapunov method is given. In the second section we present in a parallel manner the two possible ways that the Lyapunov method can be applied, taking into account the structure of the systems under consideration ie. interconnected systems. Without entering into details here, of these two ways, termed in the literature scalar and vector, a certain advantage appears to emerge in favour of the first one. This simple intuitive idea is tempered in a third section by means of a rapid examination of two particular problems, namely : the examination of stability subject to structural perturbations and the estimation of the areas of stability. In the fourth section is discussed a potential application of the vector Lyapunov function approach to the synthesis of robust, multilevel control structures, ie. capable of overcoming certain types of structural perturbations. Finally, the last section of the chapter tackles the basic problem of the choice of Lyapunov functions and some elements of response are given in the case of interconnected linear systems, finishing with a critical discussion of the results obtained highlighting certain limitations inherent in the approach under consideration.

4.1. The Lyapunov method (recapitulation)

The object of this section is to give a brief recapitulation of the principal definitions and results concerning the second Lyapunov method, also called the direct Lyapunov method. We shall exclude from our presentation those results which concern the determination itself of the Lyapunov functions, but shall mention certain works of the Russian school with Lure, Kraskovski, Zubov. A well documented presentation of works in this area can be found in the work of W. Hahn, 1967 (Stability of motion, Springer Verlag. 1967).

One of the fundamental advantages of the Lyapunov method is that it does not require knowledge of the solutions of a mathematical model to be able to study their stability. This explains the qualifier "direct" which has been given to it. The principle of the Lyapunov method is relativrly simple and can be introduced - as did Kalman (1960) [KAL - 60] , - by considerations of energy.
Let us consider an isolated physical system for which it has been possible to express the energy E in terms of the state X, and let us suppose that this system has only one point of equilibrium. Then if $\frac{dE}{dt} < 0$, $\forall X$, the point of equilibrium is stable. The energy will therefore decrease along the trajectories to reach its minimum value at the point of equilibrium. $\frac{dE}{dt} < 0$ explains the fact that the system is dissipating.
This energy function is very difficult to express in cases where one is faced with an "abstract" mathematical model. The Lyapunov functions are, therefore, to some extent a generalisation of energy. Some authors have also talked, in this connection, of "generalised distance".
If the principle of the method is relatively simple, its application occasionally comes up against the delicate problem of determination of the functions termed Lyapunov V(X). As we shall deal later with the examination of the stability of an isolated fixed point, we shall limit ourselves to that case for the definitions and results which follow.
We shall concern ourselves first with the case of invariant systems :

$$\overset{\circ}{X} = F(X) \; , \qquad X(0) = X_o \qquad t \geq 0 \tag{4.1}$$

in which $X \in R^n$, $F(.) : R^m \to R^n$ is a continuous function satisfying the Lipschitz conditions (uniqueness of the solu
We shall note $\Phi(t, X_o)$ as the unique solution to (4.1) with the origin X_o at $t = 0$, and $\|X\|$ as the Euclydian norm of X. Without loss of generality we shall suppose that the origin is the point of equilibrium.

4.1.1. Some definitions

Def. 4.1 : the equilibrium point 0 of (4.1) is stable if $\forall \epsilon > 0$ $\exists \delta(\epsilon)$ such that

$$\|X_o\| < \delta(\varepsilon) \quad \Rightarrow \|\Phi(t, X_o)\| < \varepsilon \;\; \forall t \geq 0$$

Def. 4.2 : O is asymptotically stable if

a) O is stable
b) $\exists \eta > 0$ and $\|X_o\| < \eta \quad \Rightarrow \quad \lim_{t \to \infty} \|\Phi(t, X_o)\| = 0$ (attractivity)

Def. 4.3 : O is exponentially stable if $\exists \sigma > 0$ such that $\forall \varepsilon > 0$ $\exists \delta(\varepsilon)$ such that

$$\|X_o\| < \delta(\varepsilon) \Rightarrow \|\Phi(t, X_o)\| \leq \varepsilon e^{-\sigma t} \quad \forall t \geq 0$$

Geometric interpretation

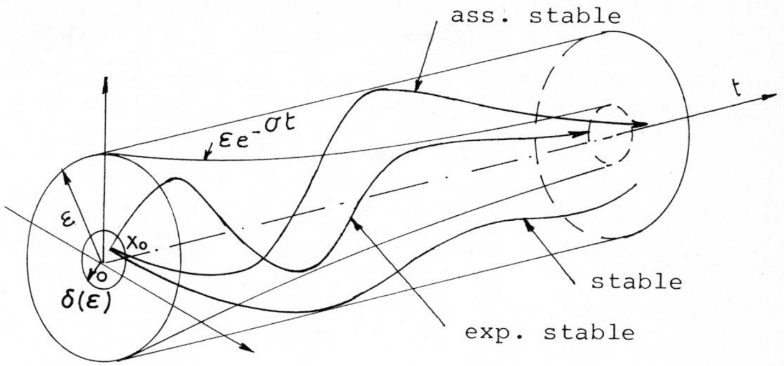

Figure 4.1

example : $F(X) = AX$, A with distinct eigenvalues

$\Phi(t,X_o) = e^{At}\, X_o = e^{T^{-1}\Lambda\, Tt}\, X_o$ (T similarity matrix, Λ diagonal matrix of the eigenvalues of A)

$$\|\Phi(t,X_o)\| \leqslant \|e^{T^{-1}\Lambda Tt}\|\ \|X_o\| \leqslant \|T^{-1}\|\|T\|\ \|X_o\|\|e^{\Lambda t}\|$$

$$= \|T^{-1}\|\ \ \|T\|\ \ \|X_o\|\quad \exp\left[(\max \lambda_i)\ t\right]$$

so then with $\delta(\varepsilon) = \varepsilon\,(\ \|T^{-1}\|\ \ \|T\|\ )^{-1}$ if

$(\max \lambda_i) = 0$, 0 is stable

$(\max \lambda_i) < 0$, 0 is exponentially stable (asymptotically, also, of course).

Further definitions :

Def. 4.4. : The continuous function $\varphi : [0,\infty) \rightarrow R^+$ is said to be of class K if $\varphi(o) = 0$ and φ strictly increasing.
Let $v(X)$ be a continuous, scalar function $v(.) : R^n \rightarrow R$.

Def. 4.5. : The function $v(X)$ is said to be positive definite (positive semi-definite) if $v(X) > 0$ $(v(X) \geqslant 0)$ $\forall X \in R^n$ and $v(o) = 0$ if and only if $X = 0$.

Remark : $v(X)$ is positive definite if $\exists\, \varphi \in K$ and $v(X) \geqslant \varphi(X)$

4.1.2. Lyapunov theorems

Application of the Lyapunov's second method consists in studying the sign of increase in relation to time of the function $v(t) = v\,[X(t)]$ along the trajectories of (4.1), namely

$$\overset{\circ}{v}_+(t) = \lim_{\Delta t \rightarrow 0^+} \left\{ \frac{v(T+\Delta t) - v(t)}{\Delta t} \right\}$$

namely

$$\overset{\circ}{v}_+(t) = \lim_{\Delta t \to 0^+} \left\{ \frac{v[X(t) + \Delta t\ F(X)] - v[X(t)]}{\Delta t} \right\} \qquad (4.2')$$

If, in addition, function v(X) is differentiable in relation to all its arguments, then

$$\overset{\circ}{v}_+(t) = \frac{dv}{dt} = \overset{\circ}{v} = \nabla v(X)'\ f(X)$$

In the following we shall adopt the unique notation $\overset{\circ}{v}$

Theorem 4.1 : If there exists a positive definite function v(X) such that

$$\begin{aligned} &a) \quad \varphi_1(\|X\|) \leq v(X) \leq \varphi_2(\|X\|) \\ &b) \quad \overset{\circ}{v}(x) \leq 0 \end{aligned} \qquad (4.3)$$

φ_1 and φ_2 being class K functions, then the point of equilibrium x = 0 of (4.1) is stable

Proof : from b) we have $v(\Phi(t,X_o)) - v(X_o) = \int_o^{t_o} \overset{\circ}{v}(\Phi(\tau,X_o))\,d\tau \leq 0$

ie. $\quad v[\Phi(t,X_o)] \leq v(X_o)$

namely $\quad \varphi_1[\|\Phi(t,X_o)\|] \leq \varphi_2[\|X_o\|]$

For every $\varepsilon > 0$ let us choose $\delta(\varepsilon)$ such that $\varphi_1(\varepsilon) \geq \varphi_2[\delta(\varepsilon)]$

Then $\forall X_o \quad \|X_o\| < \delta(\varepsilon), \quad \varphi_1[\|\Phi(t,X_o)\|] \leq \varphi_2[\delta(\varepsilon)] \leq \varphi_1(\varepsilon)$

namely $\|\Phi(t, X_o)\| \leq \varepsilon$. q.e.d.

Theorem 4.2 : If there exists a positive definite function v(X) such that

$$\begin{aligned} &a) \quad \varphi_1(\|X\|) \leq v(X) \leq \varphi_2(\|X\|) \\ &b) \quad \overset{\circ}{v}(X) \leq -\varphi_3(\|X\|) \end{aligned} \qquad (4.4)$$

in which $\varphi_1, \varphi_2, \varphi_3 \in K$, the point of equilibrium X = 0 is asymptotically stable.

Proof : $\overset{\circ}{v} \leq 0$ certainly and therefore 0 is stable. On the other hand

$$v[\Phi(t,X_o)] - v(X_o) = \int_o^t \overset{\circ}{v}[\Phi(\tau,X_o)]\,d\tau \leq -\int_o^t \varphi_3(\|\Phi(\tau,X_o)\|)\,d\tau$$

ie.

$$\varphi_1(\|\Phi(t,X_o)\|) \leq \varphi_2(\|X_o\|) - \int_o^t \varphi_3(\|\Phi(\tau,X_o)\|)\,d\tau$$

Now $\varphi_1(.) \geq 0$, therefore

$$\varphi_2(\|X_o\|) \geq \int_o^t \varphi_3(\|\Phi(\tau,X_o)\|)\,d\tau$$

Let $\|X_o\| \leq \eta$ positive, therefore

$$\lim_{t\to\infty} \int_o^t \varphi_3(\|\Phi(\tau, X_o)\|)\,d\tau \leq \varphi_2(\eta) < \infty$$

wich implies

$$\lim_{t \to \infty} \| \Phi(t, X_o) \| = 0$$

Theorem 4.3 : If there exists a positive definite function such that :

$$\text{a)} \quad \mu_1 \varphi(\|X\|) \leq v(X) \leq \mu_2 \varphi(\|X\|) \tag{4.5}$$

$$\text{b)} \quad \overset{\circ}{v}(X) \leq -\mu_3 \varphi(\|X\|)$$

$\mu_1, \mu_2, \mu_3 > 0, \varphi \in K$, the equilibrium point $X = 0$ of (4.1) is exponentially stable (conditions (4.5) are the same as (4.4) with the condition $\varphi(\|X\|) = \varphi_1/\mu_1 = \varphi_2/\mu_2 = \varphi_3/\mu_3$.
We then say that $\varphi_1, \varphi_2, \varphi_3$ are of the same order of magnitude).

Proof :

$$v[\Phi(t, X_o)] - v[X_o] = \int_0^t \overset{\circ}{v}[\|\Phi(\tau, X_o)\|] d\tau \leq -\mu_3 \int_0^t \varphi[\|\Phi(\tau, X_o)\|] d\tau$$

$$\leq -\frac{\mu_3}{\mu_2} \int_0^t v[\Phi(\tau, X_o)] \, d\tau$$

a possible solution to this integral inequality is :

$$v(\Phi(t, X_o)) \leq e^{-\sigma t} v(X_o) \quad \text{where } \sigma = \mu_3/\mu_2$$

ie. $$\varphi(\|\Phi(t, X_o)\|) \leq \frac{\mu_2}{\mu_1} \varphi \|X_o\| e^{-\sigma t} \quad \text{that is}$$

$$\|\Phi(t, X_o)\| \leq \frac{\mu_2}{\mu_1} \|X_o\| \; e^{-\sigma t}$$ q.e.d.

4.1.3. Time varying systems

Here we shall briefly give a few results and definitions concerning time varying systems, that is systems described by non-autonomous differential equations of the type

$$\overset{\circ}{X} = F(X, t) \; ; \tag{4.6}$$

The basic difference in comparison with invariant systems is the introduction of the notion of uniformity, uniformity in relation to the initial instant. Moreover, we know that in the case of invariant systems, the only thing that counts is the duration $t - t_o$.

$$\Phi(t, X_o, t_o) = \Phi(t - t_o, X_o) = \Phi(\tau, X_o)$$

this is no longer the case for time varying systems $\Phi(t, X_o, t_o)$.
For example, the point of equilibrium 0 may be attractive for a couple (X_o, t_o') and unstable for a couple (X_o, t_o'') of initial conditions.

Def. 4.6 : The point of equilibrium 0 of (4.6) is stable if $\forall \, \varepsilon > 0$ and $t_o \in R^+$

$\exists\, \delta(\varepsilon, t_o)$ such that $\|\Phi(t, X_o, t_o)\| < \varepsilon$ for every $t \geq t_o$

Def. 4.6' : The point of equilibrium 0 of (4.6) is uniformly stable if $\forall\, \varepsilon > 0 \;\; \exists\, \delta(\varepsilon)$ and $\|\Phi(t, X_o, t_o)\| < \varepsilon \;\; \forall\, t_o \in R^+$, $\forall t \geq t_o$.

Def. 4.7 : The point of equilibrium 0 of (4.6) is asymptotically stable if it is stable and if there exists a number $\eta(t_o) > 0$ such that $\|X_o\| < \eta(t_o) \Rightarrow \lim_{t \to \infty} \|\Phi(t, X_o, t_o)\| = 0$

Def. 4.7' : There is uniform, asymptotic stability when there is uniform stability and uniform attractivity, not dependent on t_o.

It is clear that in practice we shall concern ourselves more particularly with the different types of uniform stability.

Theorem 4.4 : If there exists a positive definite function $v(X,t)$ such that

a) $\varphi(\|X\|) \leq v(X,t); \;\; \forall t \geq 0$

b) $\mathring{v} \leq 0$

$\varphi \in K$, the point of equilibrium $X = 0$ of (4.6) is stable.

Proof : from b), by integration, we have

$$v[\Phi(t, X_o, t_o), t] \leq v[X_o, t_o] \qquad \forall t \geq t_o$$

that is

$$\varphi(\|\Phi(t, X_o, t_o)\|) \leq v[X_o, t_o]$$

where

$$\|\Phi(t, X_o, t_o)\| \leq \varphi^{-1}[v(X_o, t_o)]$$

It is possible to find a $\delta(\varepsilon, t_o)$ s.t $\|X_o\| < \delta(\varepsilon, t_o) \Rightarrow \|\Phi\| < \varepsilon$.

Let $\rho(t_o) = \sup_{\|X_o\| = \delta} v(X_o, t_o)$, choose $\varphi^{-1}[\rho(t_o)] \leq \varepsilon \ldots$

Theorem 4.5 : If there exists a positive definite function $v(X,t)$ such that

a) $\varphi_1(\|X\|) \leq v(X,t) \leq \varphi_2(\|X\|)$

b) $\mathring{v} \leq 0$

$\varphi_1, \varphi_2 \in K$, the point of equilibrium $X = 0$ of (4.6) is uniformly asymptotically stable.

The proof is identical to that of theorem 4.1.. It should be noted that it is the existence of an upper limit φ_2 of class K for $v(X,t)$ which gives the property of uniformity. This condition makes the solution of $\rho = \inf_{t_o \geq 0} \rho(t_o)$ different from zero.

In general, theorems 4.1 and 4.2 provide the conditions which guarantee the uniformity of the stability results. It is also to be noted that the concept of exponential stability is necessarily linked to uniformity since its proof requires the existence of an upper limit for $v(X,t)$.

4.1.4. Corduneanu theorem (1960)

In this section we are going to set out a theorem which generalises the conditions stated in the Lyapunov theorems particularly at the level of condition b).

Theorem 4.6 : If there exists a positive definite function $v(X,t)$ such that :

a) $\varphi(\|X\|) \leq v(X,t)$

b) $\overset{\circ}{v} \leq h(v,t)$

and for $\dot{w} = h(w,t)$, the point of equilibrium 0 is stable (asymptotically stable), then the point of equilibrium 0 of (4.6) is stable (asymptotically stable).
The proof can be written immediately by observing that if $v(X_o,t_o) \leq w(t_o)$ then $v(X,t) \leq w(t)$ $\forall t \geq t_o$.
The property of uniformity is obtained by adding the hypothesis of an upper limit ($v \leq \varphi_2$), for the function v and the hypothesis of uniformity of the results for the system $\dot{w} = h(w,t)$.
We can in fact say that this result generalises to some extent the preceding results as it is not necessary to assume the semi-negativity or the negativity of $\dot{v}$.
The differential equation $\dot{w} = h(w,t)$ is called the equation of comparison for system (4.6) and we generalise this result in the case of vector Lyapunov equations by introducing a system of comparison.

Note : it must be noted that the preceding theorems give results of a local nature. We shall return in the following sections to the problem of estimation of areas of stability. If we wish to guarantee overall results (in the whole space) we must add a further condition : Lyapunov functions not limited at infinity ie.

$$\lim_{\|X\| \to \infty} \varphi(\|X\|) = \infty \qquad \text{th. 4.1 and 4.4}$$

$$\lim_{\|X\| \to \infty} \varphi_1(\|X\|) = \infty \qquad \text{th. 4.2 and 4.3}$$

4.2. Stability of interconnected systems - Vector Lyapunov functions

The term "vector Lyapunov function" can be attributed to Bellman (1962) but the work which has been at the origin of extensive literature on this subject is attributable to Bailey (1966). The aim of this section is to present as broadly as possible the basic results obtained in stability analysis of interconnected systems. In order to avoid a bibliographical list of exaggerated length we shall limit ourselves to mentioning two books which have been published in this field, each of which provides a complete bibliography : Michel (1977), Siljak (1978). Finally, we should like to mention the works of Matrosov in the USSR (1971).

To avoid writing too much, we shall deal with non-linear, invariant systems of the interconnected type, ie. described by a system of equations of the form

$$S = \bigcup_{i=1}^{N} S_i; \quad (S_i) : \dot{x}_i = g_i(x_i) + h_i(X), \qquad i=1,2,\ldots N \tag{4.7}$$

where N represents the number of sub-systems

x_i, X are respectively the state vector of the ith sub-system and of the global system,

$x_i \in R^{n_i}$; $X \in R^n$: $n = \sum_{i=1}^{N} n_i$

$g_i(x_i) : R^{n_i} \rightarrow R^{n_i}$, vector of non-linear functions characterising the dynamics of the isolated sub-systems.

$$(\hat{S}_i) \quad \overset{\circ}{x}_i = g_i(x_i) \tag{4.8}$$

$h(X) : R^n \rightarrow R^{n_i}$, characterises the couplings between sub-systems.

The results which will follow concern essentially the examination of the stability properties of an equilibrium point which, without loss of generality will be taken at the origin. We shall assume, moreover, that the origin is also a point of equilibrium for each isolated sub-system.

Without giving details, we shall assume in the following pages that the diverse conditions and inequalities which we shall write about are globally true and that consequently the results obtained will be conditions for overall stability. (We shall talk of stability of the system instead of stability of the equilibrium point).

In the next section we shall return to the problem of the estimation of areas of stability which arises in the case of non-linear systems when the hypotheses made on the Lyapunov functions are only valid locally.

In the examination of the stability properties of interconnected systems using vector Lyapunov functions, we can distinguish two approaches. The first consists of looking for a scalar Lyapunov function for the overall system starting from the Lyapunov functions determined for each isolated sub-system. The second approach consists of the determination of a real differential system called the comparison system (an extension for the vectorial case of the theorem of Corduneanu stated in the scalar one).

4.2.1. Approach : Scalar Lyapunov function

To each isolated sub-system (4.8) is associated a positive definite function $v_i(x_i)$ satisfying the following inequalities :

$$\begin{cases} \varphi_{i1}(\|x_i\|) \leq v_i(x_i) \leq \varphi_{i2}(\|x_i\|) & i=1,2,\ldots N \\ \dot{v}_{i\hat{S}_i} = (\nabla v_i)' \, g_i(x_i) \leq \mu_i \, \varphi_{i3}(\|x_i\|) \end{cases} \tag{4.9}$$

a) let us suppose that the terms of interconnection satisfy

$$(\nabla v_i)' \, h_i(X) \leq \sum_{j=1}^{N} \xi_{ij} \, \varphi_{j3} \, (\| x_j \|), \quad i=1,2,\ldots N \tag{4.10}$$

then

$$\dot{v}_{iS_i} = \dot{v}_{i\hat{S}_i} + (\nabla v_i)' \, h_i(X) \leq \mu_i \varphi_{i3} \, (\| x_i \|) +$$

$$+ \sum_{j=1}^{N} \xi_{ij} \, \varphi_{j3} \, (\| x_j \|) \quad i=1,\ldots N$$

This can be written in matrix form :

$$\overset{\circ}{V} \leq A \, W \tag{4.11}$$

in which $V = [v_1, v_2 \ldots v_N]'$ is termed vector Lyapunov function

$$W = [\varphi_{13}, \varphi_{23}, \ldots \varphi_{N3}]'$$

and $\quad A = \{a_{ij}\} \quad a_{ij} = \mu_i \, \delta_{ij} + \xi_{ij} \quad$ (δ_{ij} Kronecker symbol)

<u>Theorem 4.7</u> : If there exists a vector $\alpha \in R^N$ with strictly positive components such that $\alpha' A \leq 0$ (< 0) (the inequality being valid component by component) then the system (4.7) is (asymptotically) stable.

Proof : we are considering the scalar Lyapunov function

$$\nu = \alpha' V = \sum \alpha_i \, v_i$$

Firstly, ν is positive definite and satisfies the limit inequalities

$$(\min \alpha_i) \sum_i \varphi_{i1} \leq (\min \alpha_i) \sum v_i \leq \nu \leq (\max \alpha_i) \sum v_i \leq (\max \alpha_i) \sum_i \varphi_{i2}$$

The time derivative of $\dot{\nu}$ is written as $\dot{\nu} = \alpha' \overset{\circ}{V} \leq (\alpha' A) \, W$ and since $W > 0$ $\dot{\nu} < 0$ if and only if all the components of the row vector are negative.

b) let us now suppose that the terms of interconnection satisfy

$$(\nabla v_i)' \, h_i(X) \leq [\varphi_{i3} \, (\| x_i \|)]^{\frac{1}{2}} \sum_{j=1}^{N} \zeta_{ij} \, [\varphi_{j3} \, (\| x_j \|)]^{\frac{1}{2}} \tag{4.12}$$

<u>Theorem 4.8</u> : If there is a vector $\alpha \in R^N$ with strictly positive definite components such that the matrix $S = \{s_{ij}\}$ defined by

$$s_{ij} = \begin{cases} \alpha_i [\mu_i + \zeta_{ii}] & i=j \\ [\alpha_i \, \zeta_{ij} + \alpha_j \, \zeta_{ji}] / 2 & i \neq j \end{cases} \tag{4.13}$$

is negative semi-definite (negative definite) then system (4.7) is stable (asymptotically stable)

(NB : a symmetrical matrix S is negative definite if

$$(-1)^k \, \Delta_k > 0, \quad \forall k = 1,2, \ldots N$$

Δ_k being the principal minor of order k of the matrix S. It is negative semi-definite if $(-1)^k \Delta_k \geq 0$, $\forall k$.

Proof : this is done as for theorem 4.7 using the scalar Lyapunov function $v = \alpha' V$. It is easy to see that the time derivative of v along the trajectories satisfies the inequality

$$\dot{v} \leq w' S w \quad \text{where} \quad w = \left[\varphi_{13}^{\frac{1}{2}}, \varphi_{23}^{\frac{1}{2}}, \ldots, \varphi_{N3}^{\frac{1}{2}}\right]'$$

therefore

$$\dot{v} \leq \lambda_{max}(S) \sum_{i=1}^{N} \varphi_{i3}(\|x_i\|)$$

Theorem 4.9 : If the functions $\varphi_{i1}, \varphi_{i2}, \varphi_{i3}$ are of the same order magnitude and if there exists a vector $\alpha > 0$ such that

say $\alpha' A < 0$

say S negative definite

then system (4.7) is exponentially stable.

Proof : this is carried out by direct application of theorem 4.3. In fact, if $\varphi_{i1}, \varphi_{i2}, \varphi_{i3}$ are of the same order of magnitude $\forall i = 1,2,\ldots N$ we can show [MIC - 77 p. 34] that

$$\Sigma \varphi_{i1}, \Sigma \varphi_{i2}, \Sigma \varphi_{i3}$$

are equally of the same order of magnitude.

Notes :

1. It is very difficult and even impossible, to say in general terms which of the preceding cases a) or b) is the better. This is not surprising in view of the fact that the Lyapunov method provides sufficient conditions of stability and that these conditions are broadly dependent on the choice of the Lyapunov functions themselves. We can see, however, that conditions (4.10) and (4.12) cannot in general be compared (except when the functions $\varphi_{i1}, \varphi_{i2}, \varphi_{i3}$ are of the same order of magnitude); this means that in fact conditions (4.10) and (4.12) correspond to a different choice of Lyapunov functions. Nevertheless, qualitatively we can say that (4.12) appears a priori to be simpler to establish than (4.10) bearing in mind that in the product $(\nabla v_i)' h_i(X)$ crossed terms $x_i x_j$ will probably appear. A condition such as (4.10) can more easily be derived in the case where one is concerned with a finite area of the state space $\|x_i\| \leq C_i$ (> 0), a problem which we shall come across later when we are dealing with the estimation of areas of stability.

2. Determination of the vectors $\alpha > 0$ of the theorems 4.7 and 4.8 is not a trivial matter, and in cases where there are a large number of sub-systems one may have to use linear programming techniques ($\alpha' A < 0$) or even non-linear techniques in cases where one wishes to test the determinants Δ_k (S negative definite).

This problem is to a very large extent simplified when one introduces the supplementary assumption that

$$\xi_{ij} \geq 0 \quad i \neq j \qquad \text{or} \qquad \zeta_{ij} \geq 0 \quad i \neq j$$

This hypothesis is, without doubt, more restrictive (formally). In order to have stability it is then necessary that

$$\mu_i + \xi_{ii} < 0 \qquad \text{or} \quad \mu_i + \zeta_{ii} < 0 \qquad \forall i$$

Nevertheless, with respect to expressions (4.10) and (4.12) one can easily show that, except for special cases, these inequalities can only be established with $\xi_{ij} \geq 0$ and $\zeta_{ij} \geq 0$.

4.2.1.1. Application of the M matrices

Let $A = \{a_{ij}\}$ be a matrix such that $a_{ij} \geq 0$, $i \neq j$ (non-negative terms outside the diagonals).

Theorem 4.10 : For all the matrices belonging to this class, the following conditions are equivalent

(i) the eigenvalues of A have a negative real part

(ii) there exists an eigenvector $U > 0$ (positive components) such that $AU < 0$
or $\bar{U} > 0$ such that $A'\bar{U} < 0$

(iii) $(-1)^k D_k > 0$, D_k being the principal minors of A (this condition is generally called Sevastyanov Kotelyanski condition)

(iv) there exists a diagonal matrix with positive elements such that $DA + A'D$ is negative definite.

Other equivalent conditions exist [FIE - 62]but we have restricted ourselves here to those which will be of direct use to us later. From now on we shall designate by M matrices the matrices with non-negative diagonal terms which meet one of the conditions given above.
Therefore with the conditions $\xi_{ij} \geq 0 \quad i \neq j$ and $\zeta_{ij} \geq 0 \quad i \neq j$ the theorems 4.7 and 4.8 can be reformulated in the following way :

Theorem 4.11 : If the matrix A (case a) or $\bar{A} = \{\mu_i \delta_{ij} + \xi_{ij}\}$ (case b) is an M matrix then system (4.7) is asymptotically stable.

Proof : for case a choose $\alpha = \bar{U}$ (condition ii) and Lyapunov function $\nu = V' \bar{U}$, then $\dot{\nu} = V' A' \bar{U} < 0$ according to (ii).
For case b, choose $D = \text{diag } \alpha_i$ then $S = D\bar{A} + \bar{A}' D$ and according to (iv) S is negative definite.

To sum up, for cases in which the matrices A or $\bar{A}$ belong to the category of matrices with non-negative off diagonal terms, stability may be analysed by testing one of the equivalent conditions of the theorem (preferably iii).
Of course it is obvious that simple stability conditions are obtained simply by replacing the strict inequalities in the theorem by inequalities in the broader sense.

4.2.2. Comparison system

Before developping this approach in the context of systems we shall recall some results attributed to Kamke and Wazeski, dealing with differential inequalities.

Def. 4.8. : the non-linear vector function $G(X) : R^m \longrightarrow R^n$ is said to be quasi monotone if $G(0) = 0$ and for every $i=1,2,\ldots n$, $g_i(X)$ satisfy

$$g_i(X) \leq g_i(Y) \qquad \forall X,Y \text{ with } \quad x_i = y_i,\ x_j \leq y_j \qquad j \neq i$$

<u>Theorem 4.12</u> : Let us suppose the differential inequality $\overset{\circ}{X} \leq G(X)$, G quasi monotone (Lipschitz), then

a) $\forall\, X_o \leq Y_o \Rightarrow X(t) \leq Y(t)$ where $Y(t)$ is the solution of the (real) differential system.

$$\overset{\circ}{Y}(t) = G(Y)$$

b) $Y_o \geq 0 \Rightarrow Y(t) \geq 0 \quad \forall t$

Proof : part a) true by hypothesis for $t = t_o$. Let us suppose that $\exists t^*$ such that

$$x_i(t^*) = y_i(t^*) \quad \text{and} \quad x_j(t^*) \leq y_j(t^*)$$

then

$$\dot{x}_i(t^*) \leq g_i[X(t^*)] \leq g_i[Y(t^*)] = \dot{y}_i(t^*)$$

therefore to

$$t^* + \Delta t \qquad x_i(t^* + \Delta t) \leq y_i(t^* + \Delta t)$$

part b) true by hypothesis for $t = t_o$. Let us suppose that $\exists t^*$ with $y_i(t^*) = 0 \quad y_i(t^*) \geq 0$, it is easy to see that $\dot{y}_i(t^*) = g_i[(Y(t^*)] \geq 0$ and therefore $y_i(t + \Delta t) \geq 0$

This theorem constitutes, in a way, the extension for the vectorial case of the theorem of Corduneanu.

Let us return to hypotheses and inequalities (4.9) and (4.10) with $\xi_{ij} \geq 0$ and let us make the further hypothesis that the functions of class K are of the same order of magnitude, namely

$$\left[\begin{array}{l} \mu_{i1}\varphi_i(\|x_i\|) \leq v_i(x_i) \leq \mu_{i2}\varphi_i(\|x_i\|) \\ \dot{v}_{i\hat{S}_i} = (\nabla v_i)' g_i(x_i) \leq \mu_{i3}\varphi_i(\|x_i\|) \\ (\nabla v_i)' h_i(X) \leq \sum_{j=1}^{N} \xi_{ij}\varphi_i(\|x_i\|) \end{array}\right. \tag{4.14}$$

Naturally the results of paragraph 4.2.1. are directly applicable and one obtains an inequality of the type $\overset{\circ}{V} \leq AW$ where $W = [\varphi_1, \varphi_2, \ldots \varphi_N]$. The system is stable if A is a (M) matrix. It is possible to transform this inequality by using the property with the same order of magnitude of the functions φ as follows

$$\begin{cases} \dot{v}_{i\hat{S}_i} \leq \dfrac{\mu_{i3}}{\mu_{i2}} \; v_i\,(x_i) & (\mu_{i3} < 0) \\ (\nabla v_i)'\; h_i(X) \leq \sum_{j=1}^{N} \xi_{ij}\, \mu_{j1}^{-1} \; v_j\,(x_j) \end{cases} \tag{4.15}$$

and we then get a differential inequality of the type

$$\overset{\circ}{V} \leq \tilde{A}V \text{ where } \tilde{A} \in R^{N\times N}, \quad \tilde{a}_{ij} = \mu_{i3}\mu_{i2}^{-1}\,\delta_{ij} + \xi_{ij}\,\mu_{j1}^{-1} \tag{4.16}$$

As by hypothesis $\xi_{ij} \geq 0$, $i \neq j$, the function $G(V) = \tilde{A}V$ is quasi-monotone (def. 4.8) and the solutions of the differential inequality (4.16) are bounded from above by those of the real differential system

$$\overset{\circ}{W} = \tilde{A}\,W \quad \text{provided that} \quad V_o \leq W_o \tag{4.17}$$

System (4.17) is called a comparison system. Therefore :

<u>Theorem 4.13</u> : If $\tilde{A}$ is a M matrix then (4.7) is asymptotically stable.
Indeed, in this case exponential stability and asymptotic stability are equivalent to the fact
1) of the hypothesis on the class K functions (same order of magnitude)
2) of the exponential stability of the system of comparison (linear).

One can see that for the same hypotheses case b) of section 4.2.1. can be reduced to the study of a system of comparison. Indeed, let us now consider the hypotheses :

$$\begin{cases} \mu_{i1}\,\varphi_i\,(\|x_i\|) \leq v_i(x_i) \leq \mu_{i2}\,\varphi_i(\|x_i\|) \\ \dot{v}_{i\hat{S}_i} \leq \mu_{i3}\,\varphi_i\,(\|x_i\|) \\ (\nabla v_i)'\,h_i(X) \leq \varphi_i^{\frac{1}{2}}\,(\|x_i\|) \sum_{j=1}^{N} \xi_{ij}\,\varphi_j\,(\|x_j\|) \end{cases} \tag{4.18}$$

in which $\xi_{ij} \geq 0$. Then comes

$$\begin{aligned} \dot{v}_{iS_i} &\leq \mu_{i3}\,\varphi_i\,(\|x_i\|) + \varphi_i^{\frac{1}{2}}\,(\|x_i\|) \sum_{j=1}^{N} \xi_{ij}\,\varphi_j^{\frac{1}{2}}\,(\|x_j\|) \\ &\leq \mu_{i3}\,\mu_{i2}^{-1}\,v_i(x_i) + v_i(x_i)^{\frac{1}{2}}\mu_{i1}^{-\frac{1}{2}} \sum_{j=1}^{N} \xi_{ij}\mu_{j1}^{-\frac{1}{2}}\,v_j^{\frac{1}{2}}(\|x_j\|) \\ &\qquad (\mu_{i3} \leq 0) \end{aligned} \tag{4.19}$$

Let us make the change of variable in (4.19) $\tilde{v}_i = 2\sqrt{v_i}$, we get, in matrix form :

$$\dot{\tilde{V}} \leq 2\,\tilde{A}\,\tilde{V} \tag{4.20}$$

where $\tilde{V} = [2\sqrt{v_1}, \ldots 2\sqrt{v_N}]'$, $\tilde{A} = \{\tilde{a}_{ij}\}$ and $a_{ij} = \mu_{i3}\mu_{i2}^{-1}\delta_{ij} + \xi_{ij}\,\mu_{i1}^{-\frac{1}{2}}\mu_{j1}^{-\frac{1}{2}}$

and therefore a condition of stability for (4.7) is that A be a M matrix.

4.2.3. Recapitulatory table. Notes

$$S = \bigcup_{i=1}^{N} S_i$$

$$S_i : \underbrace{\dot{x}_i = g_i(x_i)}_{\hat{S}_i} + h_i(X) \qquad \left[\begin{array}{l} g_i(0)=0 \\ h_i(0)=0 \end{array}\right.$$

<u>hyp</u>. $\varphi_{i1}(\|x_i\|) \le v_i(x_i) \le \varphi_{i2}(\|x_i\|)$ $\qquad \dot{v}_{i\hat{S}_i} \le \mu_{i3}\varphi_{i3}(\|x_i\|)$

$$V = [v_1, v_2, \dots v_N]'$$

a) $(\nabla v_i)' \, h_i(X) \le \sum_{j=1}^{N} \xi_{ij}\, \varphi_{j3}(\|x_j\|)$

$\mathring{V} \le AW, \quad W = (\varphi_{13}, \dots \varphi_{1N})'$

$A = \{a_{ij}\} \quad a_{ij} = \mu_{i3}\,\delta_{ij} + \xi_{ij}$

b) $(\nabla v_i)' \, h_i(X) \le \varphi_{i3}^{\frac{1}{2}}(\|x_i\|) \sum_{j=1}^{N} \zeta_{ij}\, \varphi_{j3}^{\frac{1}{2}}(\|x_j\|)$

$\bar{A} = \{\bar{a}_{ij}\}$

$\bar{a}_{ij} = \mu_{i3}\,\delta_{ij} + \zeta_{ij}$

Scalar Lyapunov function : $V = \alpha' V, \alpha \in R_N^+$

$\alpha' A \le 0$ (≤ 0) : stable (ass. stable) $\qquad S = \frac{1}{2}[\operatorname{diag}(\alpha_i)\,\bar{A} + \bar{A}' \operatorname{diag}(\alpha_i)] \le 0$ (<0) stable (ass. stable)

$\xi_{ij} \ge 0 \quad ; \quad \zeta_{ij} > 0$

<u>Hyp</u>.

A is a M matrix $\Rightarrow$ stability $\qquad$ $\bar{A}$ is a M matrix $\Rightarrow$ stability

<u>hyp.</u> $\quad \varphi_{i1}, \varphi_{i2}, \varphi_{i3}$ same order of magnitude

$$\varphi_i = \varphi_{i1}/\mu_{i1} = \varphi_{i2}/\mu_{i2} = \varphi_{i3}$$

$\mathring{V} \le \tilde{A}\, V \qquad \tilde{a}_{ij} = \mu_{i3}\mu_{i2}^{-1}\,\delta_{ij} + \xi_{ij}\,\mu_{j1}^{-1}$.

$\Downarrow$

$\mathring{W} = A\, W$

System of Comparison

$\tilde{A}$ is a M matrix $\Rightarrow$ stability

Notes :

1. It is clear from the table that the vector Lyapunov functions comparison system approach requires, for its application, more restrictive hypotheses than the scalar Lyapunov function approach. This has led some writers to criticise relatively severely the vectorial approach in comparison with the scalar approach[SAN - 78]. Nevertheless, as we shall see in the next section, this statement must be moderated, particularly in the case of estimation of areas of stability and structural perturbations. We shall see that within the framework of linear systems the vectorial Lyapunov function approach makes it possible to pose a problem of search of "optimal" stability test conditions, ie. search for optimal Lyapunov functions.

2. In the preceding pages we have obtained a linear system of comparison by construction. It is equally possible to consider the existence and the determination of non-linear systems of comparison [BIT - 78].

Let us make the following assumption :

$$\dot{v}_{iS_i} \leq f_i \; (v_1, v_2, \dots v_i, \dots v_N) \quad i=1,2,\dots N \tag{4.21}$$

and let us suppose that $F = [f_1, \dots f_N]'$ is quasi-monotone, then the system

$$w_i = f_i [w_1, \dots w_i, \dots w_N] \quad i=1,2,\dots N; \quad \overset{\circ}{W} = F(W) \tag{4.22}$$

constitutes a comparison system for (4.21).(4.22) is a real non linear differential system for which all the known conventional methods for the stability study (Popov, Lyapunov...) can be applied. Because F is quasi-monotone it is possible to determine for (4.22) tests of an algebraic kind.

For example :

a) if there exists a point $W_o \in R_+^N$ such that $F(W_o) \leq 0$ then

$$V(t) \leq W_o \quad \forall t \quad \text{(bounded solutions)}$$

demonstrated by $W(t) \leq W_o \;\; \forall t$

b) if there exists a point $W_o \in R_+^N$ such that $F(\lambda W_o) \leq 0$, $\forall \lambda \in [0,1]$ then the origin is stable

proof : $\delta(\varepsilon) = \varepsilon = \|x_o\|$, x_o : $\sup [v_i(x_{io})] \leq \min w_{oi}$

c) if there exists a point $W_o \in R_+^N$ such that $F(\lambda W_o) < 0$, $\forall \lambda \in [0,1]$ then the origin is asymptotically stable

Proof : using the scalar function, that of Rosembrock

$$v = \max_i \frac{w_i}{w_{oi}}$$

if $W_o \to \infty$ in the preceding theorems, then the stability is global.

Example : this is taken from [MIC - 77 p. 69] and concerns the controlled longitudinal movement of an aircraft

$$\dot{x}_i = -\rho_i x_i + \sigma$$

$$\dot{\sigma} = \sum_{k=1}^{4} \beta_k x_k - p\sigma - f(\sigma)$$

with $\rho_k > 0$, $p > 0$, β_k constants

$$f(0) = 0 \text{ and } \sigma f(\sigma) > 0$$

Let us adopt a decomposition in two sub-systems S_1 and S_2 respectively associated with the state spaces

$$S_1 : \{x_1, x_2, x_3, x_4\}$$

$$S_2 = \{\sigma\}$$

a) the Lyapunov functions associated with each sub-system are

$$v_1 = x_1^2 + x_2^2 + x_3^2 + x_4^2 \quad ; \quad v_2 = \sigma^2$$

then comes

$$\dot{v}_1 \leq -2 \min(\rho_i)\, v_1 + 4 v_1^{\frac{1}{2}} v_2^{\frac{1}{2}}$$

$$\dot{v}_2 \leq 2 \left(\Sigma \beta_k^2\right)^{\frac{1}{2}} v_1^{\frac{1}{2}} v_2^{\frac{1}{2}} - 2 p v_2$$

$$\bar{A} = 2 \begin{bmatrix} -\min(\rho_i) & 2 \\ (\Sigma \beta_k^2)^{\frac{1}{2}} & -p \end{bmatrix}$$

as $\bar{A}$ has positive off diagonal elements a sufficient condition of stability is that A be an M matrix, namely the Sevastyanov Kotelyanski condition gives.

$$p \min \rho_i > 2 (\Sigma \beta_k^2)^{\frac{1}{2}}$$

b) with $\tilde{v}_1 = \sqrt{v_1}$ $\tilde{v}_2 = \sqrt{v_2}$ there comes

$$\begin{bmatrix} \dot{\tilde{v}}_1 \\ \dot{\tilde{v}}_2 \end{bmatrix} \leq \bar{A} \begin{bmatrix} \tilde{v}_1 \\ \tilde{v}_2 \end{bmatrix}$$

, therefore, the same condition of stability.

4.3 Areas of stability and structural perturbations

The grouping together in one single section of two distinct problems is justified by the fact that we are dealing with two cases in which the comparison system approach may turn out to be better than the scalar function approach.

4.3.1. Areas of stability

In this sub-section we shall presume that the hypotheses made on the Lyapunov functions and the terms of interconnection are only valid in a finite area surrounding the origin. This area will be defined thus

$$\mathcal{D} : \left\{ X : \quad v_i \ (x_i) \quad \leqslant \ v_{oi}, \ i = 1,2 \ldots N \right\} \tag{4.23}$$

In this area the following inequalities are satisfied

$$\left[\begin{array}{l} \varphi_{i1} \ (\|x_i\|) \ \leqslant \ v_i \ (x_i) \leqslant \varphi_{i2} \ (\|x_i\|) \\ \dot{v}_{i\hat{S}_i} \ \leqslant \mu_{i3} \quad \varphi_{i3} \quad (\|x_i\|) \\ (\nabla v_i)' \ h_i(X) \ \leqslant \ \sum_{j=1}^{N} \ \xi_{ij} \quad \varphi_{j3} \quad (\|x_i\|) \end{array}\right. \tag{4.24}$$

which provides the differential inequality

$$\overset{\circ}{V} \ \leqslant \ A \ W \qquad a_{ij} = \mu_i \ \delta_{ij} + \xi_{ij} \tag{4.25}$$

<u>Theorem 4.14</u> : If there exists a vector $\alpha = [\alpha_1, \ldots \alpha_N]'$ such that

$$\alpha' \ A \leqslant 0 \quad (<0)$$

then the area $\mathcal{D}_s$ defined by

$$\mathcal{D}_s : \left\{ X : \Sigma \alpha_i \ v_i(x_i) \leqslant \min_j \alpha_j \quad v_{oj} \right\}$$

is an area of stability (asymptotic stability) for the point of equilibrium at the origin.

Proof : it must be shown that $\forall t$, $X(t) \in \mathcal{D}$. Let us consider the Lyapunov function $\quad \upsilon = \alpha' V.$

then $\quad \dot{\upsilon} \leqslant \alpha' A \quad W < 0$

so $\quad \upsilon(t) \ \leqslant \ \upsilon \ (t_o)$

ie. $\quad \sum_i \ \alpha_i \quad v_i \ [x_i(t)] \leqslant \sum_i \alpha_i \ v_i \ [x_{io}] \leqslant \min_j \alpha_j \ v_{oj}$

therefore $\quad \forall_i \quad v_i(x_i) \leqslant \dfrac{\min \alpha_j \quad v_{oj}}{\alpha_i} \leqslant v_{oi}$ q.e.d.

For the system of comparison approach let us assume that

$$\xi_{ij} \geqslant 0 \quad \text{and} \ \varphi_{i3} = \varphi_{i1} / \mu_{i1} = \varphi_{i2} / \mu_{i2}$$

then (4.25) is changed to

$$\overset{\circ}{V} \leqslant \bar{A} \ V \quad \text{in which } \bar{a}_{ij} = \mu_{i3} \mu_{i2}^{-1} \ \delta_{ij} + \xi_{ij} \ \mu_{i1}^{-1}$$

with the associated system of comparison

$$\overset{\circ}{W} = \bar{A}W \tag{4.26}$$

<u>Theorem 4.15</u> : If the matrix $\bar{A}$ is a M matrix, ie. if there exists a vector $\beta > 0$ such that $\bar{A}\beta \le 0$ (< 0) then the area $\mathcal{D}_s$ defined by

$$\mathcal{D}_s : \left\{ X : v_i(x_i)\,\beta_i^{-1} \le \min_j \; v_{jo}\,\beta_j^{-1},\ i=1,2,\dots,N \right\}$$

is an area of stability (asymptotic stability) for the equilibrium point

Proof : this is done from the Lyapunov function known as Rosembrock's function

$$\vartheta = \max_i \; v_i(x_i)\,\beta_i^{-1}$$

$$\dot{v} \le \frac{1}{\beta_i} \sum_j \bar{a}_{ij} v_j \le \frac{1}{\beta_i^{2}} \Big[\sum_j a_{ij}\beta_j\Big] v_i \le 0 \quad \text{therefore}$$

$$\forall t \quad v(t) \le v(o) \ \text{ ie. } \max_i v_i[x_i(t)]\beta_i^{-1} \le \min_j v_{jo}\beta_j^{-1} \le v_{io}\beta_i^{-1} \quad \forall i$$

therefore

$$v_i(t) \le v_{io} \qquad \forall i = 1,2,\dots N \qquad\qquad \text{q.e.d.}$$

Of course it is possible to apply theorem (4.14) to (4.26) but this is of little interest since $\bar{A}$ is derived from A after further majorations.

It is impossible to say a priori which of the theorems (4.14) and (4.15) will provide the better estimate of the area of stability : scalar approach or vectorial approach. An obvious advantage of the area given in theorem (4.15) is that it provides an estimate on each sub-system independently of the other sub-systems [WEI - 73].

In the figure below we are considering a case with N = 2

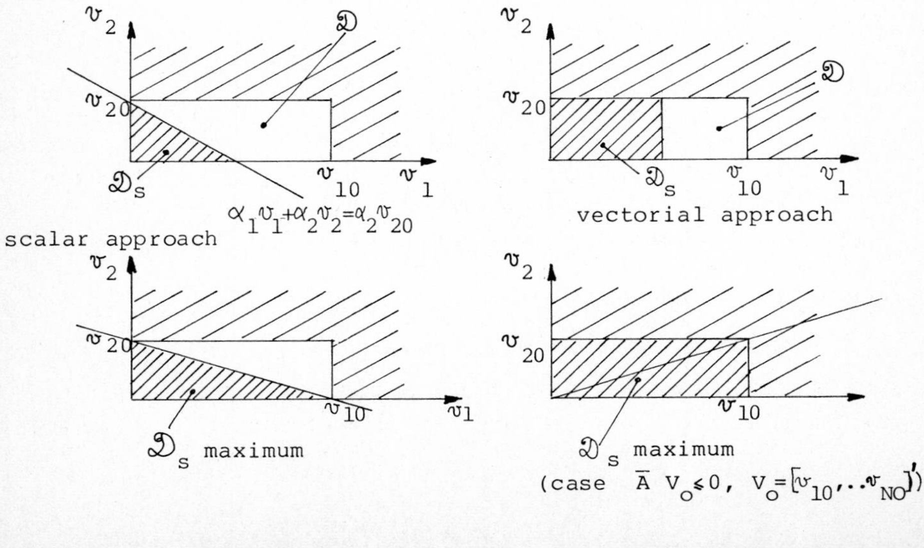

Example : let us consider the system

$$\begin{cases} \dot{x}_1 = A_1 x_1 + b_1 f_1 (c_1' x_2) \\ \dot{x}_2 = A_2 x_2 + b_2 f_2 (c_2' x_1) \end{cases}$$

Where b_1 c_1 b_2 c_2 are appropriate vectors and f_1, f_2. $R \to R$ are such that $0 \le \sigma f_1(\sigma) \le k_1 \sigma^2$ if $|\sigma| < \rho_1$

$$0 \le \sigma f_2(\sigma) \le k_2 \sigma^2 \text{ if } |\sigma| < \rho_2$$

Let us suppose that A_1 and A_2 are asymptotically stable so that there exists two symmetrical positive definite matrices P_1 and P_2 such that

$$A_1' P_1 + P_1 A_1 = -Q_1; \quad A_2' P_2 + P_2 A_2 = -Q_2$$

where Q_1 and Q_2 are positive definite.
Let us take as Lyapunov functions for each sub-system $v_1 = \sqrt{x_1' P_1 x_1}$ and $v_2 = \sqrt{x_2' P_2 x_2}$

$$\lambda_{min}^{\frac{1}{2}}(P_1)\|x_1\| \le v_1 \le \lambda_{max}^{\frac{1}{2}}(P_1)\|x_1\|$$

then

$$\dot{v}_{1\hat{S}_1} = -\frac{1}{2}\frac{x_1' Q_1 x_1}{v_1} \le -\frac{1}{2}\lambda_{min}(Q_1)\lambda_{max}^{-\frac{1}{2}}(P_1)\|x_1\|$$

and

$$(\nabla v_1)' b_1 f_1(c_1' x_2) = \frac{(P_1 x_1)'}{v_1} b_1 f_1(c_1' x_2) \le \frac{\|P_1 x_1\|}{v_1}\|b_1\|\|k_1\|\|c_1\|\|x_2\|$$

$$\le \lambda_{max}^{\frac{1}{2}}(P_1)\|b_1\|\|x_1\|\|k_1\|\|x_2\|$$

with the same thing for (S_2), we then get

$$\overset{\circ}{V} \le \begin{bmatrix} -\frac{1}{2}\lambda_{min}(Q_1)\lambda_{max}^{-\frac{1}{2}}(P_1) & \lambda_{max}^{\frac{1}{2}}(P_1)\|b_1\|\|c_1\| k_1 \\ \lambda_{max}^{\frac{1}{2}}(P_2)\|b_2\|\|c_2\| k_2 & -\frac{1}{2}\lambda_{min}(Q_2)\lambda_{max}^{-\frac{1}{2}}(P_2) \end{bmatrix} \begin{bmatrix} \|x_1\| \\ \|x_2\| \end{bmatrix} = AW$$

where, taking it a bit further and replacing x_1 x_2 by v_1 and v_2 we get

$$\overset{\circ}{V} \le \begin{bmatrix} -\frac{1}{2}\lambda_{min}(Q_1)\lambda_{max}(P_1) & \lambda_{max}^{\frac{1}{2}}(P_1)\lambda_{min}^{-\frac{1}{2}}(P_1)\|b_1\|\|c_1\|k_1 \\ \lambda_{max}^{\frac{1}{2}}(P_2)\lambda_{min}^{-\frac{1}{2}}(P_2)\|b_2\|\|c_2\|k_2 & -\frac{1}{2}\lambda_{min}(Q_2)\lambda_{max}(P_2) \end{bmatrix} V = \bar{A}V$$

A and $\bar{A}$ both have positive off diagonal elements. By applying the conditions of Sevastyanov Kotelyanski one can easily see that $\bar{A}$ provides more restrictive conditions of stability (A may

be stable and not $\bar{A}$). If we now consider the case where $\bar{A}$ and A are stable, then we can examine the problem of the estimation of the area of stability.
The following differential inequalities are valid for

$$\| c_1 \; x_2 \| < \rho_1 \qquad \text{and} \qquad \| c_2 \; x_1 \| < \rho_2$$

Let us transpose these inequalities in v_1 and v_2

$$\| x_1 \| \leq \rho_2 \; \| c_2 \|^{-1} \quad \Longrightarrow \quad v_{10} \leq \rho_2 \; \| c_2 \|^{-1} \lambda_{min}^{-\frac{1}{2}} \; (P_1)$$

$$\text{and} \qquad v_{20} \leq \rho_1 \; \| c_1 \|^{-1} \lambda_{min}^{-\frac{1}{2}} \; (P_2)$$

Let us continue on a numerical application $A_1 \equiv A_2$, $P_1 \equiv P_2$, $Q_1 \equiv Q_2$ ...

$$\lambda_{min} \; (P_1) = \frac{1}{2}, \lambda_{max}(P_1) = 1 \, , \lambda_{min}(Q_1) = 1, \| b_1 \| = \| c_1 \| = 1,$$

$$k_1 = k_2 = 0{,}3 \; , \qquad \rho_1 = \rho_2 = 1$$

The scalar approach provides :

$$\overset{\circ}{V} = \begin{bmatrix} -\frac{1}{2} & 0{,}3 \\ 0{,}3 & -\frac{1}{2} \end{bmatrix} W, \; (v_1, \; v_2 < \frac{1}{2}) \quad \alpha = [1,1] \text{ is such that}$$

$\alpha ' A \leq 0$ therefore a possible Lyapunov function is $(v_1 + v_2)$
The associated area of stability is given by :

$$v_1 + v_2 \leq \min \; \alpha_i \; v_{io} = \frac{1}{2}$$

With the vectorial approach :

$$\overset{\circ}{V} = \begin{bmatrix} -\frac{1}{2} & 0{,}3\sqrt{2} \\ 0{,}3\sqrt{2} & -\frac{1}{2} \end{bmatrix} \qquad \beta = [1, \; 1]' \text{ is such that } \bar{A} \, \beta \leq 0 \text{ there-}$$

$$\text{fore } \mathfrak{D}_s = \left\{ X, \; v_1, v_2 \leq \min \frac{v_{oi}}{\beta_i} = \frac{1}{2} \right\}$$

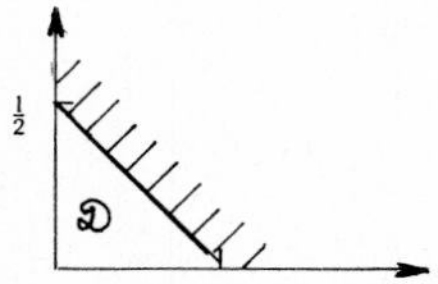

scalar approach

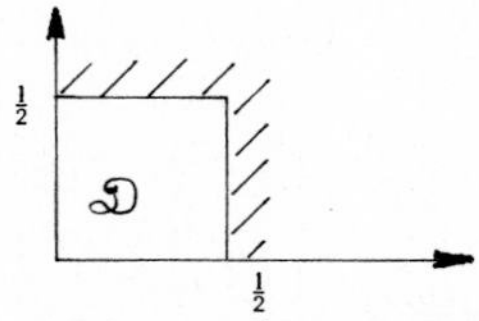

system of comparison approach

Further details concerning this problem can be found in [WEI - 73],[BIT - 76].

4.3.2. Structural perturbations

We shall assume, in this sub-section, that structural type perturbations act on the system. The term "structural" is used to specify that we are concerned not with perturbations "external" to the system but with changes in the very dynamics of these systems. One type of structural perturbation may, for example,

be a line rupture between sub-systems. This type of perturbation may be modelled by a change in the mathematical model which describes the system.
We shall assume that the number of perturbations is finite. These hypotheses are physically realistic in the case, for example, of cutting of transmission lines between sub-systems [BUR - 78]. The problem can be formulated in the following way :

$$S = \bigcup_{i=1}^{N} S_i$$

$$(S_i) \quad \dot{x}_i = g_i(x_i) + h_i^k (X) \qquad k=1,2,\dots p \tag{4.27}$$

in which k is an index specifying a perturbation. Let p be the maximum number of these perturbations. We propose to establish conditions which guarantee the stability of (4.27) whatever the sequence of perturbations to which the system may be subject.
For each isolated sub-system there exists a function $v_i(x_i)$ such that

$$\varphi_{i1} (\|x_i\|) \leq v_i(x_i) \leq \varphi_{i2} (\|x_i\|) \tag{4.28}$$

and in addition, $\forall k = 1,2,\dots p$; $\quad v_{i\hat{S}_i} \leq \mu_i^k \varphi_{i3} (\|x_i\|)$

$$(\nabla v_i)' h_i^k (x_i) \leq \sum_{j=1}^{N} \xi_{ij}^k \varphi_{j3} (\|x_i\|) \tag{4.29}$$

So for each perturbation of index k, we obtain

$$\mathring{V} \leq A^k W \quad \text{where} \quad A^k = \left\{ a_{ij}^k = \mu_i^k \delta_{ij} + \xi_{ij}^k \right\} \tag{4.30}$$

Theorem 4.16 : If for every k=1,2,...p there exists a vector α' such that

$$\alpha' A^k \leq 0$$

then system (4.27) is stable under structural perturbations.

Proof : obvious from the scalar Lyapunov function $\nu' = \alpha' V$

Notes :
1) one must determine a unique vector α for all the matrices A^k k=1,2,...p. In the case where $\xi_{ij}^k \geq 0$ it is no longer enough to test stability of the matrix A^k using, for example, Sevastyanov Kotelyanski conditions. The problem of the determination of the vector α may be undertaken using linear programming tools.
2) In his concept of connective stability Siljak avoids this danger by determining a differential inequality which is greater than the k differential inequalities (4.30). In fact

$$\mathring{V} \leq A^k W \leq A_M W \quad \text{if} \quad A_M = \left\{ a_{ijM} = \sup_k (a_{ij}^k) \right\}$$

Of course this causes a more restrictive condition and it may happen that theorem 4.16 is verified while no vector α exists for A_M such that

$$\alpha' A_M \leq 0$$

Let us now suppose that the inequalities (4.28) and (4.29) are verified with the further condition that $\xi_{ij}^{\;k} \geq 0$ and that the functions $\varphi_{i1}, \varphi_{i2}, \varphi_{i3}$ are of the same order of magnitude.

$$\varphi_{i1}\ \mu_{i1}^{\;-1} = \varphi_{i2}\ \mu_{i2}^{\;-1} = \varphi_{i3}$$

Then for each perturbation k = 1, 2, ...p, there comes

$$\mathring{V} \leq \bar{A}^k\ V \text{ with } \bar{A}^k = \left\{ \bar{a}_{ij}^{\;k} = \mu_i^{\;k}\ \mu_{i2}^{\;-1}\delta_{ij} + \xi_{ij}^k\ \mu_{i1} \right\}$$

Theorem 4.17 : If for every k = 1, 2,... p there exists

a) either a vector $\alpha > 0$ such that $\alpha'\ \bar{A}^k \leq 0$

b) or a vector $\beta > 0$ such that $(A^{k})'\beta \leq 0$

then system (4.27) is stable for the structural perturbations defined above.

Proof : a) with $\nu = \alpha'\ V$

b) with $\nu = \max_i\ (v_i\ \beta_i^{\;-1})$

When p > 1, a and b are not equivalent, ie. a) cannot be satisfied while b) is satisfied. There again the system of comparison approach may in certain cases turn out to be more efficient than the scalar approach.

example :

$$\bar{A}_k^{\;1} = \begin{bmatrix} -1 & 4 \\ -0,1 & -2 \end{bmatrix} \qquad \bar{A}_k^{\;2} = \begin{bmatrix} -1 & 0,1 \\ 1 & -5 \end{bmatrix}$$

The vector $\alpha > 0$ with $\alpha'\ \bar{A}_k^{\;1} \leq 0$ and $\alpha'\ \bar{A}_k^{\;2} \leq 0$ does not exist whereas $\beta = [1,5]'$ exists such that $\bar{A}_k^{\;1}\ \beta \leq 0$ and $\bar{A}_k^{\;2}\ \beta \leq 0$

Conclusion : we have just shown two cases in which the vector approach may prove to be more efficient than the scalar approach. We shall now examine another case where the vectorial approach may be justified; the case of interconnected linear systems.

4.4. On the problem of the robustness of multi-level control

The object of this section is to give a rapid presentation of a potential application of vectorial Lyapunov functions in the context of the analysis and the synthesis of multi-level control for interconnected systems. Although in its principal it can be considered for all types of system (non linear, non autonomous ...) this approach will be presented in the case of linear time invariant systems described by :

$$(S)\quad \mathring{x}_i = A_i\ x_i + B_i\ u_i + \sum_{j=1}^{N} A_{ij}\ x_j;\ i=1,2...N, \qquad (4.31)$$

and on the hypothesis that the control u_i applied to each sub-system is the sum of two terms :

u_{ie} : local component generated by a local controller associated with the sub-system i.

u_{ig} : component of co-ordination generated by an organ of supervision.

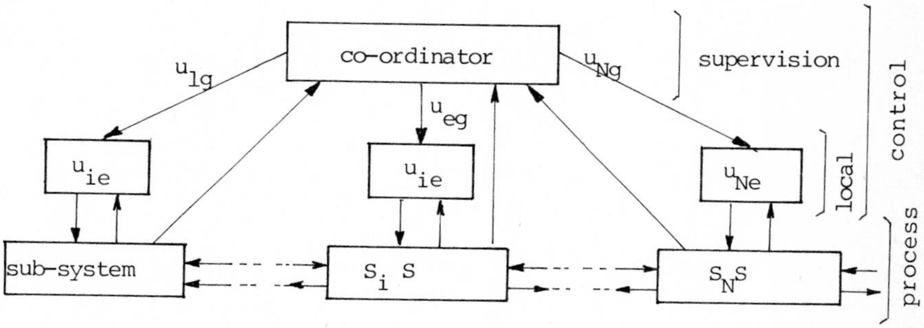

We are going to deal principally with the case in which the structural perturbations consist in a link cut between the higher level and the sub-systems. Two types can be considered :

Case 1 - Perturbations of "control" : a rupture of descending liaisons in the figure (absence of u_{ig})

Case 2 - Perturbations of observation which correspond to a rupture of ascending liaisons, deterioration of the signals u_{ig} coming from loss by the co-ordinator of all information on sub-system k.

With this hypothesis of perturbation it is clear that the robustness of the controlled system will depend essentially on the local components of control. To test the robustness of the structure, one method consists of using the vectorial Lyapunov function tool previously presented on the model :

$$\mathring{x}_i = A_i x_i + B_i\ u_{ie} + \sum_{j=1}^{N} A_{ij}\ x_j + \overline{u_{ig}} \quad ; \ i=1,2 \ldots N \qquad (4.32)$$

where $\overline{u_{ig}}$ represents the control from the supervisor after a given perturbation.

The principle consists of looking for (iteratively) a set of controls u_{ie} making it possible to guarantee the stability of the perturbed system.

Let us consider the case in which the controls are of the feedback type :

$$\begin{cases} u_{ie} = -\ K_i\ x_i \\ u_{ig} = -\displaystyle\sum_{\substack{j=1 \\ \neq i}}^{N} T_{ij}\ x_j \end{cases} \qquad (4.33)$$

For better characterisation of the structural perturbations let us introduce in the model a matrix E with binary elements

$$E = \left\{ e_{ij} \;/\; e_{ij} = 0 \text{ or } 1,\ i,j = 1,\dots N \right\}$$

So then (4.31) is written as :

$$\overset{\circ}{x}_i = (A_i - B_i\,K_i)\,x_i + \sum_{j=1}^{N} (A_{ij} - e_{ij}\,T_{ij})\,x_j \quad i=1,2,\dots N \tag{4.34}$$

A perturbation between the level of supervision and the sub-system i finds expression as :

$$e_{ij} = 0 \quad \forall j = 1,\dots N \text{ for case no. 1}$$

$$e_{ji} = 0 \quad \forall_j = 1,\dots N \text{ for case no. 2}$$

The problem of the synthesis of a robust structure is that of the determination of a local control $u_i = - K_i x_i$ which guarantees stability in every possible case of perturbation ($\forall$ E). Let us make the assumption (not very restrictive in practice) that each isolated sub-system is asymptotically stable; (A_i - $B_i\ K_i$ asymptotically stable for every i).
To each of the sub-systems i is associated the Lyapunov function

$$v_i\ (x_i) = (x_i'\ P_i\ x_i)^{\frac{1}{2}} \tag{4.35}$$

in which P_i is the matrix solution of the Lyapunov equation

$$(A - B_i\ K_i)'\ P_i + P_i\ (A_i - B_i\ K_i) + D_i'\ D_i = 0$$

With $[A_i - B_i\ K_i\ ,\ D_i]$ observable, P_i is positive definite.
The Lyapunov function v_i satisfies the following inequalities

$$\begin{cases} \lambda_{mi}^{\frac{1}{2}} \|x_i\| \leq v_i\ (x_i) \leq \lambda_{Mi}^{\frac{1}{2}} \|x_i\| \\ \|\nabla v_i\| \leq \lambda_{Mi}^{\frac{1}{2}} \end{cases} \tag{4.36}$$

where λ_{mi}, λ_{Mi} are the minimum and maximum eigenvalues of P_i respectively.
Evaluating the time derivative of v_i along the trajectories of each sub-system i, we get

$$\overset{\circ}{v}_i = -\frac{1}{2}\ \frac{x_i'\ D_i'\ D_i x_i}{x_i'\ P_i\ x_i}\ v_i + (\nabla v_i)' \sum_j (A_{ij} - e_{ij}\ T_{ij})\ x_j \tag{4.37}$$

Majoring the second term of the equation, we can write down the following inequality

$$\overset{\circ}{v}_i \leq -\alpha_i\ v_i + \lambda_{Mi}^{\frac{1}{2}}\ \lambda_{mj}^{-\frac{1}{2}}\ h_{ij}\ v_j \tag{4.38}$$

where :

$$\begin{cases} \alpha_i = \min\limits_{x_i} \dfrac{x_i' D_i' D_i x_i}{x_i' P_i x_i} \\ h_{ij}(e_{ij}) = \| A_{ij} e_{ij} T_{ij} \| \end{cases}$$

With $V = \{ v_1, v_2 \dots v_N \}'$, we get the comparison system :

$$\overset{\circ}{V} = L(E)\, V$$

in which $L(E) = \{ l_{ij}(E) = -\alpha_i \delta_{ij} + \lambda_{Mi}^{\frac{1}{2}} \lambda_{mi}^{-\frac{1}{2}} \| A_{ij} - e_{ij} T_{ij} \| \}$

(4.39)

The problem of robust control considered here is posed as being that of synthesis of a decentralised gain [u = - Kx, K = block diag [Ki]] such that the matrix L(E) is stable for every matrix E, which characterises the structural perturbations.
The first problem is that of finding the most unfavorable case of perturbations, ie. the one which would provide the "least" stable comparison system (4.39).
It is not at all easy to determine the worst (in the Lyapunov sense) matrix L(E). The problem is one of a combinatorial nature with a large number of combinations to be tested (as large as the number of sub-systems is high). One way of avoiding this problem is to maximise independently each off diagonal term of the matrix L(E), in order to obtain in the end a unique system of comparison.

$$\overset{\circ}{V} = L_M V$$

where :

$$L_M = \{ \ell_{ij} = -\alpha_i \delta_{ij} + \lambda_{Mi}^{\frac{1}{2}} \lambda_{mi}^{-\frac{1}{2}} \max_{e_{ij}} \| A_{ij} - e_{ij} T_{ij} \| \} \qquad (4.40)$$

and L_M stable $\Longrightarrow$ $L(E)$ stable, $\forall E$.

This matrix L_M may not have any physical significance, ie. may not correspond to any possible case of physical structural perturbation. It may moreover, impose relatively severe conditions for the stability tests and therefore prove to be very restrictive for the application of the algorithm which follows.
For a more accurate study, it is necessary, then to determine "the worst" system of comparison and to escape from a too large combination of the associated problems the hypothesis can be made (which is physically realistic) of non-multiplicity of structural perturbations simultaneously. However this problem, which is a problem of preliminary analysis is only of little interest for what follows and we shall still need the determination of decentralised gains K_i guaranteeing the stability of the system of comparison (4.40).
Considering the matrix L_M, it is clear that L_M is stable if the diagonal terms are sufficiently negative, ie. if α_i is big enough. The coefficient α_i is a measure of the degree of sta-

bility, estimated by way of the function v_i of the sub-system i. Therefore, a strategy for the synthesis of a robust control which can be accepted is based on finding controls which guarantee a sufficient degree of stability to each sub-system. This strategy is expressed by the following general algorithm :

step 1 : preliminary analysis : determination of the most unfavourable case L_M.

step 2 : test : L_M stable?
yes - stop
no - step 3

step 3 : determine K = bloc. diag. $[K_i]$ such that α_i increases
i=1,2,... N
go to step 2.

In addition to the problem posed by the choice of Lyapunov functions v_i associated with each sub-system (we shall return to this later), is posed the problem of the increase of the coefficient α_i, the stability degree of the sub-system i. Certain works [SIL - 76] dealing with the problem of robust control have chosen to search uniform degrees of stability ($\alpha_i = \alpha, \forall i$) which can only be justified a priori in the case of ident cal sub-systems. It is more reasonable not to impose such a constraint, and much more reasonable to allow oneself to adapt the increases on the α_i in terms of the characteristics of the problem, by exploiting, as the evolution of the algorithm occurs, information which can be given by the previous iterations. Indeed it may happen that for such a sub-system an increase of α_i results only in an enormous increase of the gains K_i whereas for some other the same increase requires only a moderate increase of the gains K_i. This may also make it possible to preserve a satisfactory degree of sub-optimality.
In order to illustrate this fact we are going to develop the preceding idea in the particular case where the synthesis of local controls is obtained by minimisation of a quadratic criterion. This choice makes it possible to determine both the local control and the associated Lyapunov function.
Nevertheless this choice remains arbitrary and any other method for the synthesis of local controls may be accepted (techniques of poles placement for example).
To each sub-system S_i is associated a quadratic criterion

$$J_i \ (u_i) = \frac{1}{2} \int_0^{\infty} (x_i' \ D_i' \ D_i \ x_i + u_i' \ R_i \ u_i) \ dt \qquad (4.41)$$

$R_i > 0$ - The local controls being written as

$$u_i = - K_i \ x_i = - R_i^{-1} \ B'_i \ P_i \ x_i \qquad (4.42)$$

where P_i is the positive definite solution (hypothesis D_i; A_i observable) of the Riccati equation :

$$A'_i P_i + P_i A_i - P_i B_i R_i^{-1} B'_i P_i + D'_i D_i = 0 \tag{4.43}$$

A lyapunov function for the sub-system S_i is of course :

$$v_i = (x'_i P_i x_i)^{\frac{1}{2}} \tag{4.44}$$

The increase of the coefficient α_i (diagonal) of the matrix of the system of comparison (4.40) L_M can be carried out as follows : to the sub-system i is applied the control

$$u_i = - \bar{K} x_i = - R_i^{-1} B'_i \bar{P}_i x_i \tag{4.45}$$

where $\bar{P}_i$ is the solution of

$$(A_i + \sigma_i I)' \bar{P}_i + \bar{P}_i (A_i + \sigma_i I) - \bar{P}_i B_i R_i^{-1} B'_i \bar{P}_i + D'_i D_i = 0 \quad \sigma_i \geq 0 \tag{4.46}$$

It is known that this control minimises the criterion :

$$\int_0^\infty e^{2\sigma_i t} (x'_i D'_i D_i x_i + u'_i R_i u_i)\, dt$$

This type of control makes it possible to position the eigen values of the matrix $(A_i - B_i R_i^{-1} B'_i \bar{P}_i)$ in the part of the complex plane situated on the left of the abscissa line $-\sigma_i$.
With the Lyapunov functions $v_i = (x'_i \bar{P}_i x_i)^{\frac{1}{2}}$ we obtain a new comparison system

$$\mathring{V} = \bar{L}_M V \tag{4.47}$$

$$\bar{L}_M = - (\bar{\alpha}_i + \sigma_i)\, \delta_{ij} + \lambda_{Mi}^{\frac{1}{2}} \lambda_{mj}^{-\frac{1}{2}} h_{ij}$$

where

$$\bar{\alpha}_i = \min_{x_i} \frac{1}{2} \frac{x'_i D'_i D_i x_i}{x'_i \bar{P}_i x_i}$$; λ_{Mi} , λ_{mi} maximum and minimum eigenvalues of the matrix $\bar{P}_i$, $h_{ij} = \max_{e_{ij}} \| A_{ij} - e_{ij} T_{ij} \|$

It is always possible to determine σ_i so that $\bar{\alpha}_i + \sigma_i > \alpha_i$, <u>but</u> it is not obvious a priori that the conditions of stability supplied by (4.47) and the matrix $\bar{L}_M$ will always be less restrictive than those obtained at the previous step with (4.40) and the matrix L_M. This is because the off diagonal terms of the matrix L change with the applied control.
The problem appears to be inextricable, taking into account the terms with the form ($\lambda_{Mi} \lambda_{mj}^{-1}$) appearing in L , which induce a coupling at the level of the Lyapunov functions associated with each sub-system. This difficulty can easily be removed by

noting that the stability properties of the matrix L_M are identical to those of the matrix $\tilde{L}_M$ where

$$\tilde{L}_M = \left\{ \ell_{ij} = -\alpha_i \, \theta_i^{-1} \delta_{ij} + h_{ij} \right\} \qquad (4.48)$$

$$\theta_i = \lambda_{Mi}^{\frac{1}{2}} \; \lambda_{mi}^{-\frac{1}{2}}$$

In fact one can easily verify that the "Savastyanov-Kotelyanski" tests are identical for the matrices L_M and $\tilde{L}_M$. The second of these matrices has a particularly interesting form (a form which will be explored in the next paragraph) for the terms which depend on the Lyapunov functions v_i are isolated on the principal diagonal and each term depends only on the Lyapunov function corresponding to it (aspect of decentralisation in calculations).
Therefore the problem now consists of determining gains K_i iteratively so that the ratios α_i / θ_i are large enough.
This task can be carried out by defining an auxiliary problem for the determination of the "minimal" degrees of stability, in such a way as to avoid moving too far away from the initial control which may be considered as an optimal or slightly sub-optimal control from which it is better to move away as little as possible.
For this the following algorithm may be used : [GER - 79] .

<u>Step 1</u> : calculation of the matrix $H_M = h_{ij}$ representative of the interconnections.

<u>Step 2</u> : determination of the "minimal" degree of stability

$\min_{\pi} f(\pi)$ (f, auxiliary function)

subject to

$[H_M - \text{diag}\,\pi_i]\; Y \leq 0$ (condition of stability)

$Y = [y_1, \dots y_N]' > 0$

$\pi = [\pi_1 \;\; \pi_2 \; \dots \; \pi_N]$

<u>Step 3</u> : for each isolated sub-system, from an initialisation on σ_i

a) solve the Riccati equation (4.46)

b) calculation of α_i and θ_i

c) test whether $(\alpha_i + \sigma_i)\,\theta_i^{-1} > \pi_i$ stop

if not go on to 3-a with new initialisation for σ_i.

<u>Note</u> : The object of the auxiliary function f (π) is to enable the determination of a vector π whose components may not be too big (useful for the test 3-c). It can be fixed a priori by translating, for example, the effect of deterioration of the global criterion (measurement of the sub-optimality of the obtained control (see appendix), or it can be modified iterati-

vely in terms of the information given by previous iterations the aim is to decrease the values π_i of the sub-systems which, because of their structure and the choice of Lyapunov functions, do not permit an efficient increase of the ratio α_i / θ_i.

4.5. Interconnected linear systems : choice of Lyapunov functions.

Just as in the scalar case the quality of the results obtained depends to a large degree on the choice made for the Lyapunov functions. In the vectorial case this choice is made even more difficult by taking into account the fact that the choice must be exercised not only in terms of the nature of the mathematical model of the isolated sub-systems but also in terms of the interconnection terms. Moreover, a Lyapunov function which would make it possible to estimate the stability degrees of the isolated sub-systems accurately, may not be satisfactory when it came to taking into account the interconnection terms. Without being able to offer an exact response to this problem, the aim of this section is to give a method of approach which makes it possible to reduce the conservativeness of the stability conditions provided by the vectorial Lyapunov functions analysis.

4.5.1. Scalar function or comparison system

That which follows is valid for interconnected linear systems of the type :

$$(S_i) \quad \overset{\circ}{x}_i = A_i x_i + \Sigma A_{ij} x_j \qquad i=1,\ldots N \tag{4.49}$$

With A_i asymptotically stable, and P_i matrix solution to the Lyapunov equation

$$A'_i P_i + P_i A_i + D'_i D_i = 0$$

$(P_i > 0 \iff (A_i\ D_i)$ observable)

we form the Lyapunov functions associated with each isolated sub-system

$$v_i (x_i) = (x'_i P_i x_i)^{\frac{1}{2}}$$

satisfying

$$\lambda_{mi}^{\frac{1}{2}} \| x_i \| \leq v_i < \lambda_{Mi}^{\frac{1}{2}} \| x_i \|$$

$$\| \nabla v_i \| < \lambda_{Mi}^{\frac{1}{2}}$$

Evaluating the time derivative of v_i along the trajectories of (S_i) we get

$$\overset{\circ}{v}_i = -\frac{1}{2} \frac{x_i' D'_i D_i x_i}{(x'_i P_i x_i)^{\frac{1}{2}}} + \frac{(P_i x_i)'}{(x'_i P_i x_i)^{\frac{1}{2}}} \sum_{j=1}^{N} A_{ij} x_j \tag{4.50}$$

After majoration of (4.50) we get the differential inequality

$$\overset{\circ}{V} \leq AW$$

where $V = \{v_1, v_2, \ldots v_N\}'$; $W = [\|x_1\|, \|x_2\| \ldots \|x_N\|]'$ and

$$A = \left\{ a_{ij} = -\frac{1}{2} \lambda_m (D'_i D_i) \lambda_M^{-\frac{1}{2}} (P_i) \delta_{ij} + \lambda_M^{\frac{1}{2}} (P_i) \|A_{ij}\| \right\}$$

δ_{ij} being the Kronecker delta symbol.

The existence of a scalar Lyapunov function (the weighted sum of the Lyapunov functions associated with each of the sub-systems) with a negative time derivative along the trajectories of the system is guaranteed when A is a M matrix (Theorem 4.10). Let us take again the equation (4.50) and write it as :

$$\overset{\circ}{v}_i = -\frac{1}{2} \frac{x'_i D'_i D_i x_i}{x'_i P_i x_i} v_i + \frac{(P_i x_i)'}{(x'_i P_i x_i)^{\frac{1}{2}}} \sum_{j=1}^{N} A_{ij} x_j \qquad (4.51)$$

This enables us to write, after majoration, a new differential inequality of the comparison system type :

$$\overset{\circ}{V} \leqslant L V$$

where

$$L = \left\{ \ell_{ij} = -\alpha_i \delta_{ij} + \lambda_M^{\frac{1}{2}} (P_i) \lambda_m^{-\frac{1}{2}} (P_j) \|A_{ij}\| \right\}; \alpha_i = \min_{x_i} \frac{x'_i D'_i D_i x_i}{x'_i P_i x_i}$$

Here again, since L has positive off diagonal elements, stability is guaranteed if L is a M matrix.
It is easy to see from the Sevastyanov Kotelyanski conditions that the tests on the previous matrices A and L can be made in the same way on the following matrices $\bar{A}$ and $\bar{L}$

$$\bar{A} = \left\{ \bar{a}_{ij} = -\frac{\lambda_m (D'_i D_i)}{\lambda_M (P_i)} \delta_{ij} + \|A_{ij}\| \right\} \qquad (4.52)$$

and

$$\bar{L} = \left\{ \bar{\ell}_{ij} = -\alpha_i \theta_i^{-1} \delta_{ij} + \|A_{ij}\| \right\} \qquad (4.53)$$

where $\theta_i = (\lambda_M (P_i) \lambda_m^{-1} (P_i))^{\frac{1}{2}}$

Indeed, every pre and post multiplication of an M matrix by diagonal matrices with strictly positive elements provides another M matrix.
The matrices $\bar{A}$ and $\bar{L}$ are written in a way which is particularly interesting owing to the fact that the variable terms, depending on the choice of the Lyapunov functions, are isolated on the principal diagonal. This at first makes it possible to attempt a comparison between the two approaches; scalar (matrix A) and vectorial (matrix L), only to realise immediately that it is useless to wish to determine which is better, for

$$\alpha_i \geqslant \frac{\lambda_m (D'_i D_i)}{\lambda_M (P_i)} \quad \text{and} \quad \theta_i \geqslant 1$$

Then this written form makes it possible to set out for the two approaches two problems of optimisation, the solution to which would provide the least restrictive conditions of stability :

for the matrix $\bar{A}$

$$p_1 : \max_{D_i} (\lambda_m (D'_i D_i) \lambda_M^{-1} (P_i) \text{ subject to}$$

$$A'_i P_i + P_i A_i + D'_i D_i = 0$$

for the matrix $\bar{L}$

$$p_2 : \max_{D_i} (\alpha_i \theta_i^{-1}) \quad \text{subject to } A'_i P_i + P_i A_i + D'_i D_i = 0$$

We have here two rather difficult problems, the complexity of which depends essentially on the non-differentiability of the objective function. In the following sub-section we propose an approximate resolution of the problem p_2.

4.5.2. For a "good" comparison system

The obtaining of an "optimal" comparison system passes through the resolution of N (one for each sub-system) problems of optimisation of the form :

$$\max_{D} \alpha \theta^{-1} \text{ subject to } A'P + PA + D'D = 0 \qquad (4.54)$$

where

$$\alpha = \min_{x} \frac{x'D'Dx}{x'P x} \quad ; \quad \theta = \left[\lambda_M (P) \lambda_m^{-1} (P)\right]^{\frac{1}{2}}$$

As we have said, this problem presents the difficulty of having a cost function which is non-differentiable in relation to D. We are going to eliminate this difficulty by proposing an approximate problem for which we obtained good results.
An early difficulty attached to solving the problem (4.54) concerns the determination of the coefficient α. This difficulty can be overcome by noting that the problem (4.54) is equivalent to :

$$\max_{0 \leq \beta \leq \sigma \ ; \ D} \beta \theta^{-1} \quad \text{subject to} \qquad (4.55)$$

$$F'P + PF + D'D = 0; \ F = A + \beta I$$

σ being the degree of stability of the matrix A

$$\sigma = \max (\beta / (A + \beta I) \text{ stable})$$

Proof : let $(\bar{\alpha}, \bar{\theta})$ and $(\hat{\beta}, \hat{\theta})$ be the optimal values obtained for (4.54) and (4.55) respectively. Let us first show that $\bar{\alpha}\bar{\theta}^{-1} \geq \hat{\beta}\hat{\theta}^{-1}$. Indeed, the restriction (4.54) is written as $A'P + PA + D'D + 2\hat{\beta} P = 0$, therefore

$$\hat{\alpha}\,\hat{\theta}^{-1} = (\beta + \min_x \frac{x'D'Dx}{x'Px})\,\hat{\theta}^{-1} \leq \bar{\alpha}\,\bar{\theta}^{-1}$$

Let us now show that $\hat{\beta}\,\hat{\theta}^{-1} \geq \bar{\alpha}\,\bar{\theta}^{-1}$. The restriction in (4.55) can be written as

$$(A'+\bar{\alpha}\,I)\,P + P\,(A+\bar{\alpha}I) + D'D - 2\,\bar{\alpha}I = 0$$

By definition of α, the matrix $(D'D - 2\alpha I)$ is at least positive semi-definitive. Therefore there exists a matrix $\bar{D}$ such that

$$\bar{D}'\bar{D} = D'D - 2\,\alpha\,I$$

and therefore

$$\bar{\alpha}\,\bar{\theta}^{-1} = \bar{\beta}\,\bar{\theta}^{-1} \leq \hat{\beta}\,\hat{\theta}^{-1}$$

The problem (4.55) does not make it possible to get rid of the difficulty of non-differentiability mentioned at the beginning, but nevertheless, allowing θ to appear as the only awkward term, it makes it possible to suggest the resolution of a problem which is practically equivalent and defined from the following considerations.
For positive definite matrix $P \in R^{n \times n}$ we have

$$1 \leq \frac{1}{n^2}\;T_r(P)\;T_r(P^{-1}) \leq \theta^2 \leq T_r(P)\;T_r(P^{-1})$$

and

$$1 = \inf \theta^2(P) = \frac{1}{n^2}\;\inf T_r(P)\;T_r(P^{-1})$$

Let us now consider the following problem

$$\max_{0 \leq \beta \leq \sigma}\;\beta\,\theta^{-1}\,(P) \qquad (4.56)$$

where P is the solution obtained by resolution of

$$\min_D J(D) = T_r\,(P)\;T_r\,(P^{-1}) \text{ subject to } F'P+PF+D'D = 0$$

There appears a problem which is split into two sub-problems :
- a problem of minimisation in relation to D which, as we shall see, may be resolved using a direct method of the gradient type
- a problem of maximisation in relation to β which is resolved using unidirectional search by successive evaluations of the function objective $\beta\theta^{-1}$.
For the first sub-problem (4.56) we therefore have to calculate the gradient matrix dJ/dD, which is done by direct application of the theorem 3.1.
We get

$$\left[\begin{array}{l} \frac{dJ}{dD} = 2\,D\Lambda \\ F\Lambda + \Lambda F' + \psi = 0 \\ \psi = T_r\,[P^{-1}]\;I - T_r\,[P]\quad P^{-2} \end{array}\right. \qquad (4.57)$$

Proof : with $L = T_r\,[P]\,T_r\,[P^{-1}] + T_r[\Lambda'\,(F'P+PF+D'D)]$

we have directly

$$\frac{\partial L}{\partial D} = 2\,D\Lambda = \frac{dJ}{dD}$$

$$\frac{\partial L}{\partial \Lambda} = F'P + PF + D'D$$

$$\frac{\partial L}{\partial P} = F\Lambda + \Lambda F' + \frac{d}{dP}\,(T_r(P)\,T_r\,(P^{-1}))$$

and

$$\frac{d}{dP}\,(T_r(P)\,T_r(P^{-1})) = T_r\,(P^{-1})\,I - T_r\,(P)\,P^{-2}$$

Example : the problem (4.55) has been resolved for

$$F = \begin{bmatrix} 0 & 1 & 0 & 0 \\ 0 & 0 & 1 & 0 \\ 0 & 0 & 0 & 1 \\ -1 & -4 & -6 & -4 \end{bmatrix}$$

For an initial condition D = I with a stop test $\left\| \frac{dJ}{dD} \right\| < 10^{-4}$ the convergence of the algorithm was obtained after 25 iterations with

$$D = \begin{bmatrix} 0{,}08 & -0{,}07 & 0{,}18 & 0{,}05 \\ -0{,}03 & 0{,}59 & 0{,}15 & 0{,}55 \\ 0{,}26 & -0{,}01 & 0{,}65 & 0{,}38 \\ 0{,}16 & 0{,}96 & 0{,}76 & 1{,}20 \end{bmatrix}$$

The figure shows how the present terms in the inequality above evolve, showing for this example a considerable reduction in the value of θ .

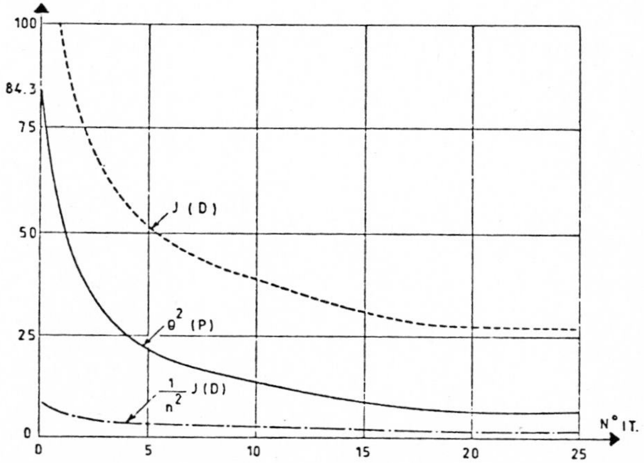

Note : the problem (4.55) is not strictly equivalent to the problem (4.54). Indeed in (4.55) appears not the minimisation of θ but of its upper limit. Nevertheless, it is a question of a systematic method to try to decrease the conservativeness of the vector Lyapunov functions approach. Other illustrative examples may be found in [GER - 79].

4.5.3. Discussion

By applying the preceding calculation to each of the sub-systems we are in a position to provide (or at least to try to do so) the least restrictive system of comparison possible.
Let us reconsider the problem brought up in paragraph (4.4) including the results of the preceding paragraph, and which concerns the synthesis of multilevel controls presenting a degree of robustness vis à vis structural perturbations. The algorithm presented may be summarised in the following way :
- sequential increase of the degree of stability of each isolated sub-system.
- obtainment at each step of the "optimal" system of comparison
- iteration until a stable system of comparison is obtained.

Is it possible to determine classes of systems for which such an algorithm would converge without fail? This is a particularly difficult problem which, as far as we know, has received no satisfactory reply. The limitations of this approach lie essentially in the use of a method of the Lyapunov type for stability tests.

To a given system of comparison one can make correspond an infinity of starting systems; for example, the interconnections in the vectorial approach are taken into account by the intermediary of their norm, which constitutes extremely aggregated information. Taking the opposite view of the question previously posed, it is conversely possible to supply a priori tests of non-convergence of the proposed algorithm. For this the idea is to look in the class of systems which will provide the same system of comparison for a system which is unstable.
This reflexion may be used for the determination of an upper bound for the ratio $\alpha\theta^{-1}$.

Theorem 4.18 :

$$\alpha\theta^{-1} < \min_{j=1,\dots n} \left\{ \|A_j\| , \|\bar{A}_j\| \right\}$$

where A_j ($\bar{A}_j$) represent respectively the jth row (column) of the matrix A (nxn).
The demonstration is done easily by constructing an interconnected system with two identical sub-systems whose global matrix is given by

$$\begin{bmatrix} A & I_j A \\ I_j A & A \end{bmatrix} \quad \text{or} \quad \begin{bmatrix} A & AI_j \\ AI_j & A \end{bmatrix}$$

then

$$\int_0^t e^{-\mu\theta} d\theta \,.\, Q \leq P(t) \leq \int_0^t e^{-\xi\theta} d \,.\, Q$$

where $\mu t = \max_x \frac{x'H(Q)\ x}{x'\ Qx}$, $\xi = \min_x \frac{x'H(Q)\ x}{x'\ Qx}$; $H(Q) = -(A'Q+QA)$

Proof : $P(t) = \int_0^t T(\theta)\ d\theta$ where $T(\theta) = e^{A'\theta}\ Q\ e^{A\theta}$

and let : $v(\theta) = x'\ T(\theta)\ x$

$\frac{dv}{d\theta} = -\ x'\ e^{A'\theta}\ A(Q)\ e^{A\theta}\ x$, which allows as limits

$$-\left(\max_x \frac{x'e^{A'\theta}\ H(Q)e^{A\theta}\ x}{x'\ H\ (Q)\ x}\right) v \leq \frac{dv}{d\theta} \leq -\left(\min_x \frac{x'e^{A'\theta}\ H(Q)e^{A\theta}x}{x'\ H(Q)\ x}\right) v$$

Let $-\ \mu v \leq \frac{dv}{d\theta} \leq -\ \xi v$

and therefore by integration

$$e^{-\mu\theta}\ x'Qx \leq v\ (\theta) \leq e^{-\xi\theta}\ x'\ Qx$$

Integrating once again, we get :

$$\int_0^t e^{-\mu\theta} d\theta\ x'Qx \leq \int_0^t v(\theta)\ d\theta = x'Px \leq \int_0^t e^{-\mu\theta} d\theta\ x'Qx,\ \forall x$$

with, for associated matrix

$$\begin{bmatrix} \alpha\,\theta^{-1} & \|A_j\| \\ \|A_j\| & \alpha\,\theta^{-1} \end{bmatrix} \quad \text{or} \quad \begin{bmatrix} \alpha\,\theta^{-1} & \|\bar{A}_j\| \\ \|\bar{A}_j\| & \alpha\theta^{-1} \end{bmatrix}$$

The matrix $\bar{L}$ cannot be a M matrix because the systems thus constructed have 2 identical rows (2 columns) and therefore, necessarily

$$\alpha\theta^{-1} \leq \|A_j\|^2 \qquad \text{or } \alpha\,\theta^{-1} \leq \|\bar{A}_j\|^2 \quad, \forall_j$$

APPENDIX

Determination of the function f (π) from the bounds of the Lyapunov equation :

Bounds of the solution of the Lyapunov equation : [GER, 79]

Let $\overset{o}{P}(t) = A'P + PA + Q;\quad P(0) = 0$

and let $\Delta A = \{ X > 0;\ A'X + AX < 0 \}$

For the algebraic equation

$$A'P + PA + Q = 0 \quad \text{we get} \quad \frac{Q}{\mu} \leq P \leq \frac{Q}{\xi}$$

Determination of f (π)

Let us return to problem (4.41). Each isolated sub-system has a control which minimises the criterion

$$J_i\,(u_i) = \frac{1}{2}\int_0 (x'_i\,D'_i D_i\,x_i + u'_i\,R_i\,u_i)\,dt$$

$$J_i = \frac{1}{2}\,x'_{i_o}\,P_i\,x_{i_o} \quad \text{where } P_i \text{ is the solution of}$$

$$A'_i\,P_i + P_i\,A_i - P_i\,B_i\,R_i^{-1}\,B'_i\,P_i + D'_i\,D_i = 0$$

The increase of the coefficient α_i of the matrix of the system of comparison is done by minimising the criterion :

$$\bar{J}_i(u_i) = \frac{1}{2}\int_0^\infty e^{2\sigma_i t}\,(x'_i D'_i D_i x_i + u'_i R_i u_i)\,dt$$

giving

$$\bar{J}_i\,(u_i) = \frac{1}{2}\,x'_i\,P_i x_{i_o} \quad \text{with } (A_i + \sigma_i I)\,P'_i + P_i\,(A_i + \sigma_i I) -$$

$$- P_i B_i R_i^{-1}\,B'_i P_i + D'_i D_i = 0$$

now (σ_i small)

$$\bar{J}_i - \overset{\circ}{J}_i \cong \frac{1}{2}\,x'_{i_o} \cdot \left[\frac{dP_i}{d\sigma_i}\right]_{\sigma_i = 0} x_{i_o} \cdot \sigma_i$$

$\dfrac{dP_i}{d\sigma_i}$ satisfies the Lyapunov equation

$$(A_i - B_i\,R_i^{-1}\,P_i^\circ + \sigma_i I)'\,\frac{dP_i}{d\sigma_i} + \frac{dP_i}{d\sigma_i}\,(A_i - B_i{}^{-1}\,B'_i\,P_{i_o} + \sigma_i I) + 2P_i^\circ = 0$$

It is easy to verify that $2\,P_i^\circ \in \Delta\,(A_i - B_i R_i^{-1}\,B'_i P_i^\circ + \sigma_i)$ and therefore

$$\frac{2P_i^\circ}{\mu} \leq \frac{dP_i}{d\sigma_i} \leq \frac{2P_i^\circ}{\xi}$$

with

$$\mu = \max_x \frac{x'_i(D'_i D_i + P_i^\circ B_i R_i^{-1}\,B'_i\,P_i^\circ)\,x}{x'_i\,P_i^\circ\,x_i}\;;\quad \xi = \min_x\,(\;.\;)$$

Therefore we get

$$\frac{\mu_i}{\sigma_i} \leqslant \frac{\bar{J}_i - J_i^{\circ}}{\bar{J}_i} \leqslant \frac{\sigma_i}{\xi_i}$$

which makes it possible to determine limits to the degree of sub-optimality of the isolated sub-systems. f (π) may be chosen in such a way as to minimise the upper limits of the degree of sub-optimality, for example

$$f(\pi) = \sum_{i=1}^{n} \xi_i^{-1} \pi_i$$

REFERENCES

BAI - 66 BAILEY F.N., "The application of Lyapunov's second method to interconnected systems", SIAM Journal of control, 3, 443-462, 1966.

BEL - 62 BELLMAN R., "Vector Lyapunov functions", SIAM Journal of control, 1, n° 6, 32-34, 1962.

BIT - 76 BITSORIS B. and C. BURGAT, "Stability conditions and estimates of the stability region of complex systems", International Journal of Systems Science, 7, 911-928, 1976.

BIT - 78 BITSORIS B., "Principe de comparaison et stabilité des systèmes complexes", Thèse d'Etat, Université Paul Sabatier, Toulouse, 1978.

BUR - 78 BURGAT C., J. BERNUSSOU, Lj.T. GRUJIC, J.C. GENTINA and P. BORNE, "Sur la stabilité des systèmes de grande dimension : les perturbations structurelles arbitraires et périodiques", RAIRO, vol. 12, n° 30, 245-267, 1978.

COR - 60 CORDUNEANU C., "Sur la stabilité asymptotique,". Revue de mathématiques pures et appliquées, 6, pp. 573-576, 1960.

FIE - 62 FIEDLER M. and V. PTAK, "On matrices with non positive off diagonal elements and positive principal minors", Czch. Math. Journal, 12, 382-400, 1962.

GER - 79 GEROMEL J.C., "Contribution à l'étude des systèmes dynamiques interconnectés : aspect de décentralisation". Thèse de Doctorat d'Etat, 903, Université Paul Sabatier, Toulouse, 1979.

GER - 79 GEROMEL J.C. and J. BERNUSSOU, "Stability of two level control schemes subjected to structural perturbations", International Journal of Control, vol. 29, 2, 1979.

GER - 79 GEROMEL J.C. and J. BERNUSSOU, "On bounds of Lyapunov's equation", IEEE Transactions on Automatic Control, A.C.-24, 3, 1979.

HAH - 67 HAHN W., "Stability of motion", Springer Verlag, 1967.

KAL - 60 KALMAN R.E. and BERTRAM J.E., "Control system analysis and design via the second method of Lyapunov - Part. I : continuous time systems", Trans. ASME J. Basic Eng., 82, 371-393, 1960.

MAT - 71 MATROSOV V.M., "Vector Lyapunov functions in the analysis of non-linear interconnected systems", Symposia Math., vol. 6, Academic Press, N.Y., 1971.

MIC - 77 MICHEL A.N. and R.K. MILLER, "Qualitative analysis of large scale dynamical systems", Mathematics in science and engineering, vol. 134, Academic Press, 1977.

SAN - 78 SANDELL N.R., P. VARAIYA, M. ATHANS, M.G. SAFONOV, "Survey of decentralized control methods for large scale systems", IEEE Transactions on Automatic Control, AC-23, 2, April 1978.

SIL - 76 SILJAK D.D. and M.K. SUNDARESHAN, "A multilevel optimization of large scale dynamic systems", IEEE Trans. Aut. Control, AC-21, 79-84, 1976.

SIL - 78 SILJAK D.D., "Large scale dynamic systems : stability and structure, Series in system science and engineering", North Holland, 1978.

WAZ - 50 WAZEWSKI T., "Systèmes des équations et des inégalités différentielles ordinaires aux deuxièmes membres monotones et leurs applications". Annales de la Société Polonaise de Mathématiques, 23, 112-166, 1950.

WEI - 73 WEISSEMBERGER S., "Stability regions of large scale systems", Automatica, 9, 653-663, 1973.

CHAPTER 5

ON THE REAL TIME MANAGEMENT OF TELEPHONE NETWORKS

5.1 Introduction

There is often cause to say that progress in technology, with the means it offers, gives encouragement and challenge to methodological studies. In the example of telephone networks, which are the subject of this chapter, the appearance of digital tools, rapid and powerful as to their capacity, makes it possible to consider dynamic management policies for the calls routing. Dynamic management signifies adaptative control which by use of real time measurements, changes its parameters in order to react to perturbations such as traffic overloads, communication links breakdowns,...

Most studies in this field have so far assumed above all a static aspect of planning. This generally meant large optimization problems for the specification and definition of the network structure evolution in order to adapt it in the best possible way to traffic forecasts and to technological developments. The accepted criteria were varied. The first to be mentionned is the economic criterion and then quality of service, safety ... [RES - 80]. In most of these studies, optimisation of the structure was done for a simple fixed policy for calls routing termed the "alternate routing policy". We shall introduce this later.

The complexity aspects for the determination of a real time management policy are manifold. Firstly this is a very large process, for it is geographically widespread on a regional or a national scale. To these large physical dimensions correspond, of course, a large mathematical dimension due to the fact that there are a large number of "goods", flows to be transported. A further characteristic which adds to this complexity is the random nature of the signals and phenomena to be dealt with.

All these reasons mean that, in spite of the development in power of data processing methods and numerical tools, it is unrealistic to think of a purely centralized management mode. The data processing and numerical tool must be used rather with the view for a shared assignment of the control over the whole network. This division, partition in the control amounts to a distribution of "intelligence" at the network nodes.
This view corresponds to the partition or decomposition approach which is a now classical way to tackle, large and complex control problems at least when the "real time" aspect is con-

cerned. This decomposition can show the two aspects of spatial decomposition of the control structure (and this to adapt as well as possible to the process structure itself) and vertical or functional decomposition, which corresponds to a time horizon hierarchisation of the different level tasks (and this to better apprehend the adaptative functions needed for a control in a random environment). Spatial decomposition is generally relatively well defined by the physical structure of the process under consideration. In the case of telephone networks, for example, the sub-systems are well identified with the commutation centers and the transmission links define the interconnection structure between these sub-systems. As regards the vertical decomposition of the control task, the problem is not as clear-cut and its solution contains a good measure of heuristics (or even common sense) based, more often than not on simplifying hypotheses and approximations. Validation of this type of approach from a mathematical point of view is relatively difficult [FOR - 78]and sometimes can only be totally achieved at the simulation, or even the experimentation test.

These two aspects of decomposition will be approached here and a two-level management structure will be proposed for the routing of calls in the telephone network. The essential aim of this study has been to show the potentiality of modern control engineering ideas and methods to tackle this type of complex problem, and not to provide a universal solution (if such exists) to the problem of the adaptive routing of calls in a telephone network.

One stage which is of prime importance in mastering complex systems is the analysis - modelling - formulation phase. It is indeed important to adapt the model and the formulation of the problem well to the aim desired and this must at the same time meet conflicting demands, which are accuracy and quality of results on the one hand and practical feasibility on the other. Here again this compromise cannot be solved generally in a mathematically accurate way, but it is rather the outcome of a series of tests, of simplifications, of "coming and going" until satisfactory results are obtained. We shall be able to illustrate this point in this chapter.

5.2 The process, the traffic, the problem

This section is devoted to a deliberately very simplified presentation of the process and of the problem of call routing in a telephone network.

5.2.1. Structure

Most national telephone networks have a hierarchic structure, with a hierarchy which makes possible the discrimination of local or regional centres (with a limited zone of influence) from those intended for the handling of inter-regional (long distance) calls. In the French system there are four hierarchical levels, the two upper levels defining the "inter-city (inter-regional) network", which is schematically shown in figure 5.1

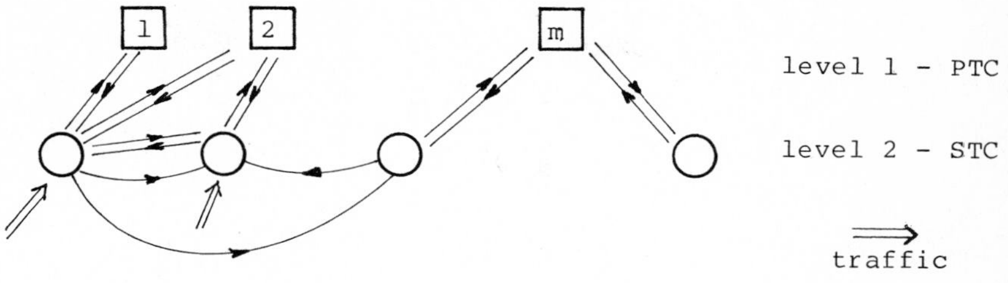

Figure 5.1

This network comprises "secondary transit centres " (STC) and "principal transit centres" (PTC). At the moment all connections are unidirectional, ie. for connections between STC and PTC the distinction can be made between "upstream" connections (STC → PTC) and "downstream" connections (PTC → STC). The traffic (flow of calls) arrives at level 2 (STC), passes if necessary through level 1, to be routed towards the STC. One practical rule stipulates that if a direct connection exists between two STC it must always be used preferentially as first choice (rule of minimisation of the number of links used by one call). Therefore no management need be made of this type of connection. This is why we shall consider a network without inter STC connection. The unidirectionality of the connections enables us then to modify diagram 5.1 in an exploded representation defining a graph where the origin, transit and destination nodes are distinct from each other.

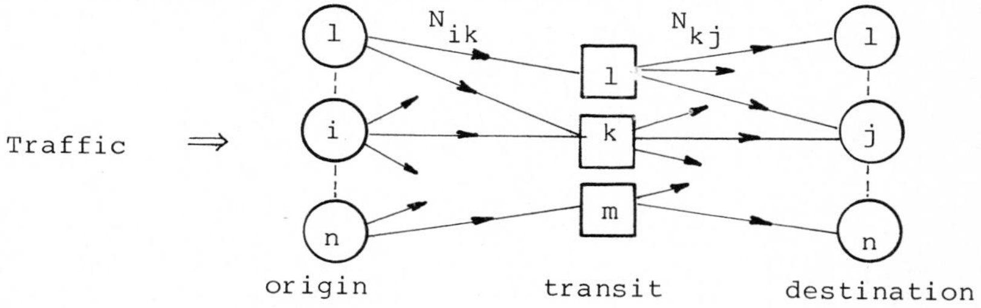

Figure 5.2

For simplicity of notation in the rest of the text the letter i will stand for the origin nodes, j the sink nodes and k the transit nodes. Thus N_{ik} and N_{kj} will represent respectively the capacities of the source links $STC_i \longrightarrow PTC_k$ and the sink links $PTC_k \longrightarrow STC_j$.

5.2.2. The traffic

To define the traffic load of the network we must characterise

two variables, which are
- the quantity, the number of calls (per unit of time)
- the duration of communication (duration of service).

The following hypotheses are widely used in the study of service networks or queuing networks [KLE - 75].

H1 : the input of calls process is likened to a Poisson process :

$$F_k\ (t, \theta) = \frac{(\lambda\theta)^k}{k!}\ \exp\ (-\lambda\theta) \tag{5.1}$$

where $F_k\ (t, \theta)$ is the probability that k call inputs occur between t and t+ θ and λ is the rate of input of the calls (parameter of the poisson law).
Some properties of this process will be used later and we recall them briefly without proof.
- The probability of input of more than one call between t and t+dt is infinitesimal in dt.
- The superposition of independent (Poisson) processes is a poisson process whose rate of input is the sum of the input rates of the elementary processes.
- The variance is equal to the average
- The superposition of a large number of "independent renewal processes" tends towards a poisson process.

H2 : The service law (duration of communication) is a negative exponential of parameter T :

$$F(t) = 1 - \exp\ (- t/T) \tag{5.2}$$

in which F(t) is the probability that the duration of conversation is less than t.
It is demonstrated that the probability for a conversation which has already lasted t to have a duration of between t and t+dt is equal to dt/T. This probability is independent of t (memoryless property). This means that the only information necessary to predict the future concerns the number of calls present at a given moment. For example the probability for any call from k present to end during an interval dt is equal to kdt/T.

In the modelling which we shall approach later we shall of course discriminate between the various call flows in terms of the origin-destination couple (i,j). Therefore to each flow (i,j) will be associated a parameter λ_{ij} (parameter of the poisson law - rate of input of calls coming from the STC_i to the destination STC_j). On the other hand, we are making the simplifying and relatively realistic hypothesis that the average duration of conversation is the same for all the pairs (i,j).

In the telephone field there exists the interesting notion of offered traffic which, by definition, is the traffic which would be dealt with by a system with infinite capacity. Thus for a poisson type input of parameter λ and an exponential duration of parameter T the traffic offered (average dealt with by a link of infinite capacity) is $A = \lambda . T$

This quantity is expressed in Erlangs (unit of traffic).
We shall speak subsequently of the matrix of traffic offered for a network with source nodes i and sink nodes j. Let us say simply that it is the matrix whose elements A_{ij} are the traffic offered between the nodes i and j.

$$A_{ij} = \lambda_{ij} \cdot T$$

5.2.3. The problem - its criteria

The control, ie. the policy of routing calls in the network, must guarantee a sufficient quality of service. The choice of a criterion for the evaluation of this quality of service is in itself a delicate problem. This is a fairly common characteristic for complex systems and it is dependent on several factors. In this sense the complete problem if it could be formulated, should be formulated in the form of a multiple criteria problem.

First of all there is the economic aspect, which is broadly taken into account at the planning level. It corresponds, crudely, to the question of the choice of minimal investments for a required performance. Of course this problem is not at all independent from that of the definition of a real time management policy. Nevertheless we shall not approach it here but shall content ourselves with mentioning in conclusion how it could be integrated in the proposed management structure.

The planning stage also covers the aspect of making the network secure vis à vis perturbations such as links ruptures, exchange breakdowns etc. In general the solution is given in terms of multi-routing (several independent routes for each pair (i,j)). The policy of "load sharing" proposed here satisfies this type of constraint.
The principal parameter which will help us for the establishment and evaluation of the control is the calls loss parameter. Here again we must distinguish between the global loss (number of calls lost over the whole network) and the loss per flow, per pair (i,j). Indeed, since this is a public service it is a good thing for the quality of service to be more or less uniform for every pair ((i,j), ie. for every subscriber. For this we talk of point to point efficiency E_{ij}, which is the ratio of the number of calls loss over the total number of calls for the flow (i,j). This is the "global losses" criterion which will be used here, the efficiencies appearing then as constraints to be verified a posteriori for the qualification of the proposed solutions.

5.3. Functional decomposition of the control

Before moving on to the parts relating to modelling and optimisation we should like to introduce here the philosophy governing, and to our mind justifying, the solution recommended.

As is stated in the introduction to the chapter, the complexity of the study is due to the dimension of the problem, both geographical and mathematical (number of explicative variables) and also to the random nature of the signals and phenomena to be controlled. These two reasons make any purely centralised solution for control unacceptable. It would be economically prohibitive to want to centralise all the information and even if it could be done, does there exist an instrument of calculation and decision-making powerful enough to digest all this information?

Nevertheless, this being a problem which, using the hypotheses made about the traffic, is Markovian we can consider the "optimal" control to be of the state "feedback" type, and even complete state feedback where all the components of the state intervene. Knowing that this ideal situation which requires knowledge of the complete system cannot be met, one can try to move away from it as little as possible (or at least reasonably so). Although detailled global knowledge of the short term of the global system cannot be obtained we can try to obtain less detailled knowledge which provides only medium term information (principal component).
Returning to the problem which concerns us here, if we look at some characteristic recordings of the traffic offered to the network, it is reasonable to consider :

- an average traffic which varies "slowly", the time scale being the hour.
- rapid variations, fleeting "perturbations" around the quasi-stationary component

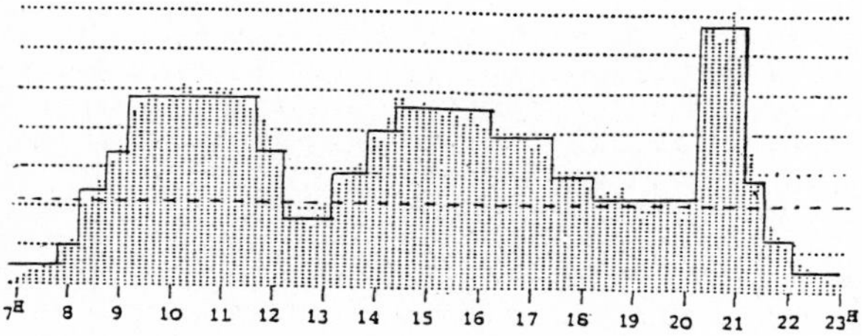

Figure 5.3

This analysis, although brief, also proposes to distinguish, for the elaboration of the control, between two types of problems. The first consists in considering only the fundamental part of the characteristic of the offered traffic which, slowly varies and can be approximated by a piecewise constant characteristic. Inside these pieces, for a constant average value of the offered traffic (hypothesis of stationarity) we shall apply ourselves to determining the optimal strategy for the distribution of the load on the network. As regards the load-sharing technique, this therefore consists in finding the routing percentages for each flow on each route.

The second problem concerns the truly real time aspect and consists of the definition of tactical actions for the control of transitory variations around the average value taken into account in the first problem. The objective is to force the system to follow, on average, the strategic indications defined by the first level problem by smoothing the transient perturbations on the traffic. From this the local call routing actions are defined, taking into account, for practical purpose, a decentralisation constraint in order to take into account the structure of the network. The type of problem which we shall deal with here will be of the regulator type.

To sum up, the proposed structure here for the real time management of call routing in telephone networks is a two-level structure which integrates :

- a strategic level which, making the assumption of quasi-stationnarity applies itself to determining the optimal asymptotic situation in relation to a criterion which evaluates the losses due to saturation on the transmission links. We shall see that a problem of this kind can be resolved in a centralised fashion.
- a tactical level which, integrating the strategic aims (co-ordination) acts in real time to remove transitory overloading as quickly as possible . It is a level of real time actions and has a decentralised structure.

Note : for a system of this sort, even local actions of real time routing cannot be the result of "calculations" and of individualised decisions for each call. In practice, routing laws are chosen a priori and the auto-adaptive, dynamic aspect of control then consists of modifying the parameters of these laws in real time.

The most widespread law at the present time is the one called the "alternate routing policy". It is almost certainly because of the electro-mechanical technology that this law was formerly imposed, since it requires simple commutation actions. The outgoing links of a commutation centre are, for a given flow classed as first choice, second choice etc. A call belonging to the flow coming into this centre will be systematically routed towards the first choice link and if it is blocked, towards the second choice etc. This grouping can be translated using routing tables for each flow (pair i-j). At the present time these tables are fixed. A dynamic management can therefore consist of an adaptation of the grouping of routes in terms of the state of the network.

More recently, and prospectively, the idea of a "load sharing policy" has been introduced. This consists simply of sharing the load on all the links possible coming out of a commutation center. The realisation of this law does not seem impossible in the perspective of digital commutation exchanges (random generators, shift registers). Of course, for this law, dynamic management consists of adapting the parameters (the percentages) of the load sharing in terms of the measurements carried out on the network.

It is this last law which interests us particularly here, whilst at the same time we shall try to make a comparison with the first one.

5.4. Modelling

5.4.1. Elementary study

As a preliminary, let us consider an isolated trunk group of capacity N to which is offered a traffic (λ ,T) and let

i be the instantaneous state of the link (number of calls présent)

and

$P_i(t)$ the probability that the state **of the link is equal to i.**

It is well known that such a process can be modelled by means of a Markov chain and even, taking into account the hypotheses made, as a birth-death process (in an infinitely small interval of time dt only the transitions between adjacent states are possible). In the notations used, a state transition rate diagram is given by figure (5.4).

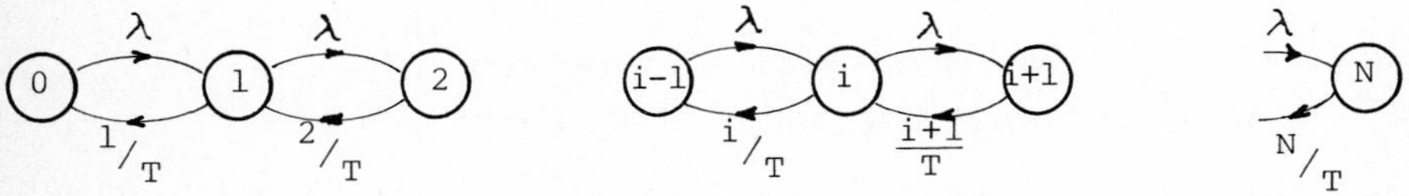

Figure 5.4

It is thus possible to write a differential system concerning the probabilities $P_i(t)$; we have

$$\mathring{P}_i(t)=\lambda P_{i-1}(t)-(\lambda^*+\frac{i}{T})P_i(t)+\frac{(i+1)}{T}P_{i+1}(t);\ i=0,\ldots N \qquad (5.3)$$

with $\lambda^* = \begin{cases} \lambda \text{ for } i = 0,1, \ldots N-1 \\ 0 \quad i=N \end{cases}$

and $P_{-1}=0$

System (5.3) is completed by the equation of normalisation

$$\sum_{i=0}^{N} P_i(t) = 1$$

Of special interest for what is to follow is the study in the stationary case, $\dot{P} = 0$. It is relatively easy to see that system (5.3) with $\dot{P} = 0$ and $\sum P_i = 1$ admits a unique solution which can be written as

$$P_i = \frac{(\lambda T)^i}{i!} P_o \quad ; \quad P_o = \left[\sum_{i=o}^{N} \frac{(\lambda T)^i}{i!} \right]^{-1} \qquad (5.4)$$

In particular we get the blocking probability in steady state as

$$P_N = \frac{(\lambda T)^N}{N!} \left[\sum_{i=o}^{N} \left(\frac{\lambda T}{i!}\right)^i \right]^{-1} = E[\lambda T, N] \qquad (5.5)$$

Let us note that here we again come across the product λT termed traffic offered and denoted by A. E $(\lambda T, N)$ is called the Erlang function. Since it will often be used later, let us repeat that it concerns the formula which, knowing the traffic offered and the capacity of the link makes it possible to evaluate the blocking probability in steady state.

$$E(A,N) = \frac{A^N}{N!} \left[\sum_{i=o}^{N} \frac{A^i}{i!} \right] \qquad (5.6)$$

A single calculation would make it possible to show that the passed traffic (average value of the state of the trunk) is written as

$$m = A \left[1 - E(A,N)\right]$$

and therefore the losses (average of the calls blocked) as A E (A,N).

5.4.2. Markovian model for a star type network

Let us consider the network in figure (5.2) and, with a load sharing type of control, having parameters α_{ikj}; α_{ikj} is the percentage of traffic between the STC i and j routed by the PTC_k. We are making the hypothesis that this flow sharing conserves the Poisson character for each portion (this is true for a shift generator with uniform distribution). This being assumed, and for given α_{ikj} we can see that it is possible to decompose the formulation by transit centre k and obtain, in a similar way to the case with only one trunk, a Markovian model bringing into probabilities of the type (x_{ikj} integers)

$$P(x_{1k1}, x_{1k2}, \dots x_{1kn}, x_{2k1}, \dots x_{ikj}, \dots x_{nk1}, \dots x_{nkn})$$

$= \text{Prob}\Big[\dots, x_{ikj}$ calls from i to j and passing through k are present at the instant t $\dots \Big]$

Thus we can write a differential system involving these probabilities

$$\begin{aligned} \dot{P}(x_{1k1} \dots x_{ikj} \dots x_{nkn}) = & \sum_{(i,j)} \lambda_{ikj} P(\dots x_{ikj}-1 \dots) \\ & - \sum_{(i,j)} \left(\lambda^*_{ikj} + \frac{x_{ikj}}{T}\right) P(\dots x_{ikj} \dots) \\ & + \sum_{(i,j)} \frac{(x_{ikj}+1)}{T} P(\dots x_{ikj}+1 \dots) \end{aligned} \qquad (5.7)$$

where $\lambda_{ikj} = \alpha_{ikj} \lambda_{ij}$

This differential system is defined for every n^2 uple (x_{1k1}, ... x_{ikj} ... x_{nkn}) which satisfies constraints linked to the capacities of the trunk groups constituting the network :

$$(x_{1k1}, \dots x_{ikj} \dots x_{nkn}) \in \mathcal{D}$$

where

$$\mathcal{D} = \left\{ (x_{1k1} \dots x_{ikj} \dots) \Big/ \sum_{j=1}^{n} x_{ikj} \leq N_{ik} \text{ (source links) and } \sum_{i=1}^{m} x_{ikj} \leq N_{kj} \text{ (sink links)} \right\}$$

Moreover

$$\lambda^*_{ikj} = \begin{cases} 0 \text{ if } \sum_{j=1}^{n} x_{ikj} = N_{ik} \text{ or } \sum_{i=1}^{n} x_{ikj} = N_{kj} \\ \lambda_{ikj} \text{ elsewhere} \end{cases}$$

This formulation already gives an idea of the formidable complexity of the Markovian model for a network of adequate size and it is obvious that it cannot be of any use for the purpose of real time control.

Let us note, however, that in the stationary case, the probabilities are provided by the product form :

$$P(x_{1k1}, \dots x_{ikj} \dots x_{nkn}) = \prod_{i,j} \frac{(\lambda_{ikj})^{x_{ikj}}}{x_{ikj}!} P(0,0,\dots 0) \quad (5.8)$$

to approach formula (5.4) concerning a single circuit. P(0,0, ...0) can (in theory!) be calculated by the expedient of the constraint of normalisation.

$$\sum_{\mathcal{D}} P(x_{iki}, \dots x_{ikj} \dots x_{nkn}) = 1$$

When the size of the network (number of links and their capacity) increases, the resolution of this combinatoring problem is a very important task if it is to be used in a control scheme. Its interest lies in the fact that it can be used for the validation of approximated models, as is done later.

5.4.3. Model of the mean for one trunk

The Markovian representation proves to be too complex to be used in practice. So one idea which then comes naturally to mind is to try to model the random process which constitutes the telephone traffic by using its first statistical moments. The simplest way is to apply oneself to working out a model of the first order, a model of the mean.

The mean value of the traffic passed by one trunk group is written as :

$$m(t) = \sum_{i=o}^{N} i \, P_i(t)$$

Using (5.3) we find the dynamic equation

$$\dot{m}(t) = - \frac{m(t)}{T} + \lambda \, (1 - P_N(t)) \qquad (5.8)$$

This equation is of little use as no simple analytical expression exists which makes it possible to determine $P_N(t)$ accurately, except in the stationary case where we have

$$P_N = E\ (A,N) \quad \text{and } m = A \left[1 - E\ (A,N)\right] \qquad (5.9)$$

The expressions (5.9) suggest a way of estimating $P_N(t)$, by extending them to the non-stationary case in the following way : for a given value of $M(t)$ a fictitious traffic offered $y\ (t)$, is determined which would give in the stationary case a passed traffic equal to $m(t)$.

$$m(t) = y(t) \left[1 - E\ (y(t),\ N)\right]$$

the estimation of $P_N(t)$ then being taken as equal to $E\ (y(t),\ N)$.
An approximation of this sort, which consists of extending an accurate expression for the stationary case to the transitory case provides a non-linear model of the first order.

$$\left[\begin{array}{l} \dot{m}(t) = - \dfrac{m(t)}{T} + \lambda \left[1 - E\ (y(t),N)\right] \\[2ex] m(t) = y(t) \left[1 - E\ (y(t),\ N)\right] \end{array}\right. \qquad (5.10)$$

Figure (5.5) presents a comparison between $P_N(t)$ obtained by integration of the Markovian model and $P^*(t) = E\ (y,\ N)$, the chosen estimator.

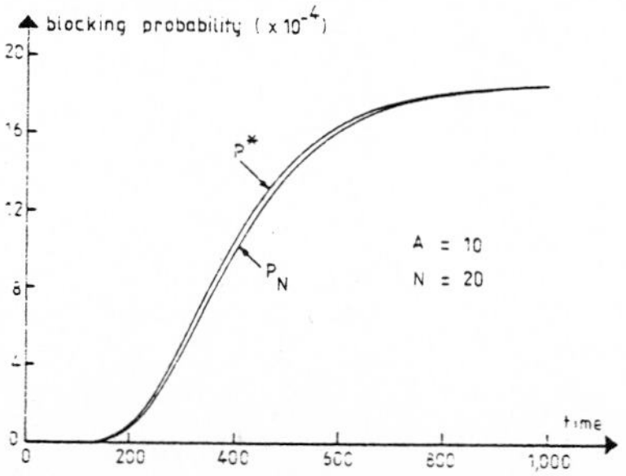

Figure 5.5

Other validations by numerical simulation have been made showing a relatively good quality of the estimator thus defined which tends towards the accurate probability of blockage for the stationary case. It is to be noted, and this is intuitively comprehensible, that this estimator makes an over-estimation of the blocking probability, which is also a positive point for what is to follow [GAR - 80].

5.4.4. Model of the mean for one network

In practically the same way, although the calculations are more delicate, by using the formulae (5.7) we can write down the dynamic equations relating to the averages of the traffic moved by the source links and the sink links [LEG - 81].

The equations obtained are similar to (5.8) and we have :

$$\begin{cases} \dot{m}_{ik} = -\dfrac{m_{ik}}{T} + \sum\limits_{j=1}^{N} \lambda_{ikj} \left[1 - P_{ikj}(t)\right] \\ \dot{m}_{kj} = -\dfrac{m_{kj}}{T} + \sum\limits_{i=1}^{N} \lambda_{ikj} \left[1 - P_{ikj}(t)\right] \end{cases} \tag{5.11}$$

where $P_{ikj}(t)$ represents the probability of blockage at instant t of the route ikj made up of the links ik (source) and kj (sink).

$$P_{ikj}(t) = \text{Prob}\ [\text{ik saturated or kj saturated.}]$$

Even more than in the case of one single trunk-group, it is unrealistic to try to calculate $P_{ikj}(t)$ accurately. Furthermore, there is no analytical formula, not even in the stationary case, which makes it possible to calculate P_{ikj}. We must therefore try to link P_{ikj} (probability of blockage of one route) to the P_{ik}, P_{kj} (probabilities of blockage of each of the trunks constituting the route). This is a very arduous task for in all strictness P_{ik} and P_{kj} are not independent. However, in the case of a large network there is a mixture of a large number of independent flows on each link. For example, in the case where the load sharing is total, n independent flows with destination the STC_j pass through the link (ik); and the same number having as their origin the STC_i pass through the link kj; one single flow passes at the same time through (ik) and (kj). This statement leads to making the approximation which consists of saying that P_{ik} and P_{kj} are independent.
We then have

$$P_{ikj} = P_{ik} + P_{kj} - P_{ik} P_{kj}$$

and the system (5.11) is written as

$$\begin{cases} \dot{m}_{ik} = -\dfrac{m_{ik}}{T} + (1 - P_{ik}) \sum\limits_{j=1}^{N} \lambda_{ikj} (1-P_{kj}) \\ \dot{m}_{kj} = -\dfrac{m_{kj}}{T} + (1-P_{kj}) \sum\limits_{i=1}^{N} \lambda_{ikj} (1-P_{ik}) \end{cases} \tag{5.12}$$

This expression makes it possible to try again here the approximation which had been made in the case of one single trunk and which consists of using the Erlang formula of the stationary case to estimate the probabilities of blockage of the trunks (ik) and (kj).
Finally we get the following non-linear system :

(5.12)

with

$$\begin{cases} P_{ik}=E(y_{ik},N_{ik}) ; \quad m_{ik}=y_{ik}\left[1-E(y_{ik},N_{ik})\right] \\ P_{kj}=E(y_{kj},N_{kj}) ; \quad m_{kj}=y_{kj}\left[1-E(y_{kj},N_{kj})\right] \end{cases} \tag{5.13}$$

An example : let us consider the simple cell of figure (5.6)

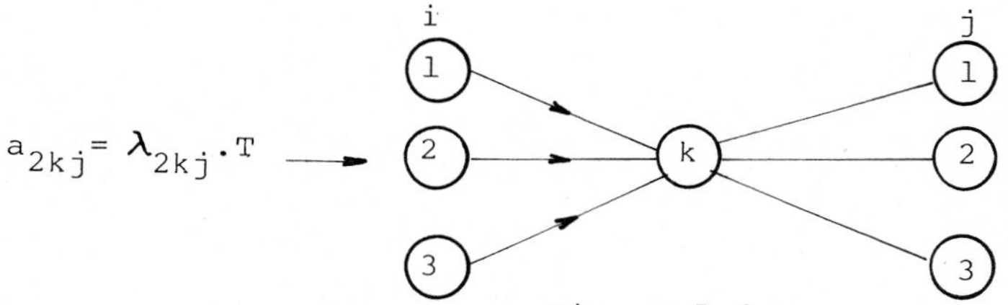

Figure 5.6

the offered traffic

i→j	1	2	3
1	0,5	0,6	1,3
2	0,5	0,8	1,4
3	0,8	1,2	2,7

and the links capacities

i	1	2	3
↓ K	2	4	6

j	1	2	3
↑ k	3	4	6

For the steady state, the two tables below provides the blocking probabilities and the mean passed traffic, obtained with the preceeding model (5.13) with $\mathring{m}=0$, and with the exact markov model (5.7)

Table 5.1 - Blocking probabilities

ikj	1k1	1k2	1k3	2k1	2k2	2k3	3k1	3k2	3k3
Markov	0,4489	0,3926	0,3817	0,2895	0,2097	0,1964	0,3225	0,2511	0,2339
model (5.13)	0,4588	0,4002	0,3924	0,2998	0,2240	0,2139	0,3369	0,2652	0,2559

Table 5.2 Mean traffic

	markov	model (5.13)
m_{1k}	1,074	1,057
m_{2k}	2,023	1,984
m_{3k}	3,998	3,901
m_{k1}	1,583	1,553
m_{k2}	1,984	1,949
m_{k3}	3,527	3,439

5.4.5. Conclusions concerning modelling

We have therefore an analytical model in the first moment (mean) to describle the traffic behaviour in telephone network. With the standard assumption of Poissonian offered traffic and from the a priori knowledge of the structure of the network and the routing laws at each switching center, it provides a relatively good approximation for the state occupancy of the transmission links and the blocking probabilities. This approximation is not only valid in the steady state but seems also to be quite accurate in the transient state. In addition, this approximation is always an upper bound for the blocking probabilities and losses. This is an important characteristic for a practical use.

The quality of the results obtained can even, at first sight, be very surprising for a method of the first moment. This can be explained by the fact that the approximations made are good in the Poissonian case and that the policy of load sharing conserves this Poissonian character which is characterised in particular by the fact that mean equals variance. It is not possible here to go into detail in this relatively technical problem of modelling, but we can say that modelling for a policy of alternate routing presents a greater complexity. The main reason lies in the fact that the traffic then begins to present a survariant character, which then poses certain problems for modelling. From existing models we can mention Wilkinson's theory of the equivalent trunk [WIL - 56]and the Interrupted Poisson Process (I.P.P.) of Kuczura [KUC - 73],[KUC - 77]. These are model which bring into play the first 2 and the first 3 moments respectively. Indeed, it can be said that their relative complexity makes them unusable in a dynamic application and that their field of utilisation is situated above all at the level of planning problems. Recently, one moment models have been derived to take into account the overflowing procedure and the overflow traffics[MAN - 79],[LEG - 82] . They provide means, in a more or less heuristic way, in order to keep in the model some informations concerning the deviations of the observed traffics from the poisson.

5.5 Determination of control

In section (5.3) we attempted a general and philosophical presentation of the proposed structure of control. Let us remember simply that we are talking about a two-level structure, the first level being applied to the quasi-static aspect for the routing of calls and based on the slow variations in average of the matrix of offered traffic (time scale of the order of the hour), the second level carrying out a type of regulation having bearing on the transitory and fleeting variations of the traffic around its average value. The aim of this section is to explain how each of the levels has been determined by explaining the associated problems (of optimisation) and mentioning the methods and algorithms used for their resolution. Here again, for obvious reasons of space, we cannot really enter into detail.

5.5.1. First level : a multiflow problem

The almost stationary nature of the offered traffic matrix leads us to define a static problem of optimisation of the multiflow type during each time interval in which it will be assumed that the matrix of offered traffic (A_{ij}) remains constant. Let us choose as the optimisation cost the global losses. Such a criterion is written (in stationary state) as

$$J = \sum_{(i,j)} \sum_{k} \alpha_{ikj} A_{ij} (1 - P_{ikj}) \qquad (5.14)$$

For $\alpha_{ikj} A_{ij}$ being the traffic offered to the route ik, kj; (a portion of the flow (i,j)); the term $\alpha_{ikj} A_{ij} (1 - P_{ikj})$ represents the losses related to this portion. Adding up in relation to k provides the losses for the flow (i,j) and finally, adding up on all the flows gives the overall losses for the network.

Going back to approximation (realistic in the case of a large network) consisting of making the independence assumption of the blocking probabilities for the source and sink trunk-groups ($P_{ikj} = P_{ik}+P_{kj}-P_{ik}P_{kj}$), the problem of optimisation is written as :

$$\min_{\alpha_{ikj}} \left\{ J = \sum_{(i,j)} \sum_{k} \alpha_{ikj} A_{ij} (1-P_{ik})(1-P_{kj}) \right\} \qquad (5.15)$$

subject to

$$\begin{cases} m_{ik} = (1-P_{ik}) \sum_j \alpha_{ikj} A_{ij} \ (1-P_{kj}) \\ m_{kj} = (1-P_{kj}) \sum_i \alpha_{ikj} A_{ij} \ (1-P_{ik}) \\ P_{ik} = E\ (y_{ik}, N_{ik});\ m_{ik} = y_{ik}\ (1-E(y_{ik}, N_{ik})) \\ P_{kj} = E\ (y_{kj}, N_{kj});\ m_{kj} = y_{kj}\ (1-E(y_{kj}, N_{kj})) \end{cases} \tag{5.16}$$

and

$$\sum_{k=1}^{m} \alpha_{ikj} = 1\ ; \quad \alpha_{ikj} \geq 0 \tag{5.17}$$

where the non-linear system (5.16) is obtained from (5.12) and (5.13) by making $\mathring{m}_{ik}=0$, $\mathring{m}_{kj}=0$ (stationary state); it makes possible the calculation of the blocking probabilities P_{ik} and P_{kj}. Although they appear to be of very great complexity at first sight, it can be said that the Erlang functions are relatively well behaved and that the resolution of (5.16) is not an impossible task using relaxation type method since efficient initial conditions (ie. not too far from the steady state solution) can be given. This point will be taken up again in the next section.

Taking into account the simple nature of the constraints (5.17) an algorithm of admissible directions of the Frank and Wolfe type has proved to be relatively effective. We recall briefly below the fundamental stages of this kind of algorithm.

Stage 1. For an admissible solution $\bar{\alpha}_{ikj}$ (ie. satisfying the constraints (5.17)), calculate the value of the criterion as well as the gradient

$$\left.\frac{dJ}{d\alpha_{ikj}}\right|_{\bar{\alpha}_{ikj}} \triangleq g_{ikj}$$

Stage 2. Resolve the auxiliary linear problem

$$\min_{\alpha_{ikj}} \sum_{(ikj)} g_{ikj}\ \alpha_{ikj} \tag{5.18}$$

subject to

$$\sum_k \alpha_{ikj} = 1, \qquad \alpha_{ikj} \geq 0$$

which makes it possible to determine a point $\tilde{\alpha}_{ikj}$ (generally unique)

State 3. Advance in the direction $(\tilde{\alpha}_{ikj} - \bar{\alpha}_{ikj})$ which is locally a direction of decrease for the criterion J. Repeat these operations until a minimum is achieved.

In an algorithm of this sort most of the work is contained in Stage 1 (cf. following paragraph). On the other hand, problem (5.18) of stage 2 is particularly simple as it can be decomposed by flows (i,j) and therefore allows the trivial solution

$$\alpha_{ikj} \begin{cases} = 1 \text{ for } k=k^* \text{ where } k^* \text{ is the index (or one of the indices) for which } g_{ikj} \text{ is minimum} \\ = 0 \text{ for } k \neq k^* \end{cases}$$

For stage 3 we mentioned that the iterative process must be continued until one minimum was obtained. Although the unicity of the minimum has not been mathematically proved it seems, intuitively and bearing in mind many digital simulations that we have undertaken that the minimum be unique.

5.5.2. First level : some results

With a network test of 10 STCs and 3 PTCs whose capacities for the source and sink links and the offered traffic matrix are given below, we get a multiflow problem with 270 variables

Table 5.3

$$N(I,K) : \begin{bmatrix} 15 & 25 & 20 \\ 10 & 20 & 15 \\ 15 & 20 & 15 \\ 20 & 20 & 20 \\ 20 & 30 & 20 \\ 30 & 30 & 30 \\ 15 & 20 & 15 \\ 15 & 15 & 15 \\ 10 & 15 & 10 \\ 20 & 15 & 20 \end{bmatrix} \qquad N(K,J) : \begin{bmatrix} 15 & 15 & 15 & 30 & 20 & 15 & 15 & 20 & 5 & 25 \\ 20 & 15 & 25 & 30 & 20 & 20 & 15 & 15 & 10 & 15 \\ 20 & 15 & 15 & 20 & 20 & 20 & 15 & 15 & 15 & 20 \end{bmatrix}$$

$$A(I,J) : \begin{bmatrix} 0 & 4 & 7 & 10 & 5 & 5 & 6 & 4 & 3 & 4 \\ 5 & 0 & 4 & 8 & 6 & 5 & 4 & 3 & 3 & 5 \\ 3 & 4 & 0 & 7 & 5 & 4 & 4 & 3 & 4 & 4 \\ 5 & 4 & 7 & 0 & 8 & 5 & 6 & 3 & 4 & 3 \\ 5 & 3 & 4 & 6 & 0 & 6 & 5 & 5 & 4 & 6 \\ 8 & 7 & 6 & 10 & 8 & 0 & 6 & 7 & 5 & 5 \\ 5 & 4 & 5 & 8 & 8 & 7 & 0 & 6 & 4 & 5 \\ 5 & 4 & 3 & 6 & 5 & 4 & 3 & 0 & 3 & 4 \\ 4 & 5 & 3 & 6 & 4 & 3 & 4 & 3 & 0 & 3 \\ 5 & 5 & 6 & 9 & 8 & 7 & 5 & 6 & 4 & 0 \end{bmatrix}$$

First of all to resolve it we used a simplified model which consists of applying the Erlang formula in chain from the origin i to the destination j. To make the calculation of losses in the network we compare the telephone traffic to a fluid which would flow from i to j with losses in each crossed link with independence (mathematically, not only statistically) of the losses on the upstream links in relation to those on the downstream links. The traffic moved by one trunk-group, which is always assumed to be Poissonian, becomes offered traffic for the following link. Thus, from the traffic offered to the (i,k) links

$$y_{ik} = \sum_j \alpha_{ikj} A_{ij} \tag{5.19}$$

we deduce the blocking probability of these links

$$P_{ik} = E\ (y_{ik}, N_{ik}) \tag{5.20}$$

as well as their "moved" traffic

$$z_{ik} = y_{ik}\ (1 - E\ (y_{ik}, N_{ik})) \tag{5.21}$$

The traffic offered to the sink links is

$$y_{kj} : \sum_i \alpha_{ikj} A_{ij}\ P_{ik} \tag{5.22}$$

and therefore their blocking probability

$$P_{kj} = E\ (y_{kj},\ N_{kj}) \tag{5.23}$$

Finally, according to these assumptions, the problem of multiflow optimisation becomes

(5.15)

subject to

$$\left[\begin{array}{l} P_{ik} = E\left[\sum_j \alpha_{ikj} A_{ij},\ N_{ij}\right] \\ P_{kj} = E\left[y_{kj}, N_{kj}\right]\ ;\quad y_{kj} = \sum_i \alpha_{ikj} A_{ij} \end{array}\right. \tag{5.24}$$

and

(5.17)

It is obvious that this is a very important simplification for the implicit non-linear problem (5.15) is replaced by (5.24) which does not really present any difficulty. In fact, (5.24) can be considered as the first iteration in a possible relaxation scheme in the resolution of (5.16).

To test the accuracy of the model (5.24) in relation to (5.16), we have used again the simple example of section (5.4.4.). In the table below we have re-written the probabilities of the routes ($P_{ik} + P_{kj} - P_{ik}P_{kj}$) obtained by Markovian resolution and by the "elaborated" model (5.16) and we have given in the last column those provided by the simplified model (5.24).

Table 5.4

ikj	1k1	1k2	1k3	2k1	2k2	2k3	3k1	3k2	3k3
Markov	0,4489	0,3926	0,3817	0,2895	0,2097	0,1964	0,3225	0,2511	0,2339
Model 5.16	0,4588	0,4002	0,3924	0,2998	0,2240	0,2139	0,3369	0,2652	0,2556
Model 5.24	0,4838	0,4320	0,4236	0,3169	0,2479	0,2368	0,3658	0,3021	0,2918

Naturally the elaborated model (5.16) makes possible for better estimations of the blocking probabilities than (5.24). It can be noted that the latter model also always provides a higher limit for the estimation of these probabilities.
The resolution of the problem (5.15), (5.24), (5.17) ie. the multiflow problem with simplified model, has provided, for the example given here, the results given in the table (5.9) in terms of $a_{ikj} = \alpha_{ikj}\, a_{ij}$.

Table 5.5

J ⟶ (columns); I ↓ (rows); K ↓ (within each cell)

0.0 K	0.879	1.449	3.108	1.127	1.109	1.453	1.076	0.479	1.340
0.0	1.664	3.496	4.448	2.279	2.111	2.531	1.538	0.827	0.873
0.0	1.457	2.055	2.443	1.594	1.780	2.015	1.386	1.694	1.786
0.708	0.0	0.842	2.346	1.478	0.973	0.908	0.830	0.403	1.226
2.641	0.0	2.090	3.565	2.488	2.226	1.734	1.284	1.291	1.417
1.651	0.0	1.068	2.090	2.034	1.801	1.359	0.885	1.306	2.357
0.864	0.935	0.0	2.507	1.758	1.059	1.010	1.360	0.650	1.551
1.383	1.912	0.0	3.318	1.539	1.709	1.681	0.845	1.419	1.106
0.753	1.153	0.0	1.174	1.703	1.232	1.308	0.795	1.931	1.343
1.584	1.820	1.897	0.0	2.943	1.344	2.665	1.263	0.650	1.415
1.469	0.991	3.585	0.0	2.458	1.378	1.385	0.877	1.398	0.655
1.946	1.190	1.518	0.0	2.599	2.277	1.950	0.859	1.952	0.930
1.187	1.000	0.828	2.146	0.0	1.450	1.827	1.787	0.638	2.530
2.769	0.973	2.563	2.593	0.0	2.564	1.420	1.874	1.406	1.646
1.044	1.027	0.610	1.261	0.0	1.987	1.752	1.339	1.956	1.824
1.652	2.666	1.393	4.617	2.812	0.0	2.296	3.322	0.828	2.398
2.372	1.910	3.111	2.760	2.462	0.0	1.742	1.740	1.490	1.154
3.976	2.425	1.495	2.623	2.726	0.0	1.962	1.938	2.681	1.448
1.546	0.825	1.133	2.773	2.289	1.877	0.0	2.758	0.652	2.135
2.110	1.727	2.198	3.597	3.264	3.002	0.0	1.483	1.524	1.336
1.344	1.447	1.669	1.630	2.447	2.121	0.0	1.759	1.824	1.529
1.428	1.512	0.777	2.627	1.758	1.123	1.089	0.0	0.487	1.929
1.476	1.255	1.534	2.083	1.437	1.244	0.930	0.0	0.953	0.867
2.096	1.234	0.689	1.290	1.805	1.633	0.981	0.0	1.560	1.203
1.188	1.251	0.911	1.757	0.997	0.937	0.842	1.210	0.0	1.192
1.583	2.116	1.443	2.307	1.659	1.286	1.892	1.144	0.0	1.162
1.229	1.634	0.646	1.937	1.345	0.777	1.266	0.646	0.0	0.646
1.364	2.057	2.288	4.219	3.551	1.948	1.898	2.450	0.658	0.0
1.268	1.190	1.644	2.465	1.771	1.867	1.424	1.465	1.324	0.0
2.368	1.752	2.068	2.316	2.678	3.185	1.678	2.085	2.018	0.0

The value of the criterion at convergence (the optimum) is practically equal to 75 Erlangs. Figure 5.7 gives an idea of the type of convergence of the algorithm described in the preceding section, showing that on the evidence the truly appreciable gains are obtained after a fairly small number of iterations (< 5).

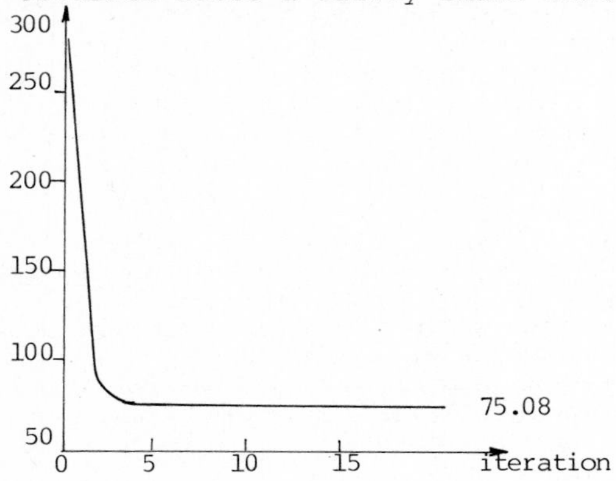

Figure 5.7

With these same values for the parameters of load-sharing, we evaluated the value of the criterion (5.15) using the complex model (5.16) and we found the value 68,2 (verification of the fact that (5.24) provides more severe upper limits than (5.15). Taking these values as initial conditions for the problem (5.15), (5.16), (5.17) and resuming the process of optimisation, we notice that the solution is not appreciably improved. For the example chosen, in 10 iterations reduction of the criterion turns on about 0,2 Erlang.

The solution obtained with the simplified model appears to be more or less the one we would obtain by using the more complex model, which requires approximately ten times more time per iteration.
We have been able to show, through this particular example, the relatively robust nature of the routing problem when it is expressed not in terms of the primal variables (the state of occupation of the trunks) but in terms of the routing variables themselves. Indeed, it is a question of relative sizes which can put up with certain inacurracies or approximations when these are, we could say, uniform and homogeneous.
It goes without saying that in a dynamic control application, for the choice of the strategic parameters of load sharing, it is the simplified model which would be used.

5.5.3. Decentralised regulation

The solution provided by the first level enables us therefore, in the interval of time in which it remains valid, ie. in which the characteristics of the traffic do not move much away from those which were retained for its calculation, to obtain an average point of function around which the system will operate. At the local level of each centre of commutation we make the hypothesis that only the local measures concerning the state of occupation of the links leaving the exchange and the solution provided by the higher level can be known. The information coming from the higher level can be used as an environment model for each exchange, and thus they make it possible to carry out a decoupling of the global problem in as many problems as there are commutation centres. With each of these problems is linked a different model constituted from the local structure of the network and from the corresponding environment model. It is then possible to return locally to a more detailed formulation of the Markovian model type and then, for the determination of the control, call on on methods and techniques derived from the study of queuing processes and Markovian decision processes [FOR - 78],[GAR - 82] and also learning automata [NAR - 77], [SRI - 82]. This kind of approach is still mainly at the level of investigation and research. One of the problems posed is in the adaptation between the detailed local model and the environment model, the latter being of necessity more or less approximate.

The approach which we shall develop a little here is more standard in control engineering and proposes regulation actions around points of operation predicted at the higher level. Thus, from the model (5.13) which, very schematically can be written as

$$\dot{M} = f(M, \alpha) \tag{5.25}$$

where M is the vector of the means of occupation of the links and α the vector of the load sharing parameters; $M \in R^{(2m.n)}$ and using values M^*, α^* corresponding to the optimal stationary regime calculated at the higher level, we carry out a linearisation around these values to obtain a "standard" linear model.

$$\begin{cases} \dot{X} = AX + Bu \\ X = M - M^* \quad , \ u = \alpha - \alpha^* \end{cases} \tag{5.26}$$

A and B being matrices obtained by derivation of F in relation to M and α the derivatives being calculated at the point M^* and α^*. We cannot here, for obvious reasons of space, enter into details of the calculations which, although of no fundamental interest, are no less relatively delicate taking into account the implicit model (5.13) and we shall content ourselves with mentioning the broad outlines of the approach presented. [BON - 81].

We are defining an auxiliary problem of parametric optimisation by consideration of a quadratic criterion and infinite horizon

$$\begin{cases} \min J = \int_0^{\infty} (X' Q X + u' Ru)dt \\ \text{subject to (5.25)} \\ \quad + \text{constraints for } u = kX \end{cases} \tag{5.27}$$

Thus we have defined a problem of the type "parametric optimisation" subject to constraints for which the algorithm described in chapter III was used after some adaptations necessitated by certain constraints specific to the problem being dealt with [BON - 81].

Firstly, regulation of the linear feedback type involves the constraint of decentralisation. On the other hand, it being a question of load sharing parameters, it is necessary that at all times they satisfy the following constraints.

$$\sum_k \alpha_{ikj}(t) = 1$$

and (5.28)

$$\alpha_{ikj}(t) \geq 0$$

Such constraints, which bring into play the instantaneous values of state, are particularly difficult to deal with and more particularly in the case where the calculation time is of prime importance for an on-line application. As the state X is itself limited by the maximum capacities of the links, it is possible to transform the constraints (5.27) into constraints which have direct bearing on the parameters of the feedback matrix k. Problem (5.26) then becomes a problem of non-linear optimisation which has been resolved using the gradient type algorithm-admissible directions described in chapter III.

5.5.4. Some results - discussion

The following tables give the characteristics of the network n=4, m=3 on which have been developped in particular certain numerical tests

Table 5.6. Offered Traffic matrix

i \ j	1	2	3	4
1	0	8	8	12
2	8	0	16	24
3	8	16	0	24
4	12	24	24	0

Table 5.7. Capacities

i \ k	1	2	3
1	18	18	18
2	8	30	40
3	36	8	18
4	36	18	36

k \ j	1	2	3	4
1	18	8	40	36
2	18	36	8	18
3	18	40	12	36

At the first level are determined the following static load sharing parameters

Table 5.8. Load sharing parameters (in %)

(i,j) \ k	1	2	3
11	-	-	-
12	2.30	85.07	12.62
13	93.33	2.67	4.00
14	24.33	18.09	57.58
21	2.30	76.52	21.17
22	-	-	-
23	20.25	30.69	49.06
24	1.64	32.22	66.15
31	75.54	5.79	18.67
32	13.71	26.70	59.59
33	-	-	-
34	88.89	1.64	9.47
41	22.82	20.59	56.59
42	1.64	30.78	67.59
43	95.47	1.81	2.72
44	-	-	-

The average states of occupation are given on table 5.9

Table 5.9.

i \ k	1	2	3
1	10.19	9.12	8.15
2	3.61	18.15	24.65
3	28.35	4.73	12.62
4	24.96	10.15	23.43

k \ j	1	2	3	4
1	8.69	2.67	32.04	23.71
2	8.96	18.05	5.00	10.15
3	9.82	26.17	8.14	24.73

It would be tedious and of little interest to give the complete table of values of the feedback gains k_{ij} and we shall content ourselves with giving below, as an example, those relating to the STC_1.

Table 5.10.

ikj \ s	1	2	3
1 1 2	-0.754	0.842	0.780
1 2 2	0.482	-5.456	3.491
1 3 2	0.272	4.614	-4.271
1 1 3	-2.183	0.950	1.033
1 2 3	0.873	-0.975	0.321
1 3 3	1.309	0.025	-1.354
1 1 4	-7.963	3.658	6.766
1 2 4	4.025	-6.612	6.121
1 3 4	3.938	2.954	-12.89
2 1 1	-1.751	0.423	0.316
2 2 1	1.130	-4.313	2.551
2 3 1	0.620	3.889	-2.863
2 1 3	-1.309	3.719	2.738
2 2 3	0.823	-5.636	3.895
2 3 3	0.486	1.917	-6.633
2 1 4	-1.242	0.300	0.221
2 2 4	0.827	-5.917	4.356
2 3 4	0.415	5.617	-4.578
3 1 1	-2.876	0.849	3.401
3 2 1	0.681	-1.829	1.530
3 3 1	2.195	0.980	-4.931
3 1 2	-1.612	0.486	2.512
3 2 2	1.289	-1.228	3.433
3 3 2	0.323	0.741	-5.946
3 1 4	-1.306	0.580	2.070
3 2 4	0.192	-1.151	0.432
3 3 4	1.114	0.572	-2.502
4 1 1	-3.047	4.210	3.246
4 2 1	1.620	-6.759	2.929
4 3 1	1.427	2.549	-6.176
4 1 2	-0.218	0.537	0.233
4 2 2	0.239	-8.533	4.378
4 3 2	-0.021	7.996	-4.611
4 1 3	-0.605	1.486	0.644
4 2 3	0.242	-0.594	-0.258
4 3 3	0.363	-0.892	-0.386

$$\Delta\alpha_{112} = -0{,}754x_{11} + 0{,}842x_{12} + 0{,}780x_{13}$$

In conclusion, an event by event simulation has shown that decentralised regulation of state would not appreciably improve the performance of the network from the point of view of the losses which remained for the network considered, around 3 Erlangs. This result is not, in fact, surprising, for the problem of regulation is not posed in terms of minimisation of losses, but rather in terms of the pursuit of a "static" solution determined at the level of supervision (first level), a solution which is optimising for a given matrix of traffic offered. However, it is stressed that the regulation actions present the advantage of smoothing the transient peaks of traffic, which are responsible to a great extent, for the overflowing traffic that has, as said before, destabilizing effects.

REFERENCES

BON - 81 BONATTI I.S., "Gestion de réseaux de service : application au réseau téléphonique interurbain", Thèse de Docteur-Ingénieur, Université Paul Sabatier, n° 767, Toulouse, 1981.

FOR - 78 FORESTIER J.P. and P. VARAIYA, "Multilayer control of large Markov chains", IEEE Trans. Aut. Control AC-23, 2, 1978.

GAR - 80 GARCIA J.M., "Problèmes liés à la modélisation du trafic et à l'acheminement des appels dans un réseau téléphonique", Thèse de Docteur-Ingénieur, Université Paul Sabatier, 733, Toulouse, 1980.

GAR - 81 GARCIA J.M., B. GOPINAH and P. VARAIYA, "Routing in telephone networks", Conference Decision and Control, San Diego, December 1981.

KLE - 75 KLEINROCK L., "Queuing Systems", Vol. 1 : Theory, John Wiley, 1975

KUC - 73 KUCZURA A., "The Interrupted Poisson process as an overflow process", Bell System Technical Journal, Vol. 52, 1973.

KUC - 77 KUCZURA A. and D. BAJAJ, "A method of moments for the analysis of a switched communication network performance", IEEE Transactions on Communication, 1977.

LEG - 81 LE GALL F. and J. BERNUSSOU, "Réseaux de Télécommunications : leur modélisation", Note Interne ASCl, n° 80.I.22, LAAS Toulouse, 1981.

LEG - 82 LE GALL F., "Contribution à la modélisation et la commande de réseaux téléphoniques", Thèse de Docteur-Ingénieur n° 796, Université Paul Sabatier, Toulouse, 1982.

MAN - 79 MANFIELD D.R. and T. DOWNS, "On the one moment analysis of telephone traffic networks, IEEE Transactions on Communications, vol. COM-27, 8, Aug. 79.

NAR - 77 NARENDRA K.S., E.A. WRIGHT and L.G. MASON, "Application of learning automata to telephone network traffic routing problems", IEEE Trans. System Man and Cybernetics, SMC-7, 1977.

RES - 80 "Reseaux, Networks, Telecommunication networks planning" Symposium. Paris, Sept. 29 th, Oct. 2nd, 1980.

SRI - 82 SRIKANTAKUMAR P.R. and I.S. NARENDRA, "A learning model for routing in telephone networks", SIAM Journal of Control and Optimization, vol. 20, 1, 1982.

WIL - 56 WILKINSON R.I., "Theories for toll traffic engineering in the USA", Bell System Technical Journal, vol. 35, 1956.

CHAPTER 6

MULTICOMMODITY NETWORK FLOW OPTIMIZATION : CENTRALIZED AND DISTRIBUTED ALGORITHMS

6.1 Communication networks structure

6.1.1. Communication networks [GRE - 80],[ZIM - 80]

This chapter is mainly devoted to the information flow routing in communication networks, constituted by geographically distributed processors and transmission links. They can be represented by means of oriented graphs $G = (N,L)$, where $N = \{1,2,\ldots n\}$ denotes the set of nodes (i.e. the processors) and $L = \{(i,j)\}$ the set of edges (i.e. the directed links). The most general topology of a meshed graph is considered. Such networks are intended to link customers located at the graph nodes for different purposes (conversation, data processing ...).
To perform the functional analysis of a communication network, it is viewed as a distributed system where each node presents a hierarchical classification of entities in different layers (fig. 6.1). At the network scale, the nth layer is the set of the n layer entities corresponding to each node. Those entities cooperate and, using the (n-1) layer services provide more elaborated services to the higher (n+1) layer. So, going up in the hierarchy is traduced by a vision becoming more and more abstract. The highest layer is the only one to see the semantic of the informations transmitted to deal with a specific application. The lower layers permit an easier analysis of the distributed system behaviour and act as service facilities for the higher layers.

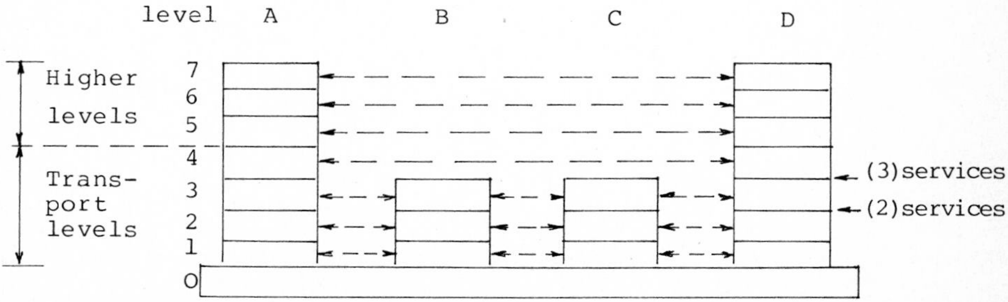

Figure 6.1

This modelling exhibits two kinds of interactions :
a) the interactions between the entities in a same layer, managed through protocols. They precisely define the cooperation, between the nth layer entities so that the (n) entities add value to the (n-1) service they get from the (n-1)layer and offer this value added service (the n service) to the (n+1) entities.

b) the interaction between adjacent levels, managed by interfaces which define the control information and data exchanges.

The network architecture is the set of protocols and interfaces to be considered for the processors communication. The ISO model [ZIM - 80] is based on a seven layer decomposition, from which two groups can be distinguished (figure 6.1).
1) the transportation functions fulfilled by the four lower layers
2) the interpreting and data processing functions fulfilled by the three higher layers.

The third layer is the one in charge of the informations routing between the customers. For instance, it defines the route to be followed between A and D (figure 6.1) and which, may be, passes through the B and C nodes. The third layer uses the 2nd layer services (information transfert between adjacent processors) to offer layer four the more elaborated service which is the routing map between terminals. The routing policies can be simple (predetermined, static) or complex (dynamic routing depending on the network states). This chapter is addressed to study the routing policies which try to minimize the average delay for the informations in the network.

6.1.2. The routing problem [SCH - 80]

The informations circulate in the network according to the store and forward technique :

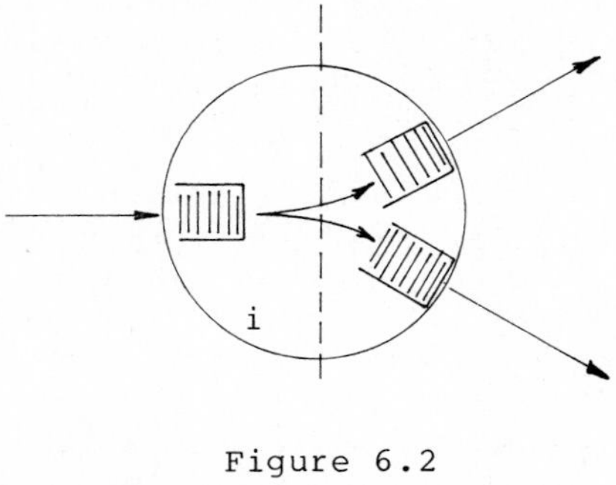

Figure 6.2

An information entering the node i is identified according to its final destination, say (n). The routing consists in choosing the next node (successor) j = v (i,n) to whom the information has to be addressed. It is buffered until the (i,j) link is idle. The delay in the buffer is considered as a link delay from i to j. The set of informations having the same n destination constitutes a kind of product (flow). An optimal routing amounts to the determination of the successors for each i node, so that the average delay is minimal. This is an optimal flow problem. When considering different kinds of products (different destinations). The coupling between them is introduced by the buffers. In this case, the optimal routing determination is a multicommodity flow problem. The routing policies can be classified between three main classes :

1) The static routing. The network presents a given structure and operates under specific given condition. The optimal routing is the solution of a classical multicommodity flow problem.
2) The dynamic routing. It takes into account instantaneous structural variations and queue length in the buffers. The optimal "real time" policy is very complex to determine and, in the case of large networks, impossible to implement in real time.
3) The quasi-static routing. From time to time, the routing policy is calculated on the basis of new data and the assumption of static problem. The dynamical aspect depends on the choice of the calculation period. This routing is well suited to the slowly varying demand problem.

The static and quasi-static cases are investigated here. Section 6.2 presents an analysis of the optimal multicommodity flow problem, which can be solved using two kinds of approaches :

1) Centralized calculations : the overall information is sent to a unique processor which determines the optimal policy and sends it back to the other processors. Different classical methods can be used in this framework, they are quickly presented in section 6.3.
2) Distributed calculations : which try to take some advantages from the specific network structure. Each processor has a partial local view of the network state, but communicates with its neighbours to provide, locally, the optimal routing policy. Two recent papers are discussed in section 6.4. The section 6.5 presents a distributed algorithm based on a dual approach of the optimal flow problem. This algorithm presents the main advantage of a really very simple distributed implementation.

6.2 Analysis of flow and multicommodity flow problems

6.2.1. Analysis of a flow problem

6.2.1.1. Some definitions

In section 6.1 we represented a communication network by a graph $G = (N,L)$, with $N = \{1, 2, \dots n\}$ the set of n nodes and $L = \{(i,j)\}$ the set of m arcs of the graph. The topology of the network is represented by the incidence matrix A (n,m) of G. If we assume the graph to be connected we know that in this case the rank of A is $\rho(A) = n-1$ [AUT - 79]. Furthermore, the columns of this matrix are relative to the arcs of the graph and if we denote by A^{ij} the column relative to the arc (i,j), we have

$A^{ij} = I^i - I^j$ where I^i and I^j are respectively the ith and j^{th} columns of the identity matrix I of dimension (n,n). Let us consider, for example, the graph in figure 6.3

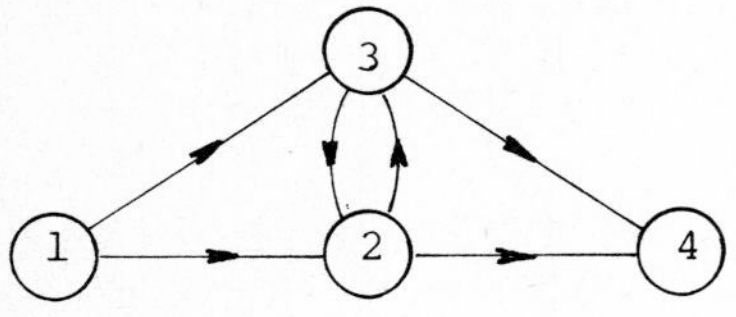

$N = \{1,2,3,4\}$

$L = \{(1,2),(1,3),(2,4),(3,4),(3,2),(2,3)\}$

Figure 6.3

Its incidence matrix is given by

$$A = \begin{bmatrix} 1 & 1 & 0 & 0 & 0 & 0 \\ -1 & 0 & 1 & 0 & -1 & 1 \\ 0 & -1 & 0 & 1 & 1 & -1 \\ 0 & 0 & -1 & -1 & 0 & 0 \end{bmatrix} \qquad (6.1)$$

The row A_i is the co-cycle of the node i. We denote by r_i the quantity of information or of material produced ($r_i > 0$) or consumed ($r_i < 0$) at node i per unit of time. The vector r (n,1) is called the network resources vector. It is assumed that $\sum_{i=1}^{n} r_i = 0$. This equation is based on the fact that the network is isolated from the outside world.

We note as x_k the flow in the arc $l_k = (i,j)$, ie. the quantity of information or material circulating in (i,j) per unit of time. The equation of conservation at note i is written as $A_i x = r_i$.

<u>Definition 6.1</u> : Flow

<u>The vector x (n,1) is called a flow for the resources vector r if Ax = r, ie. if the conservation equation is satisfied at each node of G.</u>

<u>Definition 6.2</u> : Feasible flow

<u>The vector x is a feasible flow for the resources vector r if $Ax = r$, $x \geq 0$. The set $S = \{x \; / \; Ax = r, \; x \geq 0\}$ is called the polyedron of feasible flows.</u>

By the geometric interpretation of linear programming it is possible to characterise the extreme points, edges and extreme rays of the polyhedron S from basic feasible solutions of the system Ax = r. We shall, therefore, introduce the linear program :

$$\min z = dx \qquad (6.2)$$

subject to $Ax = r$, $x \geq 0$

where d is a row vector of dimension (1,n). d^{ij} can be considered as the cost of transport of one unit of information on the arc (i,j). We begin by putting the system Ax = r into canonical form. A basis of Ax = r is a set A^B constituted by n-1 linearly independent column vectors since $\rho(A) = n-1$. It is well known that the sub-graph H = (N,B) is a maximal tree of G [AUT - 79]. If we call $\bar{B}$ the complementary set of B ($B \cup \bar{B} = L$), the sub-graph $F = (N, \bar{B})$ is the co-tree of G associated to the tree H. In the example of figure 6.3, if we choose

$B = \{(1,3),(3,2),(2,4)\}$, we get the tree and the co-tree represented on figure 6.4

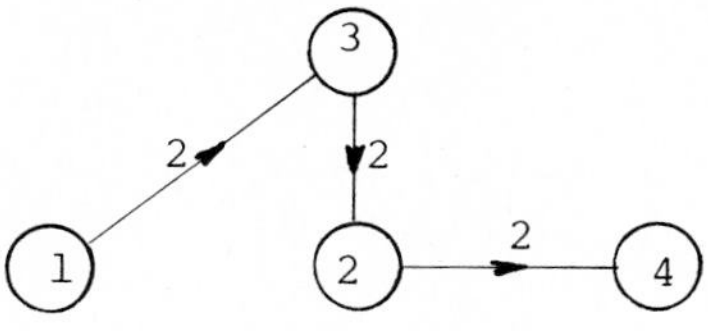

Figure 6.4a $H=(N,B)$

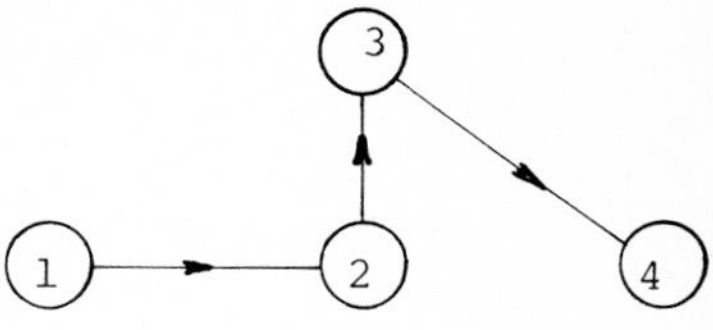

Figure 6.4b $F=(N,\bar{B})$

The system $Ax = r$ can be partitioned for a choice of the basis A^B and written in the form :

$$A^B x_B + A^{\bar{B}} x_{\bar{B}} = r$$

An arbitrary choice of the flow $x_{\bar{B}}$ on the co-tree (independent or non basic variables) fixes the flow x_B on the tree (dependent or basic variables).

<u>Definition 6.3</u> : Basic solution
<u>We call basic</u> solution of the system $Ax = r$ with respect to a basis A^B, the solution of the system $Ax = r$ resulting from the choice $x_{\bar{B}} = 0$ for the flow on the co-tree. The basic solution is said to be feasible if the corresponding flow is feasible. For the example in figure 6.3, if we suppose that $r = [2\ 0\ 0\ -2]'$, the choice of the basis of figure 6.4, gives the feasible basic solution $x_{13} = x_{32} = x_{24} = 2$, $x_{12} = x_{34} = x_{23} = 0$.

For a feasible basic solution we propose to write the criterion in terms of the only non basic variables. Program 6.2 can be written in the form :

$$d^B x_B + d^{\bar{B}} x_{\bar{B}} = z \text{ (min)} \tag{6.3a}$$

subject to

$$A^B x_B + A^{\bar{B}} x_{\bar{B}} = r, \qquad x \geqslant 0 \tag{6.3b}$$

Let p (1,n) be the vector of the dual variables, called the vector of potentials since p^i is a variable associated to a node. The criterion (6.3.a) can be written as :

$$(d^B - pA^B)\, x_B + (d^{\bar{B}} - p\, A^{\bar{B}})\, x_{\bar{B}} = z \text{ (min)} - pr \tag{6.4}$$

Let $\hat{d} = d - pA$. To write z only in terms of the non basic variables, p must be chosen such that :

$$\hat{d}^B = d^B - p\, A^B = 0 \tag{6.5}$$

This system of n-1 equations with n unknowns defines the potentials where one node potential can be chosen arbitrarily. The corresponding solution is very easily deduced from the tree associated with the basis A^B [AUT - 79] . In fact, for an arc (i,j) we have

$$\hat{d}^{ij} = d^{ij} - p\, A^{ij} = d^{ij} - (p^i - p^j)$$

and as we must have $p^i - p^j = d^{ij}$ for the arcs of the tree, the potentials can be computed by degrees from an initial node to which is allocated an arbitrary potential. p is the multiplier vector relative to basis A^B. From p we deduce the relative cost vector of the non basic variables

$$\hat{d}^{\overline{B}} = d^{\overline{B}} - p\,A^{\overline{B}}$$

It is easy to prove that the feasible basic solution corresponding to the basis A^B is optimal if, and only if [LUE - 73]

$$\hat{d}^{\overline{B}} \geqslant 0 \tag{6.6}$$

Using this condition it is possible to test easily if a basic solution is optimal.

6.2.1.2. Analysis of the polyhedron S for a single product flow

We presume that in the network there is only one producer node (origin) and only one consumer node (destination). Without loss of generality, we shall presume that these are the nodes 1 and n respectively, ie. that $r_i = 0$ for $i \neq 1, n$, $r_1 > 0$ and $r_n = - r_1$.

Analysis of the extreme points of polyhedron S

The extreme points of S correspond to the feasible basic solutions of the system $Ax = r$. We have seen that the basis of this system are trees of G, and a conventional property of the trees [AUT - 79] is that there exists a unique chain between two arbitrary nodes. In particular, the chain from 1 to n will be noted as μ_{1n}. We denote by $(i,j) \in \mu^+_{1n}$ an arc oriented in the direction of reference 1 to n and by $(i,j) \in \mu^-_{1n}$ in the opposite case. The basic solution is only feasible if the chain μ^+_{1n} defines a path from 1 to n, ie. if no arc $\in \mu^-_{1n}$ exists, because the flow on these arcs would be negative. The path from 1 to n is, of course, elementary (without loop).

A tree is constituted by n-1 arcs of G, but more often a path from the origin to the destination is constituted by a number of arcs which is distinctly inferior to n-1, so that the feasible basic solutions are generally highly degenerate. Conversely, any elementary chain from 1 to n can be considered as a sub-set of a basis of A, whence :

<u>Proposition 6.1</u> : To any extreme point of the polyhedron S there corresponds an elementary path between the origin and the destination and vice versa.

<u>Definition 6.4</u> : As any feasible basic solution is optimal for a certain metric d we shall say that the extreme points of S are shortest route flows from 1 to n.

Let us consider again the example in figure 6.3. The tree in figure 6.4 a defines an elementary path from 1 to 4. If we choose $B = \{(1,3), (3,4), (2,3)\}$, the corresponding tree in figure 6.5 also defines an elementary path from 1 to 4 but in this case the basic solution is degenerate because $x_{23} = 0$.

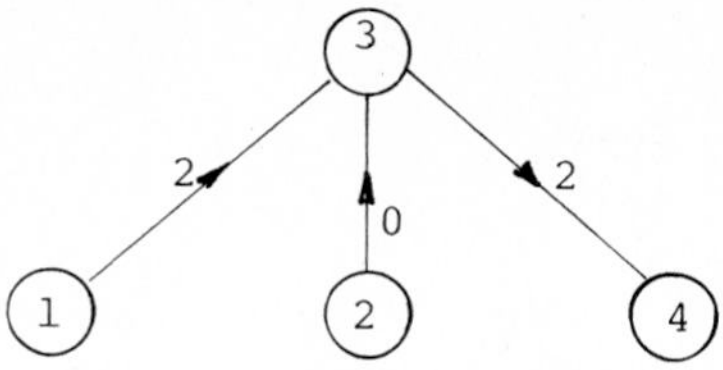

Figure 6.5

Analysis of the edges of polyedron S.

We have just said that an extreme point of S corresponds to a feasible basic solution of Ax = r. We also know [AUT - 79] that the introduction of a non basic variable in the basis is equivalent to the introduction of an arc of the co-tree in the set of basic arcs and that this introduction then constitutes a unique elementary loop μ. In vector form an elementary loop is a vector of dimension (m,1) such that :

$$\mu_{ij} = \begin{cases} +1 & \text{if} \quad (i,j) \in \mu^+ \\ -1 & \text{if} \quad (i,j) \in \mu^- \\ 0 & \text{if} \quad (i,j) \notin \mu \end{cases} \tag{6.7}$$

The direction of reference is chosen so that the new arc $\in \mu^+$. Let x_{ij} be the non basic variable which enters the basis. If we make $x_{ij} = \varepsilon > 0$ small, we can show that in the case of a non degenerate basic solution, the variations of the basic variables are given by the expression [AUT - 79].

$$x_B = \gamma_B + \varepsilon \mu_B \tag{6.8}$$

with γ_B the value of the basic variables relative to the basis B. From the geometrical interpretation of linear programming we know that the solution moves along an edge of the polyhedron having one endpoint at the current vertex. Three cases must be considered :

(1) the new arc (i,j) generates an elementary circuit μ , i.e. an elementary loop in which all arcs are oriented in the same direction μ^+.

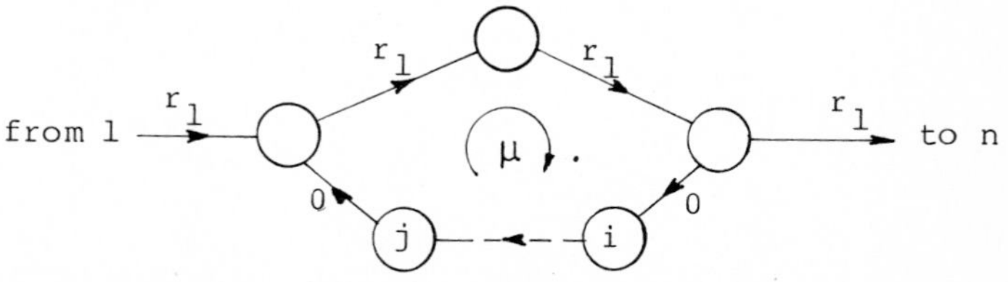

Figure 6.6

We can increase the flow on this circuit indefinitely, i.e. when ε grows in (6.8), the constraints $x_B \geq 0$ are never violated. The solution moves on an extreme ray of the polyhedron S and the circuit created defines the direction of this ray, whence :

Proposition 6.2 : To every extreme ray of the polyhedron S corresponds an elementary circuit. Conversely, every elementary circuit is the direction of a sub-set of extreme rays of S.

In the tree in figure 6.4 a the introduction of the arc (2.3) of the co-tree constitutes a unique elementary circuit passing through the nodes 2 and 3. The direction of the extreme ray is [0 0 0 0 1 1]'. The introduction of arc (3,2) in the tree in figure 6.5 creates the same elementary circuit. The extreme ray having end point at this node has the same direction as the previous one.

The potentials of the nodes of G must be such that $d^{kl}=p^k-p^l$ for the basic arcs; thus p^j-p^i is the path length from j to i which uses the tree arcs (figure 6.6). Consequently $\hat{d}^{ij} = d^{ij} - (p^i - p^j)$ is the elementary circuit length generated by the introduction of (i,j) in the tree. According to the optimality conditions (6.6), when solving a program like (6.2), the solution can move along an extreme ray of S only if the circuit corresponding to this extreme ray has a negative length. In the rest of this study we shall assume the following :

Assumption 6.1 : Graph G has no negative length circuit for the metric d. In this case we can restrict the search for an optimal solution of (6.2) to the exploration of the extreme points of S.

Assumption 6.2 : We shall only have to consider the case of a metric $d \geqslant 0$. The assumption 6.1 is therefore verified

(2) the arc introduced (i,j) creates an alternative path between two nodes A and B of the path from 1 to n associated with the node S. It is called bifurcation from A to B (figure 6.7).

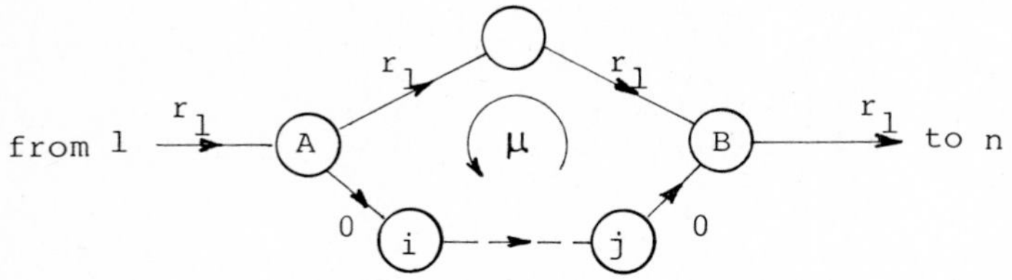

Figure 6.7

The flow on this bifurcation can be increased to the value r_1, in which case the flow on the arcs of the old path from A to B will vanish. By making any one of these arcs leave the basis a new path from 1 to n is created. The movement along the edge of the polyhedron S leads to an adjacent extreme point, whence :

Proposition 6.3 : To every edge of the polyhedron S coming from a given extreme point, there corresponds an elementary loop.

In the tree in figure 6.4.a the introduction of the arc (3,4) of the co-cycle generates the elementary loop $[0\ 0\ -1\ 1\ -1\ 0]'$ on which the flow can be increased to the value $r_1 = 2$. A and B are the nodes 3 and 4 respectively. If $r_1 = 2$ is made to pass along the bifurcation (3,4), x_{32} and x_{24} vanish simultaneously and any one can be chosen to leave the basis.
In this case, let us note that $\hat{d}^{ij}$ is the path length increment from A to B caused by the introduction of arc (i,j), i.e. by the introduction of the bifurcation. In an optimisation problem of type (6.2) a bifurcation will be introduced if $\hat{d}^{ij} < 0$. An adjacent extreme point will be reached for which the path length from 1 to n is less than that known up to now.

(3) the arc introduced (i,j) generates neither circuit nor alternative path. It is then a degenerate case, i.e. several bases generate the same basic solution. From the graphic point of view,

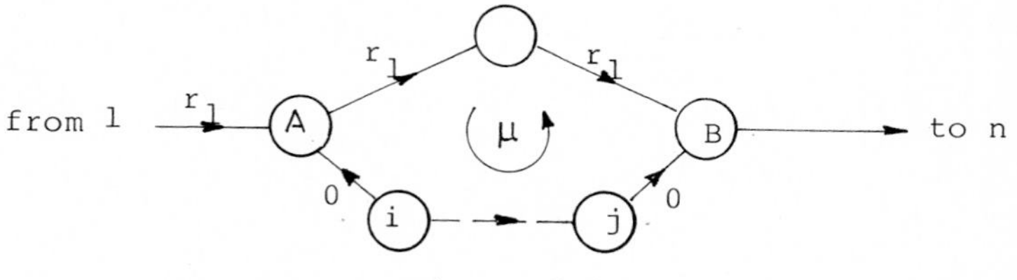

Figure 6.8

this implies that there exists at least one arc $(k,l) \in \mu^-$ such that $x_{kl} = 0$. Two or more trees (according to the degree of degeneracy) correspond to the same extreme point of the polyhedron and generate the same elementary path from 1 to n. On the example in figure 6.8 the introduction of (i,j) in the basis and the elimination of (i,A) in no way changes the elementary path from 1 to n. We have a degenerate problem.

The degeneracy of the primal problem generally complicates the resolution of the linear problem for although it is very unlikely, there are risks of cycling. In fact, we shall see in the next section that it is possible to get rid of problems of degeneracy.
In summary, we have just shown that the extreme points of S are minimal path flows for a given metric, that the extreme rays generated during the resolution of (6.2) are negative length circuits, and that the edges coming from an extreme point of S are loops constituted by a bifurcation on the elementary path from 1 to n associated with this extreme point.
In the following section we shall generalise these results in the case where each $i \neq n$ sends a product to destination n.

6.2.2. Multicommodity flow. Aggregation by destination

6.2.2.1. Multicommodity flow and aggregated flow

We now consider the case in which several products k = 1,... p circulate on the same network. We denote by x (k) the flow re-

resulting from the circulation of the product k and by r(k) the product k resources vector.

<u>Definition 6.5</u> : Multicommodity flow
The vector $[x'(1)\ x'(2)\ \dots\ x'(p)]'$ is a multicommodity flow for a resources vector $[r'(1)\ r'(2)\ \dots\ r'(p)]'$ if $Ax(k) = r(k)$ for $k = 1, \dots p$.

<u>Definition 6.6</u> : Feasible multicommodity flow
The vector $[x'(1)\ x'(2)\ \dots\ x'(p)]'$ is a feasible multicommodity flow for a resources vector $[r'(1)\ r'(2)\ \dots\ r'(p)]'$ if $Ax(k) = r(k)$, $x(k) \geq 0$ for $k = 1, \dots p$.

We denote by $S(k) = \{x(k) \,/\, Ax(k) = r(k),\ x(k) \geq 0\}$ the set of feasible flows of the product k. We denote by $\Sigma = S(1) \otimes S(2) \dots \otimes S(p)$ (where $\otimes$ represents the cartesian product) the set of feasible multicommodity flows. ($S(k)$ is a polyhedron defined in the Euclidean space E^m, Σ is therefore a polyhedron defined in E^{pxm}. Let us consider the vector

$x = \sum_{k=1}^{p} x(k)$ defined in E^m which we will call aggregated flow

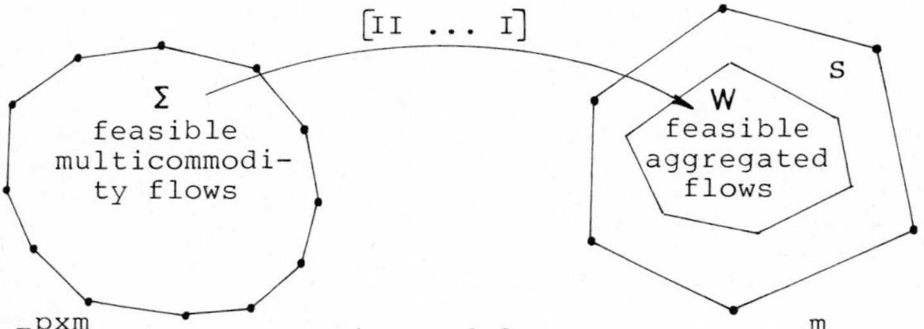

Figure 6.9

The linear operator [II ... I] which defines an aggregated flow, constituted by identity matrices I (m,m), makes the polyhedron W defined in E^m correspond to the polyhedron Σ defined in E^{pxm}, and we shall call this the polyhedron of the feasible aggregated flows.

<u>Definition 6.7</u> : Aggregated flow
The vector x is an aggregated flow for the resources vector $[r'(1)\ r'(2)\ \dots\ r'(p)]'$ if there exists a multicommodity flow $[x'(1)\ x'(2)\ \dots\ x'(p)]'$ such that

$$x = \sum_{k=1}^{p} x(k)$$

<u>Definition 6.8</u> : Feasible aggregated flow
The vector $x \geq 0$ is a feasible aggregated flow for the resources vector $[r'(1)\ r'(2)\ \dots\ r'(p)]'$ if there exists a feasible multicommodity flow $[x'(1)\ x'(2)\ \dots\ x'(p)]'$ such that

$$x = \sum_{k=1}^{p} x(k).$$

Let $S = \left\{ x \,/\, Ax = r,\ x \geq 0 \right\}$ with $r = \sum_{k=1}^{p} r(k)$ which defines a polyhedron in E^m.

Proposition 6.4 : $W \subset S$

According to definition 6.8, for $x \in W$ there exists vectors $x(k)$ which satisfy $Ax(k) = r(k)$, $x(k) \geq 0$ $\forall\, k = 1,\ldots p$ and which are such that

$\sum_{k=1}^{p} x(k) = x$. By summing on k the constraints $Ax(k) = r(k)$ we obtain $Ax = r$, $x \geq 0$ whence $x \in S$. On the other hand, $x \in S$ is a condition which is far from being sufficient to guarantee that x is a feasible aggregated flow for the resources vector $[r'(1)\ r'(2)\ \ldots\ r'(p)]'$ since aggregation reverts to the consideration that all the products are identical. The equation of conservation aggregated in a node is insufficient to guarantee the equation of conservation per product.

In general, when an aggregation of variables is made there is a loss of information, except in special cases such as the one we are going to examine in the following section.

6.2.2.2. Aggregation of products by destination

In this section we are assuming that there exists only one consumer node, all the others being producers. Without loss of generality, we shall suppose that all the products are destined for n. We call product k the product coming from k, destined for n. A resource vector $r(k)$ is therefore such that $r_i(k) = 0$ for $i \neq k, n$, $r_k(k) = r_k > 0$, $r_n(k) = -r_k$. We call $x^i(k)$ the flow resulting from the circulation of the product k. Of course there must exist in the graph at least one path to go from $k \neq n$ towards n. Since n is chosen arbitrarily, the following assumption must be satisfied :

Assumption 6.3 : The graph $G = (N,L)$ is highly connected, i.e. there exists a path from i to j $\forall i, \forall j \in N$, which must be the case in communication network.

In the rest of section we shall suppose that assumption 6.3 is verified.

Analysis of the nodes of the polyhedron Σ

The nodes of Σ correspond to the feasible basic solutions of the system $Ax(k) = r(k)$, $x(k) \geq 0$ for $k = 1,\ldots n-1$ which is decomposed into n-1 independent systems which were analysed in Section 6.2.1., whence, by generalisation of proposition 6.1 :

Proposition 6.5 : To every node of Σ one can make correspond a set of elementary paths μ_{kn} for $k = 1,\ldots n-1$, where μ_{kn} is an elementary path from k to n.

Let us note that it is always possible to make a set of cost vectors $d(k)$ for $k = 1,\ldots n-1$ correspond to a set of elementary paths μ_{kn} for $k=1, \ldots n-1$ such that the solution of the program

$$\min_{x(k) \in S(k)} \sum_{k=1}^{n-1} d(k)\, x(k) \qquad (6.9)$$

generates this set of elementary paths
Since the unit cost of transport is generally independent of the type of product in communication networks, the problem of the search for the optimal multicommodity flow $[x'(1)\ x'(2)\ \ldots\ x'(n-1)]'$ for the metric $d(k) = d\ \forall k$ is formulated as follows :

$$\min_{x(k) \in S(k)} d \sum_{k=1}^{n=1} x(k) \qquad (6.10)$$

This problem is decomposed into n-1 independent problems of search for optimal flow for the metric d

$$\min_{x(k) \in S(k)} d\, x(k) \quad \text{for } k=1,\ldots\ n-1 \qquad (6.11)$$

whence :

Definition 6.9 : An optimal multicommodity flow, solution of (6.10), is called a multicommodity flow with minimal path for d, for each of the flows of which it is composed has a minimal path for d.

Property 6.1 : A multicommodity flow with minimal path for d, solution of (6.10), is such that the set of minimal paths for d which converge towards n form an oriented tree with root n.

In fact, since (6.10) is decomposed into n-1 elementary problems (6.11) whose solution is a minimal path flow, we can therefore link to the solution of (6.10) a set of minimal paths for the metric d. On the other hand, we can observe that if the minimal paths did not form an oriented tree. There would exist one bifurcation beween 2 nodes A and B with a product passing along each branch of the bifurcation. Aside from the case of degeneracy where the length of the 2 branches is identical, the criterion could be made to decrease by causing the 2 flows to pass along the branch of minimum cost. If the length of the branches were identical, one could make the two flows pass along one single branch without changing the cost.
These results enable us to state the following proposition :

Proposition 6.6 : Let σ be the set of nodes of Σ which can be generated by the resolution of a problem of type (6.10). To every node$\in\sigma$ can be linked an oriented tree of minimal paths with root n for a given metric d.

The formulation of problem (6.10) in terms of aggregated flow becomes

$$\min_{x \in W} dx \qquad (6.12)$$

the solution of which corresponds to a node of W subject to assumption 6.2.

Analysis of the nodes of the polyhedron W

Since (6.12) is identical to (6.10)

the nodes of W are in one to one correspondence with the nodes $\in \sigma$. A node of W is therefore also characterised by an oriented tree of minimal path with root n for a given metric d.

Definition 6.10 : An aggregated flow, solution of (6.12) is called aggregated flow with minimal path for d.

Let us now characterise the nodes of the polyhedron S.

Proposition 6.7 : Every node of S is a node of W and vice versa. The nodes of S correspond to the feasible basic solutions of the system $Ax = r$, $x \geq 0$. We know (i) that a basis of $Ax = r$ is a tree of G and (ii) that a basic solution is only feasible if there exists a path from k to n for k=1,...n-1. From (i) and (ii) we deduce that to every node of S there corresponds a set of elementary paths μ_{kn} for k=1,...n-1 forming an oriented tree with root n.
This proposition if fundamental since problem (6.12) can be converted into (6.13) subject to assumption 6.2.

$$\begin{aligned} &\min \quad dx \\ &\text{subject to} \quad A_x = r, \quad x \geq 0 \end{aligned} \tag{6.13}$$

The resolution of this problem consists of finding the oriented tree of the minimal paths with root n for the metric d, a problem which is resolved in E^m when the resolution of the problem of multicommodity flow would have had to be done in $E^{(n-1)xm}$

Resolution of the problem of aggregated flow (6.13)

We have presumed $r_k > 0$ for $k \neq n$. A basic solution of (6.13) consists of making $r_k > 0$ circulate on the path μ_{kn} of an oriented tree of root n. The basic solutions are therefore non-degenerate. On the other hand, if $r_k = 0$ for at least one element $k \in \{1,...n-1\}$, the basic solution could be degenerate and may not necessarily have property 6.1. However, among the basis which generate the same degenerate basic solution, there is just one (supposing the problem to be dual non-degenerate) which has property 6.1, whence :

Proposition 6.8 : In the general case $r_k \geq 0 \quad \forall k = 1,...n-1$, to every node of W one can cause to correspond an oriented tree of minimal paths of root n. For the resolution of (6.12) we can therefore restrict the search for the optimal solution to the basis which have the property of being oriented trees of root n. One can thus eliminate the problem of the degeneracy of the primal.

Degeneracy of the dual problem.

Again we presume $r_k > 0$ for $k \neq n$. If the optimal basic solution is such that $\hat{d}^B > 0$, the optimal oriented tree is unique. There exists a unique path from every k towards n. We say that the solution is a set of paths without bifurcation. On the other hand, if $\hat{d}^{ij} = 0$ for at least one arc $(i,j) \in \bar{B}$, the optimum is no longer attained in one single node. The optimal solution is composed of paths with bifurcation (where the number of bifurcations corresponds to the degree of degeneracy of the dual). Let us suppose that the figure 6.7 represents an optimal basis with $\hat{d}^{ij} = 0$. The two branches of the bifurcation have identical costs and therefore the product on these branches can be oriented indifferently. If we call $0 \leq \varphi_{Ai} \leq 1$ the proportion of the product passing along the derivation, when φ_{Ai} is made to vary from 0 to 1 an edge of the polyhedron W is described. All the solutions of this edge are optimal. For a degeneracy of order > 1, i.e. when there are several possible bifurcations in an optimal solution, the set of optimal solutions describes a polyhedron.

We have just seen that when a network has to move products which are distinguished by their origin and their destination, as is the case in communication networks, with aggregation by destination it is possible to treat all the products destined for the same node as one single product. This aggregation does not involve any loss of information and it simplifies considerably the resolution of a multicommodity flow problem. We use it systematically in the remaining part of this chapter.

6.2.3. Other formulations of a multicommodity flow problem

In the preceding section we have shown how aggregation by destination makes it possible to reduce a multicommodity flow problem to a flow problem when concerned with the sub-set of products destined for the same node of the graph. In reality this problem can be formulated in three different ways :

(1) the nodes-arcs formulation which uses the incidence matrix. This is the one we have been considering until now. It is the most conventional.
(2) the arcs-paths formulation, closely linked to the previous one. Sometimes used in optimisation methods (particularly heuristic ones), it is interesting for its economic interpretation.
(3) formulation by the routing variables. This is less conventional, but is fundamental for the study of the distributed algorithms in section 6.4.
We are dealing here with formulations of type (2) and (3).

6.2.3.1. Arcs-paths formulation

The formulation of the nodes-arcs problem is :

$$\min_{x \in S} \; dx \tag{6.13}$$

with

$$S = \left\{ x \;/\; Ax = r, \quad x \geqslant 0 \right\}$$

According to the assumption (6.2), the resolution of (6.13) cannot generate an extreme ray. Every point $x \in S$ can therefore be written in the form of a convex combination of the nodes of S.

$$x = X\alpha \;, \quad \mathbb{1}'\alpha = 1, \quad \alpha \geqslant 0 \tag{6.14}$$

with X the matrix whose columns are the nodes of S. To each node of S is associated an oriented tree of root n. Let us put $\pi = dX$ as the cost row vector of the oriented trees. The problem (6.13) can be re-written as :

$$\begin{aligned} &\min \ \pi\alpha \\ &\text{subject to} \quad \mathbb{1}'\alpha = 1, \quad \alpha \geqslant 0 \end{aligned} \tag{6.15}$$

One can always re-arrange the indices in the basis in such a way that α_0 is the basic variable. π° is the cost of the associated oriented tree. The optimality conditions are written as :

$$\hat{\pi}^\ell = \pi^\ell - \pi^\circ \geqslant 0 \tag{6.16}$$

If (6.16) is satisfied with a strict inequality, the optimal oriented tree is unique. All the material is run on the minimal paths, which are without bifurcation. If $\hat{\pi}^\ell = 0$, there exists at least one other oriented tree of identical cost, the dual of (6.15) is degenerate. In the optimal oriented tree a bifurcation can be introduced without changing the cost of routing. In the general case one will be able to have a set of optimal oriented trees which will generate paths with bifurcations. The optimality conditions are :

$$\begin{aligned} \alpha_\ell > 0 \quad &\Rightarrow \quad \hat{\pi}^\ell = \pi^\ell - \pi^\circ = 0 \\ \hat{\pi}^\ell = \pi^\ell - \pi^\circ > 0 \quad &\Rightarrow \quad \alpha_\ell = 0 \end{aligned} \tag{6.17}$$

This is the complementary slackness theorem whose economic interpretation is :

<u>Proposition 6.9</u> : An optimal solution of problem (6.15) is such that for any couple (i,n) the cost of the paths used to move the products from i to n is identical to and lower than the cost of the non-used paths.

Test to check the optimality of a solution of problem (6.13)

Let x° be a solution of problem (6.13) which, in the general case will correspond to a set of paths with bifurcation (non-basic solution). The resolution of (6.13) generates an optimal basis solution x^*.

If $dx^\circ = dx^*$ then x° is an optimal solution (it belongs to the polyhedron of optimal solutions).

Note on this formulation. Many heuristic methods carry off one part of the flow on the high cost paths used and bring this flow back on the lowest cost path for each couple (i,n) until the (6.17) conditions are more or less satisfied.

6.2.3.2. Formulation by the routing variables

In the nodes-arcs and arcs-paths formulations, the solution of the flow problem generates the optimal flow on the arcs of the network. For application on a network these variables are not control variables. For a message coming from i, destined for n, the control variables could consist of a vector of the successive nodes to be crossed to reach n. However, since the optimal solutions are oriented trees (or combinations of oriented trees) of minimal paths, to move the products from a node i it is not necessary to know their origin. The control variables can be distributed by node, i.e. in every node i it is sufficient to know the successor node on the optimal path towards n. In the general case the solutions are paths with bifurcation. From a node i there can be several successors towards which the products can be guided. The control variables are the proportions of orientation towards the various successors, as we shall now describe in more detail.

Without losing generality let us consider the aggregated flow for destination n. Let r(n) be the resources vector, such that

$$r_i(n) \geqslant 0, \; r_n(n) = - \sum_{i=1}^{n-1} r_i(n).$$

We call t(n) the total traffic vector for the node n, i.e. $t_i(n)$ is the total flow of product n which leaves i for destination n. We obviously have $t_n(n) = 0$.
We call $\varphi_{ij}(n)$ the proportion of the traffic $t_i(n)$ using arc (i,j) and we put :

$$\Phi_{(n)} = \begin{bmatrix} \varphi_{11}(n) & \varphi_{12}(n) & \ldots & \varphi_{1n}(n) \\ \varphi_{n1}(n) & \varphi_{n2}(n) & \ldots & \varphi_{nn}(n) \end{bmatrix}$$

where $\varphi_{ij}(n) = 0$ if $(i,j) \notin L$. The total traffic to the node i is the sum of the resource in i and of all the traffic travelling through i, namely :

$$t(n) = r(n) + \Phi'(n)\, t(n) \tag{6.18}$$

which expresses implicitly the equations of conservation in each node.

<u>Definition 6.11</u> : $\Phi(n)$ is a routing matrix for the destination n if its elements $\varphi_{ij}(n)$ satisfy the following conditions :

C1 : $\varphi_{ij}(n) \geqslant 0 \quad \forall_{i,j} \quad ; \quad \varphi_{ij}(n) = 0$ if $(i,j) \notin L$

C2 : $\sum_j \varphi_{ij}(n) = 1$

C3 : $\forall\, i \neq n$ there exists a path from i to n, i.e. there exists a sequence i,j,...n such that $\varphi_{ij} > 0, \ldots \varphi_{j.} > 0, \ldots \varphi_{.n} > 0$.

C4 : $\varphi_{ni}(n) = 0 \quad \forall i \neq n \; ; \; \varphi_{nn}(n) = 1$

A matrix $\Phi(n)$ which satisfies properties C1 - C4 is a stochastis matrix to which can be associated a Markov chain whose representative graph is composed of the nodes of G and of all the arcs $(i,j) \in L$ such that $\varphi_{ij}(n) > 0$. The conditions C3 and C4 imply that n is the only recurrent state of the Markov chain

and $\phi(n)$ is therefore completely ergodic. The rank of the differential matrix $I - \phi(n)$ is therefore n-1 [GAN - 59].
The equation (6.18) can be re-written in the form :

$$[I - \phi'(n)]\ t(n) = r(n) \qquad (6.19)$$

For a routing matrix and a given resources vector, t(n) can be calculated by resolving (6.19) which is a system of dimension n with n-1 degrees of freedom. By noting that $t_n(n) = 0$ and that $[I - \phi(n)]_n = 0$, we can eliminate the component n of the vectors t and r, together with the raw and the column n of $\phi(n)$. We then obtain a system with rank n-1.

$$\left[I - \tilde{\phi}'(n)\right] \tilde{t}(n) = \tilde{r}(n) \qquad (6.20)$$

whose solution is unique :

$$\tilde{t}(n) = \left[I - \tilde{\phi}'(n)\right]^{-1} \tilde{r}(n) \qquad (6.21)$$

whence :

<u>Theorem 6.1</u>

Let there be a network for which we know the resources vector r(n) and a routing matrix $\phi(n)$ which meets the conditions C1 - C4. (i). The system (6.18) has a unique solution $t_{(n)}$. (ii) Furthermore $t_i(n)$ is non-negative and continously differentiable as a function of r(n) and $\phi(n)$.

For proof of the second part of the theorem we shall refer to [GAL - 77].

<u>Note</u>. To every matrix $\phi(n)$ satisfying the conditions C1, C2, C4, we can associate the partial graph $G^+ = (N, L^+)$ with $L^+ = \{(i,j) / \varphi_{ij} > 0\}$. The uniqueness of the solution of (6.18) is guaranteed only if condition C3 is met, i.e. G^+ is connected and if there exists at least one path from i to n in G^+. If this is not the case, two situations must be analysed :

(1) G^+ is not connected and there exists a connected component $G_o^+ = (N_o, L_o^+)$ constituted by a set of nodes $\ell \in N_o$ such that $r_\ell(n) = 0$. The system (6.18) then has an infinity of solutions as an arbitrary flow can be made to circulate on all the loops of G_o^+. Physically a routing control which would consist of making information circulate in a sub-graph of G has no sense.

(2) There exists a recurrent chain $G_o^+ = (N_o, L_o^+)$ with $n \notin N_o$, supplied by traffic. This means that the recurrent chain is not isolated or that in the event of it being isolated (constituting a connected component) we would have $r_\ell(n) > 0$ for at least one node $\ell \in N_o$. In this case there would be product accumulation in

the recurrent chain. The system (6.18) would therefore have no solution, i.e. the routing control would not be feasible. These two cases must be set aside and it is therefore natural to limit the routing controls to the matrices which also satisfy condition C3.
This formulation presents advantages which will be exploited in section 6.4

6.2.4. Total aggregation of products

In communication networks products are distinguished by their origin and their destination. In section 6.2.2. we showed that products can be aggregated by destination when they use the same metric for a given destination. The solution of the aggregated problem (6.13) provides the solution of each elementary flow problem (6.11). If we now consider each aggregated flow for a given destination as an elementary flow, we can ask ourselves to what extent it is possible to aggregate them themselves.
Let us therefore say :
$r(k)$ is the resources vector of the product destined for k which is such that $r_i(k) > 0$ for $i \neq k$, $r_k(k) = - \sum_{i \neq k} r_i(k)$

$x(k)$ is the flow generated by the circulation of the product destined for k.

In the most general case there are as many flow aggregated by destination as there are nodes in the network. The set of feasible solutions of this multicommodity flow problem defines a polyhedron Σ whose nodes can be characterised.

Analysis of the nodes of the polyhedron

The nodes of Σ correspond to the feasible basic solutions of the system $Ax(k) = r(k)$, $x(k) \geq 0$ for $k=1,\dots n$ which is decomposed into n independent systems which have been analysed in section 6.2.2. whence, by using proposition 6.5 :

<u>Proposition 6.10</u> : To every node of Σ corresponds a set of elementary oriented trees $\mathcal{A} = \{\mathcal{A}_k$ for $k=1,\dots n\}$ where $\mathcal{A}_k$ is an oriented tree of elementary paths which converge towards k.

This means that we can always find a metric $d(k)$ for $k=1,\dots n$ such that the solution of the program

$$\min_{x(k) \in S(k)} \sum_{k=1}^{n} d(k)\, x(k) \tag{6.22}$$

generates this set of oriented trees of elementary paths $\mathcal{A}$.

Since the unit cost of transport is independent of the product type, the search for the optimal multicommodity flow for the metric $d(k) = d \ \forall k$ s'écrit :

$$\min_{x(k) \in S(k)} d \sum_{k=1}^{n} x(k) \tag{6.23}$$

which is decomposed into n independent problems of flow aggregated by destination. The aggregated flow is defined by

$$x = \sum_{k=1}^{n} x(k)$$

The set of feasible aggregated flow defines a polyhedron W. In terms of aggregated flow, (6.23) can be written in the form :

$$\min_{x \in W} \ dx \tag{6.24}$$

the optimal solution of which corresponds to a node of W.

Analysis of the nodes of the polyhedron W

<u>Proposition 6.11</u> : To every node of W corresponds (i) a set of oriented trees (ii) forming a set of minimal paths between all the nodes of the graph for a certain metric d. (i) follows from proposition 6.10 and (ii) can be proved by reasoning analogous to that used to justify property 6.1.

<u>Definition 6.12</u>: An aggregated flow, solution of (6.24), is called aggregated flow with minimal path for d, i.e. to the optimal solution of (6.24) there corresponds a set of minimal paths between all the nodes of the graph for the metric d.

Let $S = \left\{ x / \ Ax=r \text{ with } x = \sum_{k=1}^{n} x(k),\ r = \sum_{k=1}^{n} r(k) \right\}$.

Contrary to the aggregation in section 6.2.2., there is no one to one correspondence between the nodes of S and the nodes of W, and so it is not possible to work exclusively with the aggregated flow. However, proposition 6.11 is important because it is exploited in certain methods approached in section 6.3.

6.2.5. Conclusion

In this section :

1) We have introduced the vocabulary particular to problems of (multicommodity) flows. We have stressed the properties of the polyhedron of feasible (multicommodity) flows which facilitate comprehension of the optimisation methods which will be taken up in the following sections.

2) We have introduced the different formulations of a problem of search for optimal flow :

   a) nodes-arcs formulation. The most conventional formulation consists of working directly on the flow vectors. Global optimisation methods are based essentially on this formulation.

   b) arcs-paths formulation. Especially used in heuristic optimisation methods (not used in this study).

   c) formulation by routing variables. Less conventional this is particularly interesting for the fact that it handles the control variables of a flow problem when in the two preceding formulations the routing policy must be deduced from the optimal solution. This formulation is used in the distributed methods examined in section 6.4.

3) For the conventional nodes-arcs formulation we have insisted on passing from a flow problem to a multicommodity flow problem by means of the aggregation of products. When several products circulate to the same destination node we have shown that the aggregation of products into one (passing from a multicommodity flow to a flow) does not involve any loss of information. When products circulate between each pair of nodes, aggregation by destination is no longer possible even though it has properties exploited in certain methods.

We have so far considered linear criteria. We now take up the examination of non-linear multicommodity flow problems and we devote the next section to conventional methods of resolution.

6.3 Conventional methods of resolution of multicommodity flow problems

6.3.1. Introduction

In a real network the total quantity of product circulating in an arc (i,j) is given an upper limit by the capacity of the arc, noted as c_{ij}. The vector c (m,1) is called the network capacity vector. In the rest of this study we make the following assumption, which is widely adopted :

<u>Assumption 6.4</u> : The transmission cost of the flow x_{ij} in the arc (i,j) is only dependent on x_{ij}. We shall denote this by $f_{ij}(x_{ij})$.

The function $f_{ij}(x_{ij})$ is generally non-linear. An interesting class of functions, which gives expression well to the phenomena of network saturation and which we adopt in the rest of this study, is that which satisfies the following assumptions:

<u>Assumptions 6.5</u> : The cost functions $f_{ij}(x_{ij})$ are such that :

a) $f_{ij}(x_{ij})$ is a convex function of class C^2

b) $f'_{ij}(x_{ij}) > 0$ for $0 \leq x_{ij} \leq c_{ij}$

c) $f''_{ij}(x_{ij}) > 0$ for $0 \leq x_{ij} \leq c_{ij}$

d) $\lim_{x_{ij} \to c_{ij}} f_{ij}(x_{ij}) = +\infty$

For the whole network it is natural to adopt an additive criterion in the form $\sum_{(i,j)\in L} f_{ij}(x_{ij})$ which supposes the non-existence of interactions between the arcs.

We are concerned with networks in which the products are aggregated by destination. $x(\ell)$ is the flow resulting from circulation of the product destined for ℓ and $r(\ell)$ is the resources vector of the product ℓ, with $r_i(\ell) \geq 0$ for $i \neq \ell$ and

$$r_\ell(\ell) = - \sum_{i\neq \ell} r_i(\ell).$$

From the set of feasible multicommodity flows, it is proposed to find the one which minimises the global criterion, i.e. the problem can be written in the form :

$$\min \sum_{(i,j) \in L} f_{ij} \left[\sum_{\ell=1}^{n} x_{ij} (\ell) \right] \tag{6.25a}$$

Subject to :

$$A\, x(\ell) = r\,(\ell), \quad x(\ell) \geqslant 0 \text{ for } \ell = 1,2,\ldots n \tag{6.25b}$$

$$\sum_{\ell=1}^{n} x(\ell) \leqslant C \tag{6.25c}$$

We have just defined a multicommodity flow problem which is constrained because of the presence of the restrictions (6.25c). By assumptions 6.5 we shall see, however, that it is possible to return to a non-constrained multicommodity flow problem. In this section we are considering four conventional methods for resolving it : one sequential method of feasible directions, one aggregated method of feasible directions, the extreme flows method and finally a projection method.

6.3.2. Sequential feasible method [FRA - 73]

To solve the multicommodity flow problem (6.25) we solve a sequence of problems of single flow. An iteration of this method consists of n sub-iterations in which the sub-iteration k is relative to the product k. Let us suppose to be known a feasible solution of (6.25) which we note as $x°(\ell)$ for $\ell = 1,2,\ldots n$. For the sub-iteration k we restrict optimisation to the sub-space of flow k by imposing $x(\ell) = x°(\ell)$ for $\ell \neq k$. The restricted problem is written as :

$$\min \sum_{(i,j) \in L} f_{ij} \left[x_{ij}(k) \right] \tag{6.26a}$$

subject to

$$Ax(k) = r(k), \quad x(k) \geqslant 0 \tag{6.26b}$$

$$x(k) \leqslant C - \sum_{\ell \neq k} x°(\ell) \tag{6.26c}$$

Let us take $S(k) = \{x(k) \,/\, Ax(k) = r(k),\ x(k) \geqslant 0\}$ and

$$\overline{S}(k) = S(k) \cap \left\{ x(k) \,/\, x(k) \leqslant C - \sum_{\ell \neq k} x°(\ell) \right\}$$

If we put $f_{ij}(x_{ij}) = +\infty$ for $x_{ij} \geqslant c_{ij}$, the constraint (6.26c) is included in criterion (6.26a) in the form of penalty function. If an initial feasible solution is known, any primal optimisation method will give rise to a sequence of solutions $x(k) \in \overline{S}(k)$ without being necessary to take into account the constraints (6.26c). The sub-iteration k is the resolution of a single flow problem, which we call a restricted problem :

$$\min f[x(k)] = \sum_{(i,j)\in L} f_{ij}[x_{ij}(k)] \qquad (6.27a)$$

subject to

$$A\,x(k) = r(k), \quad x(k) \geqslant 0 \qquad (6.27b)$$

for which an initial feasible solution $x^{\circ}(k)$ is known.

Optimality condition

Let $\hat{x}(k)$ be a feasible flow for program (6.27). We make an expansion to the first order of $f[x(k)]$ around $\hat{x}(k)$:

$$f[x(k)] = f[\hat{x}(k)] + \nabla f[\hat{x}(k)][x(k) - \hat{x}(k)] + 0\,[x(k) - \hat{x}(k)]$$

For $\hat{x}(k)$ to be a local minimum we must have

$$\nabla f[\hat{x}(k)][x(k) - \hat{x}(k)] \geqslant 0 \quad \forall x(k) \in S(k)$$

such that $x(k) - \hat{x}(k)$ is a feasible direction. Since $S(k)$ is convex, every direction $x(k) - \hat{x}(k)$ with $x(k) \in S(k)$ is feasible and the necessary optimality condition is therefore written as :

$$\nabla f[\hat{x}(k)]\,[x(k) - \hat{x}(k)] \geqslant 0 \quad \forall x(k) \in S(k) \qquad (6.28)$$

On the other hand, according to the assumptions 6.5 on the cost functions $f_{ij}(x_{ij})$, the criterion (6.27a) is convex, so that (6.28) is a necessary and sufficient condition for $\hat{x}(k)$ to be the global minimum.

Resolution of the restricted problem

As is conventional in non-linear programming, any iteration for the resolution of program (6.27) proceeds in two stages :

(1) Search for a direction to cause the criterion to decrease. A point $\hat{x}(k)$ being given, we find whether the optimality condition (6.28) is satisfied. For this we look to see if there exists a point $x(k) \in S(k)$ such that $\nabla f[\hat{x}(k)][x(k) - \hat{x}(k)] < 0$ by resolution of the linear program (6.29) in which we put $d = \nabla f[\hat{x}(k)]$:

$$\begin{aligned} &\min\ d\,x(k) \\ &\text{subject to} \\ &A\,x(k) = r(k), \quad x(k) \geqslant 0 \end{aligned} \qquad (6.29)$$

According to section 6.2 we know that the solution of this program, noted as $x^*(k)$, corresponds to an oriented tree of minimal paths of root k for the metric d.
If $d\,x^*(k) < d\,\hat{x}(k)$, a movement in the direction $x^*(k) - \hat{x}(k)$ makes it possible to decrease the criterion. The direction generated is, in a certain way, the best, because it results from a minimisation program, but it is not the direction in which the directional derivative is minimum.

(2) Search for the minimum of $f[x(k)]$ in the direction $x^*(k) - \hat{x}(k)$.

Every solution with the form $x(k) = (1-\lambda)\,\hat{x}(k)+\lambda\, x^*(k)$ is a feasible flow for $\lambda \in [0,1]$. We resolve the line search problem :

$$\min_{\lambda\in[0,1]} \quad f\left[(1-\lambda)\,\hat{x}(k) + \lambda\, x^*(k)\right] \tag{6.30}$$

which generates a new feasible solution $\hat{x}(k)$ for (6.27). We return to stage (1)

If we get $d\, x^*(k) = d\, \hat{x}(k)$ at stage (1), then $\hat{x}(k)$ is the optimal solution of the restricted problem.

Interpretation of the optimality condition

The optimality condition (6.28) is identical to the condition $d\, x^*(k) = d\, \hat{x}(k)$. Let us eliminate the very special case where $\hat{x}(k)$ would be a node of $S(k)$. We would then get $x^*(k) = \hat{x}(k)$ and the optimal solution would correspond to the tree of the minimal paths of root k for the metric d. The products would use one unique path to get to their destination. In general $\hat{x}(k)$ is not a node of $S(k)$, ie. the optimal solution corresponds to a set of paths with bifurcation. In this case we get $x^*(k) \neq \hat{x}(k)$ but the optimality condition $d\, x^*(k) = d\, \hat{x}(k)$ means that $\hat{x}(k)$ is optimal if it belongs to the polyhedron of the optimal solutions of linear program (6.29) which must necessarily be degenerate. The optimal solution $\hat{x}(k)$ therefore corresponds to a set of minimum paths with bifurcation, with destination k for the metric d. $x^*(k)$ is one of the oriented trees of minimal paths which can be generated by the resolution of program (6.29) whose dual is degenerate for $\hat{x}(k)$ optimal.

Then we can state that each product coming from i and destined for k uses paths whose marginal cost is identical to and lower than the marginal cost of non-used paths.

The most conventional use of this method consists of working sequentially on the products. In section 6.3.3. we shall show that application of this method leads to the optimum of the multicommodity flow problem (6.25). Indeed, more refined variants of the method can be used, as the Aitken double sweep method, or the method can be applied to a sub-set of products before moving to another sub-set when the optimum has been reached on the first (hierarchisation in the method of dealing with the problem).

6.3.3. Aggregated feasible method [FRA - 73]

Instead of working sequentially on the products it is possible to treat them simultaneously by considering the aggregated flow $x = \sum_k x(k)$.

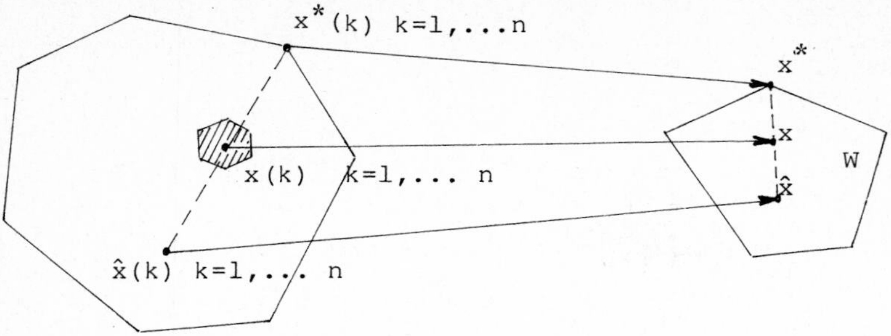

Figure 6.10

Let x° be a feasible aggregated flow for which a feasible multicommodity flow is known $x^\circ(k)$ k=1,2,...n. Formulated in terms of aggregated flow, the problem (6.25) becomes :

$$\min_{x \in W} \quad f(x) = \sum_{(i,j) \in L} f_{ij}(x_{ij}) \tag{6.31a}$$

$$x \leq C \tag{6.31b}$$

For the same reasons as those given in section 6.3.2. the capacity constraints (6.31.b) can be eliminated because they are included, in the form of penalty functions, in the cost functions. The problem is therefore the same as in (6.31.a).

Optimality condition

Let $\hat{x}$ be a feasible aggregated flow of (6.31.a). By making an expansion to the first order of $f(x)$ around $\hat{x}$, we get :

$$f(x) = f(\hat{x}) + \nabla f(\hat{x})(x-\hat{x}) + 0(x-\hat{x})$$

Since W is convex, the necessary optimality condition is written as :

$$\nabla f(\hat{x})(x - \hat{x}) \geqslant 0 \qquad \forall x \in W \tag{6.32}$$

And since, according to the assumptions 6.5, $f(x)$ is convex, the condition (6.32) is also a sufficient condition for $\hat{x}$ to be the global minimum.

Resolution of the aggregated problem

The program (6.31.a) is non-linear. Any iteration for its resolution proceeds in two stages :

(1) Search for a direction in W to make the criterion decrease. Given a point $\hat{x} \in W$ for which a multicommodity flow $\hat{x}(k)$ k=1,... n is known, we find whether the optimality condition (6.32) is verified. For this we resolve the following linear

program in which we put $d = \nabla f(\hat{x})$:

$$\min_{x \in W} \; d\,x \tag{6.33}$$

which generates the extreme point of W for the metric d. This point is the image of the extremal multicommodity generated by the resolution of the n problems :

$$\left.\begin{array}{l} \min \; d\,x(k) \\ \text{subject to} \\ Ax(k) = r(k), \quad x(k) \geqslant 0 \end{array}\right] \quad k=1,\ldots n \tag{6.34}$$

Let $x^*(k)$ be the solution of each flow problem. It corresponds to the oriented tree of the minimal paths of root k for the metric d. The set of problems generates the set of minimal paths between all the nodes for the metric d. In this sense the problem can be resolved in an aggregated way as there exist matrix methods which give the set of minimal paths between all the nodes for a given metric. From this set of minimal paths we can easily deduce x^* and $x^*(k)$ for $k=1,\ldots n$ without it having been necessary to resolve the n programs (6.34).
If $d\,x^* < d\,\hat{x}$, a movement in the direction $x^* - \hat{x}$ will produce a decrease of the criterion.

(2) Search for the minimum f(x) in the direction $x^* - \hat{x}$. Every aggregated flow of the form $x = (1-\lambda)\,\hat{x} + \lambda x^*$ is feasible for $\lambda \in [0,1]$. The following line search problem must be resolved :

$$\min_{\lambda \in [0,1]} \; f\left[(1-\lambda)\,\hat{x} + \lambda\, x^*\right] \tag{6.35}$$

There again we work with the aggregated flow. Let us note, however, that knowledge of the optimal aggregated flow alone would not be of interest, because the search for a feasible multicommodity flow whose image is a given aggregated flow is a problem as complex as the search for an optimal multicommodity flow. In general an aggregated flow is the image of a polyhedron of feasible multicommodity flows (see figure 6.10). It is however possible to alleviate easily the disadvantage of this method for at each iteration one can determine a multicommodity flow whose image is the aggregated flow resulting from (6.35). We have, in fact, assumed that to $\hat{x}$ we know how to link a feasible multicommodity flow $\hat{x}(k)$ for $k=1,\ldots n$, of which $\hat{x}$ is the image. We have also indicated that the search for the set of minimal paths between all the nodes for the metric d makes it possible to determine $x^*(k)$ for $k=1,\ldots n$. If $\tilde{\lambda}$ optimises (6.35) it is easy to see that $\tilde{x}(k) = (1-\tilde{\lambda})\,\hat{x}(k) + \tilde{\lambda}\, x^*(k)$ for $k=1,\ldots n$ has as its image $\tilde{x} = (1-\tilde{\lambda})\,\hat{x} + \tilde{\lambda}\, x^*$. From the feasible multicommodity flows whose image is $\tilde{x}$, we know how to find one by convex linear combination of already known feasible multicommodity flows (figure 6.10).
We go successively through stages (1) and (2) until we get $d\,x^* = d\,\hat{x}$, in which case $\hat{x}$ is the optimal solution of problem (6.31.a).

Interpretation of the optimality condition

The optimality condition (6.32) is identical to the condition $d\,x^* = d\,\hat{x}$. Putting aside the special case in which the optimal solution of (6.31.a) would be an extreme point of W, $\hat{x}$ is optimal if it belongs to the polyhedron of the optimal solutions of (6.33). To every node of this polyhedron there corresponds a set of minimal paths for the metric d between the nodes of the graph, so that the optimal paths are paths with bifurcation. The vector d defines the marginal cost. Thus at the optimum for each origin-destination couple, all the paths used have the same marginal cost which is lower than the marginal cost of the non-used paths.

Equivalence between the optimality conditions (6.32) and (6.28)

Proposition 6.12 : The sequential resolution and the aggregated resolution of the multicommodity flow problem (6.25) are equivalent. In fact :

(1) (6.28) for $k=1,\dots n \Longrightarrow$ (6.32). Since $\hat{x} = \Sigma\, \hat{x}(k)$, we have

$$\nabla f\,[\hat{x}(k)] = \nabla f\,(\hat{x}) \qquad \forall k=1,\dots n$$

$$(6.28) \quad \Longrightarrow \quad \nabla f\,(\hat{x})\,[x(k) - \hat{x}(k)] \geqslant 0 \qquad \forall k,\ \forall x(k) \in S(k)$$

whence, by summing on k :

$$\nabla f(\hat{x})[\sum_k x(k) - \sum_k \hat{x}(k)] \geqslant 0 \qquad \forall\ x(k) \in S(k)$$

This expression is identical to (6.32) since x describes W when the multicommodity flows describe $\Sigma = \bigotimes_k S(k)$

(2) (6.32) $\Longrightarrow$ (6.28) for $k=1,\dots n$ can be proved by contradiction. Let us suppose that (6.28) is not satisfied for $k=\ell$. By resolving the program :

$$\min\ d\,x(\ell) \qquad \text{with } d = \nabla f(\hat{x}) = \nabla f(\hat{x}(\ell))$$

subject to

$$A\,x(\ell) = r(\ell),\ x(\ell) \geqslant 0$$

we would get a solution $x^*(\ell)$ such that a movement in the direction $x^*(\ell) - \hat{x}(\ell)$ would make it possible to decrease f(x). By making :

$$\begin{cases} x(\ell) = \hat{x}(\ell) + \varepsilon\,[x^*(\ell) - \hat{x}(\ell)] \\ x(k) = \hat{x}(k) \qquad \text{for } k \neq \ell \end{cases} \tag{6.36}$$

the direction $x^*(\ell) - \hat{x}(\ell)$ would be a feasible direction for the aggregated flow from $x = \hat{x}$ in which f(x) could be decreased. Consequently (6.32) would not be satisfied.

6.3.4. Extremal flows method [CAN - 74]

The extremal flows method is a decomposition method which works on the aggregated flow. Let us recall that the problem is :

$$\min_{x \in W}\ f(x) \tag{6.31.a}$$

for which a necessary and sufficient condition for $\hat{x}$ to be the optimal solution is written as :

$$\nabla f(\hat{x})\ (x - \hat{x}) \geqslant 0 \qquad \forall\ x \in W \tag{6.32}$$

As in the Dantzig-Wolfe decomposition principle, this method consists of working with a particular definition of the polyhedron W. The expression $x \in W$ is equivalent to :

$$x = X\alpha \quad , \mathbb{1}'\alpha = 1, \quad \alpha \geqslant 0 \tag{6.37}$$

where X is a matrix whose vector columns are the nodes of W. We have shown in section 6.2.4. that these nodes correspond to sets of minimal paths between all the nodes of the graph for a certain metric. (6.37) expresses that any feasible flow can be written as a convex combination of flows with minimal paths, called extremal flows. Reformulated in α the problem (6.31.a) is written as :

$$\min \quad f(\alpha)$$
subject to
$$\mathbb{1}'\alpha = 1 \quad , \alpha \geqslant 0 \tag{6.38}$$

<u>Proposition 6.12</u> : Any aggregated flow can be expressed in the form of a convex linear combination of at the most m+1 aggregated extremal flows, with m the number of arcs of the graph.

Let x be an aggregated feasible flow. It can be written in the form $X\alpha = x$, $\mathbb{1}'\alpha = 1$, $\alpha \geqslant 0$. Therefore a feasible solution exists for the system of m+1 linear equations of rank $\rho \leqslant m+1$. According to the fundamental theorem of linear programming, there exists, a feasible basic solution with at most ρ non-zero variables.
In particular, if an aggregated flow is expressed by a combination of m+2 extremal flows, then by means of a pivot operation one can always return to a combination of m+1 extremal flows.

Principle of the method.

Let $\hat{x}^\circ$ be a feasible aggregated flow which is a solution of the program :

$$\min_{x \in V^\circ} \quad f(x) \tag{6.39}$$

where V° is a polyhedron composed of at most m+1 nodes. $\hat{x}^\circ$ can be expressed in the form of a convex combination of the nodes of V°. As in the methods examined in sections 6.3.1. and 6.3.2., the optimality condition (6.32) can be tested by resolving the program :

$$\min_{x \in W} \quad dx \qquad \text{with } d = \nabla f(\hat{x}^\circ) \tag{6.33}$$

Let X^* be an optimal basic solution of (6.33). We know that it corresponds to a set of minimal paths between all the nodes for the metric $d = \nabla f(\hat{x}^\circ)$.

(a) If $d X^* = d \hat{x}^\circ$, it is not possible to find a direction in which the criterion can decrease. $\hat{x}^\circ$ belongs to the polyhedron of the optimal solutions of (6.33). The conditions of optimality (6.32) are satisfied and $\hat{x}^\circ$ is therefore the optimal solution of (6.31.a).

(b) If $d X^* < d \hat{x}^\circ$, by moving in the direction $X^* - \hat{x}^\circ$ the criterion can be made to decrease. Let us therefore add X^* to the polyedron V° to constitute a new polyhedron $\tilde{V} = V_\circ \cup \{X^*\}$ composed of at most m+2 nodes. Resolution of the program :

$$\min_{x \in \tilde{V}} \quad f(x) \tag{6.40}$$

generates a solution $\tilde{x}$ such that $f(\tilde{x}) < f(\hat{x}^\circ)$. If $\tilde{x}$ is a combination of m+2 nodes, according to the proposition (6.12) one can always make a node go and thus return to a new polyhedron V^1, with $\hat{x}^1 = \tilde{x}$ the optimal solution in this new polyhedron.

Extremal flow method.

<u>Step 0</u> : Initialisation. We presume known an initial aggregated flow x°, a convex combination of $b \leq m+2$ extremal flows $X^\circ = [X^{\circ 1} \; X^{\circ 2} \; \ldots \; X^{\circ b}]$ which constitute the extreme points of the polyhedron V°. We have $x^\circ = X^\circ \alpha^\circ$, $\mathbb{1}'\alpha^\circ = 1$, $\alpha^\circ \geq 0$.

<u>Step 1</u> : Resolution of the master program. Find the optimal aggregated flow in V° by resolving :

$$\min \quad f(\alpha) \tag{6.38}$$

subject to

$$\mathbb{1}'\alpha = 1, \qquad \alpha \geq 0$$

for which we know the initial solution α°. The authors propose to use a gradient projection method. A dual method would seem more appropriate to us. Let $\hat{\alpha}^\circ$ be the solution of (6.38). If it is a combination of m+2 nodes, we can make a node of V° go, which constitutes a new polyhedron $\hat{V}^\circ$.

<u>Step 2</u> : Resolution of the sub-problem and optimality test. Calculate $d = \nabla f(\hat{x}^\circ)$ and resolve the program (6.33) (search for minimal paths between all the nodes for the metric d). Let X^* be the corresponding extremal flow.

(a) If $d X^* = d \hat{x}^\circ$, stop. The optimality conditions (6.32) are satisfied. $\hat{x}^\circ$ is the optimal solution of (6.31.a).

(b) If $d X^* < d \hat{x}^\circ$, the set of extreme points $[\hat{X}^\circ \; X^*]$ constitutes a new polyhedron. $\hat{x}^\circ$ can be written in the form :

$$\hat{x}^\circ = [X^\circ \; X^*]\begin{bmatrix}\hat{\alpha}^\circ \\ 0\end{bmatrix}$$

Return to step 1.

6.3.5. Gradient projection method [LUE - 73],[SCH - 76]

Together with the last two methods, this can be considered as working on the aggregated flow, because the flows of the different products are modified simultaneously. In the previous methods a linear program must be resolved at each iteration in order to carry out the optimality test. The idea of the gradient projection method consists of projecting the gradient on the active constraints to generate a direction in which the criterion can decrease. In the case where the gradient is orthogonal to the sub-space defined by the active constraints, if the optimum is not reached, relaxing a constraint it is possible to improve the criterion.

Principle of the method. Consider the problem :

$$\min \quad f(x) \qquad (6.41)$$

subject to

$$Ax = r , \quad x \geq 0$$

where A is a full rank matrix. Let x° be an initial feasible solution. We define the set of active inequality constraints by

$$\bar{J}(x^\circ) = \{u \,/\, x^\circ_u = 0\}$$

and we call the complementary set of $\bar{J}$, J. We look for a direction s such that $\nabla f(x^\circ)\, s < 0$ to make the criterion decrease. We first look for s maintaining active those constraints which were active, by doing :

$$\begin{cases} As = 0 \\ x_{\bar{J}} = 0 \end{cases} \qquad (6.42)$$

In the sub-space defined by (6.42) we choose the direction given by the projection of $-\nabla f(x^\circ)' = -g$. The components of S_J can be obtained by $s_J = -P\, g_J$ where P is a projection matrix given by

$$P = \left\{ I - [A^J]' \, [A^J (A^J)']^{-1} \, A^J \right\} \qquad (6.43)$$

with A^J the matrix constituted by the set of columns A^u of A for $u \in J$. Two cases must be considered :

(1) $s_J \neq 0$. We can show that $[\nabla f(x^\circ)]_J \cdot s_J < 0$. A movement in the direction s_J makes it possible to decrease the criterion. We carry out the line search :

$$\max_{0 \leq \alpha \leq \bar{\alpha}} \quad f(x^\circ_J + \alpha s_J) \quad . \qquad (6.44)$$

with $\bar{\alpha} = \max \left\{ \alpha \,/\, x^\circ_J + \alpha s_J \geq 0 \right\}$, which generates a new point x^1.

(2) $s_J = 0$. The vector of the Kuhn-Tucker parameters at the point under consideration is given by :

$$\beta = g_{\bar{J}} - [A^{\bar{J}}]' \, [A^J (A^J)']^{-1} \, A^J g_J \qquad (6.45)$$

(a) If $\beta \leqslant 0$, x° is optimal because the Kuhn-Tucker conditions are satisfied.

(b) In the opposite case, $\bar{J}$ is updated by taking away from the set of active inequalities constraints the one for which β_j is maximum. The gradient must then be projected in the sub-space defined by the new set of active constraints.

Resolution of a multicommodity flow problem

Let us recall that the problem to be solved is :

$$\min f \left[\sum_k x(k)\right] \tag{6.46.a}$$

subject to

$$Ax(k) = r(k) \qquad x(k) \geqslant 0 \quad \text{for } k=1,\ldots n \tag{6.46.b}$$

If the graph is connected, we know that $\rho(A) = n-1$ (see section 6.2.1.). In order to have a full rank matrix for each type of product k, we eliminate the equation of conservation relative to node k. Let A(k) be this matrix.
Let $x^\circ(k)$ for $k=1,\ldots n$ be an initial feasible solution. Let us call $\bar{J}_k(x^\circ(k)) = \{u \,/\, x^\circ_u(k) = 0\}$ and $\bar{J} = \bigcup_k \bar{J}_k$. The complementary set of $\bar{J}_k$ is denoted by J_k. To this feasible multicommodity flow corresponds the feasible aggregated initial flow $x^\circ = \sum_k x^\circ(k)$.

We first find a direction $[s'(1)\ s'(2)\ \ldots\ s'(n)]'$ such that $\nabla f\,(x^\circ(k)).\, s(k) \leqslant 0$ for $k=1,\ldots n$ keeping the same set of active constraints, namely :

$$\begin{bmatrix} A(1) & & \\ & A(2) & \\ & & \ddots \; A(n) \end{bmatrix} \begin{bmatrix} s(1) \\ s(2) \\ \vdots \\ s(n) \end{bmatrix} = \begin{bmatrix} 0 \\ 0 \\ 0 \end{bmatrix} \quad s_{\bar{J}_k}(k)=0 \text{ for } k=1,\ldots n \tag{6.47}$$

Let $\lambda_k = \text{card}\,(J_k)$ and let $\lambda = \sum_k \lambda_k$. Since $\nabla f\,(x^\circ(k)) = \nabla f(x^\circ)$ and since (6.47) has an angular block structure, $\nabla f(x^\circ)$ must be projected on complementary sub-spaces of E^λ. The projection is decomposed into n problems of projection of $-\nabla f(x^\circ)$ on the sub-spaces defined by

$$A(k)\, s(k) = 0 \qquad s_{\bar{J}_k}(k) = 0 \quad \text{for } k=1,\ldots n \tag{6.48}$$

The projection matrix therefore has an angular block structure

$$P = \begin{bmatrix} P(1) & & \\ & P(2) & \\ & & \ddots \; P(n) \end{bmatrix} \tag{6.49}$$

where the dimension of P(k) is (λ_k, λ_k). Although one is indu-

ced to work with each type of product, the method must be considered as manipulating an aggregated flow, for if the decomposition appears when projecting the gradient, all the flows will be updated simultaneously.
Let us note that

$$\sum_k \nabla f(x^\circ(k)).\ s(k) = \sum_k \nabla f(x^\circ).s(k) = \nabla f(x^\circ).\sum_k s(k)$$

By putting $s = \sum_k s(k)$, if one knows how to find $s(k)$ such that $\nabla f(x^\circ).\ s(k) < 0$ one will then have $\nabla f(x^\circ).\ s < 0$. A movement in the direction s in the set of aggregated flows makes it possible to decrease the criterion.
The projection of $-\nabla f(x^\circ)$ on each sub-space defined by (6.48) generates the direction $s_{J_k}(k)$ for $k=1,\ldots n$. Two cases must be considered :

(1) $s_{J_k}(k) \neq 0 \quad \forall k$. For each product we calculate

$$\bar{\alpha}_k = \min \left\{ \alpha_k \ / \ x^\circ_{J_k}(k) + \alpha_k \, s_{J_k}(k) \geq 0 \right\} \qquad (6.50)$$

which defines the limit in the direction s(k) beyond which the flow x(k) is no longer feasible. Let us put $\bar{\alpha} = \min_k \bar{\alpha}_k$

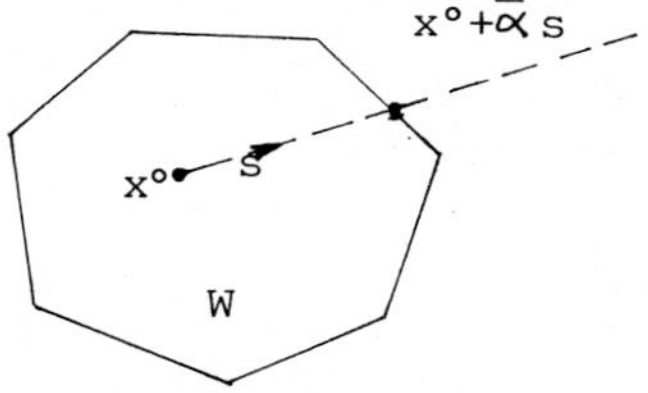

Figure 6.11

In the aggregated flows space, $s = \sum_k s(k)$ defines a direction in which the criterion can decrease. $\bar{\alpha}$ is the limit beyond which the aggregated flow is no longer feasible (figure 6.11)
Resolution of the program :

$$\min_{0 \leq \alpha \leq \bar{\alpha}} \quad f(x^\circ + \alpha\, s) \qquad (6.51)$$

generates a new aggregated flow $x^1 = x^\circ + \alpha^* s$ where α^* is the solution of (6.51). Each elementary flow is updated by

$$x^1(k) = x^\circ(k) + \alpha^* s(k)$$

(2) $s_{J_k}(k) = 0$ for at least one element $k \in \{1,\ldots n\}$. By putting $g = \nabla f(x^\circ)$ the Kuhn-Tucker parameters are given by an expression which is analogous to (6.45) :

$$\beta(k) = g_{\bar{J}_k} - \left[A(k)^{\bar{J}_k}\right]' \left[A(k)^{J_k}\left[A(k)^{J_k}\right]'\right]^{-1} A(k)^{J_k} g_{J_k} \qquad (6.52)$$

(a) If $\beta(k) \leq 0$ $\forall k$ such that $s_{J_k}(k) = 0$ then $x°$ is the optimal solution of (6.46) because the Kuhn-Tucker conditions are satisfied.

(b) In the opposite case, for every k such that $s_{J_k}(k) = 0$ one must update the set of active constraints by eliminating the inequality j such that $\beta_j(k)$ is maximum. One will then proceed with a projection on the new set of active constraints.

6.3.6. Conclusion

The feasible method in sequential or aggregated form is the best known method for resolving accurately a multicommodity flow problem with a convex criterion. Known as the "flow deviation method" (FD), it often serves as a reference to estimate the efficiency of another method. The gradient projection method (GP) seems to be more efficient when there is a small number of products ($p << n$) circulating in the network, as its convergence rate is higher than that of the FD method. On the other hand the latter becomes more efficient than the GP method for a large number of products circulating in the network. The extremal flows method (EF) seems to have a convergence rate similar to that of the GP method. It has been essentially compared to heuristic methods.
It is rather difficult to make a classification of the methods described in this section. We felt it was useful to describe them as they are the best known accurate methods acting as a basis for the development of very many heuristic methods of search for extremal multicommodity flows. They are all centralised methods. Next we will study distributed methods.

6.4 Distributed algorithms based on feasible methods of optimal routing

6.4.1. Introduction

The methods we have just described are centralised, in the sense that their application requires the centralisation of information relating to the whole network in the same processor. In geographically distributed systems it would be particularly interesting to decentralise the calculations and make the processors co-operate by an exchange of information in order to obtain the optimal routing policy.
Distribution of an optimisation problem is not always possible and it is often necessary to be satisfied with sub-optimal policies in the area of decentralised control. However, some problems have a structure which is such that distribution of the calcula-

tions does not involve any loss of information. Gallager [GAL - 77] was the first to show that a certain distribution of calculations is possible for the problem of optimal routing. His work was completed by those of Bertsekas [BER - 78]and Segall [SEG - 79]. Distributed algorithms give rise to the notion of parallel calculation and synchronisation between the processors, which we shall talk about after we have described the fundamental works of Gallager and Bertsekas.
Formulation of a distributed algorithm ends in a protocol of calculation whose correct operation must be proved. Gallager defined his distributed algorithm by assuming that certain information could be measured locally. We relax this assumption and propose a protocol of calculation for his algorithm.

6.4.2. Gallager's method [GAL - 77]

6.4.2.1. Problem formulation

This method uses the routing variables formulation described in section 6.2.3.2.. It is a sequential feasible method which consists of satisfying the optimality conditions sequentially on each product until they are satisfied on all the products. The originality of this method lies in the fact that the optimality conditions can be reached without it being necessary to centralise the information relating to the network in the same processor. The calculations can be distributed in the sense that every node of the graph can work with local information provided by its adjacent nodes and thus contribute to the calculation of the global optimum.

Given a node i, we adopt the following notations :

$I(i) = \{j \,/\, (j,i) \in L\}$ set of immediate predecessors of i

$O(i) = \{j \,/\, (i,j) \in L\}$ set of immediate successors of i

$H(i) = I(i) \cup O(i)$ set of neighbours of i

Assumption 6.6 : If $(i,j) \in L$ then $(j,i) \in L$ ie. the links are two-directional.

Since the method consists of working sequentially on the products, we shall consider, without losing generality, the product type n (destination node n). In the formulation by routing variables let us remember that $t_i(n)$ is the total flow of product n leaving i for destination n and $\varphi_{ij}(n)$ is the proportion of the traffic $t_i(n)$ which goes through the arc (i,j). We have shown that the equations of conservation are :

$$t(n) = r(n) + \Phi'(n)\, t(n) \tag{6.18}$$

where $\Phi(n)$ is a routing matrix for the destination n (definition 6.11). Theorem 6.1 establishes that (6.18) has a unique solution if $\Phi(n)$ meets conditions C1 - C4 of definition 6.11. This result is important for in the distributed algorithm which we shall analyse each node i calculates the elements $\varphi_{ij}(n)$ of the routing matrix on the basis of local information. The total traffic $t(n)$ can be calculated in terms of $r(n)$ and of $\Phi(n)$ by means of (6.21) and one could calculate $t(k)$ in terms of $r(k)$ and of $\Phi(k)$ for $k=1,\dots n-1$. in the same way. The aggregated flow in each arc is given by :

$$x_{ij} = \sum_{k=1}^{n} t_i(k)\ \varphi_{ij}(k) \tag{6.53}$$

The aggregated flow is therefore unique for $r(k)$ and $\Phi(k)$ given, $k=1,\dots n$.
Since $r(n)$ is independent of $\Phi(n)$, the problem of the search for the optimal flow of product n consists of calculating the routing matrix $\Phi(n)$ which minimises the criterion

$$f = \sum_{(i,j)\in L} f_{ij}\ (x_{ij})$$

To simplify the notations the index n will be omitted and instead of considering the routing variables φ_{ij} as the elements of a matrix, we are considering them in the form of vectors defined in the following way :

(a) for every node $i \neq n$, we fix an order of the divergent arcs (i,j) for $j \in O(i)$ and we define the vector $\varphi_i = \{\varphi_{ij} \ / \ j \in O(i)\}$. A routing policy is defined by the vector

$$\varphi = [\varphi_1' \quad \varphi_2' \quad \dots \quad \varphi_n']'$$

(b) We put $\Phi = \{\varphi_i \ / \ \varphi_{ij} \geq 0, \ \sum_{j\in O(i)} \varphi_{ij} = 1 \text{ for } i=1,\dots n-1\}$

Φ is the set of elements φ_{ij} which satisfy conditions C1 and C2 of definition 6.11. We put $\Phi^* = \{\varphi_i \in \Phi$ which also satisfy condition C3$\}$.

The elements of Φ^* constitute a routing policy in the sense of definition 6.11. We have, of course $\Phi^* \subset \Phi$. We put $t_{ij}(x_{ij}) = +\infty$ for $x_{ij} \geq c_{ij}$. The problem of search for the optimal flow is written as :

$$\min_{\varphi\in\Phi^*} \sum_{(i,j)\in L} t_{ij}\,[x_{ij}\ (\varphi)] \tag{6.54}$$

Let us analyse to what extent (6.54) is equivalent to the following problem

$$\min_{\varphi\in\Phi} \sum_{(i,j)\in L} t_{ij}\,[x_{ij}\ (\varphi)] \tag{6.55}$$

In the formulation (6.55), the condition C3 has been relaxed. In fact, every policy $\varphi \in \Phi$ belongs to one of the following three categories :

(1) $\varphi \in \Phi^*$

(2) $\varphi \in \Phi^*$ and we are in situation (2) analysed in the note in section 6.2.3.2. φ would then be a non-feasible policy which cannot be optimal.

(3) $\varphi \in \Phi^*$ and we are in situation (1) analysed in the same note. Let us suppose that φ is an optimal policy, according to the assumptions 6.5 the minimum of the criterion is obtained if the flow on the loops of G_o^+ are zero. Let us consider such a loop, for example $\{j,k,l,m\}$ on figure 6.12. It is possible to obtain another optimal policy without the existence of such a loop. Indeed, the loop can be opened and connected to the rest

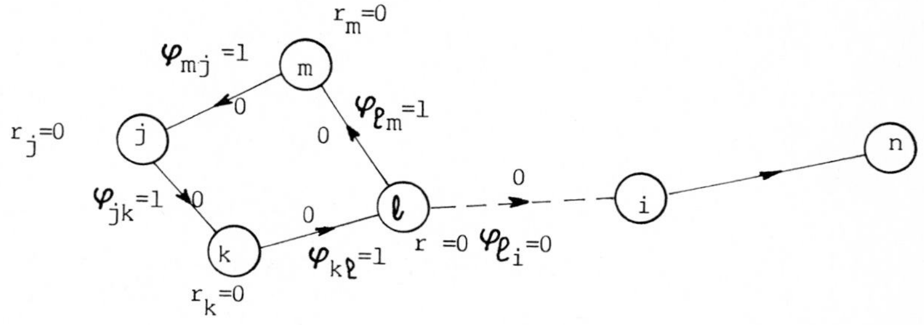

Figure 6.12

of the graph making $\varphi_{lm} = 0$ and $\varphi_{li} = 1$ for $i \notin N_o$. As there is a path from i to n, there is then a path from every node of the loop towards n. Consequently there also exists an optimal policy $\varphi \in \Phi^*$ which can be obtained simply from the optimal solution of (6.55). We shall see, in fact, that the resolution of (6.55) from an initial policy $\varphi^\circ \in \Phi^*$ generates a sequence of solutions $\varphi^\ell \in \Phi^*$ provided that some precautions are taken.

The problem (6.55) can be re-written in the form :

$$\min \quad f\,[\varphi_1, \ \varphi_2, \ \ldots \ \varphi_{n-1}] \tag{6.56.a}$$

subject to

$$\left.\begin{aligned} \sum_{j\in O(i)} \varphi_{ij} &= 1 \quad (6.56.b)\\ \varphi_i &\geqslant 0 \quad (6.56.c)\end{aligned}\right] \quad \forall \ i=1,\ldots n-1$$

6.4.2.2. Optimality conditions

Necessary optimality conditions. Let the scalar λ_i and the vector $\mu_i = \{\mu_{ij} \;\; j \in O(i)\}$ be the dual variables relative to the constraints (6.56.b) and (6.56.c). Let us put

$$\lambda = \{\lambda_i, \ i = 1,\ldots n-1\} \text{ and } \mu = [\mu_1' \ldots \mu_{n-1}']$$

We construct the Lagrangian :

$$L[\varphi,\lambda,\mu] = f\,[\varphi] + \sum_{i=1}^{n-1} \lambda_i\Big[1- \sum_{j\in O(i)} \varphi_{ij}\Big] + \sum_{i=1}^{n-1} \mu_i' \varphi_i \tag{6.57}$$

If the vectors φ_i are optimal, there exists a vector λ and a vector μ such that for every i=1,... n-1.

$$\nabla L_{\varphi_i} = 0 \qquad (6.58.a)$$

$$\mu_i' \varphi_i = 0 \qquad (6.58.b)$$

$$\mu_i \leq 0 \qquad (6.58.c)$$

Using (6.57) we can re-write (6.58.a) in the form :

$$\frac{\partial f}{\partial \varphi_{ij}} - \lambda_i + \mu_{ij} = 0 \qquad (6.59)$$

If $\varphi_{ij} > 0$, (6.58.b) $\Rightarrow \mu_{ij} = 0$, (6.59) $\Rightarrow \frac{\partial f}{\partial \varphi_{ij}} - \lambda_i = 0$

If $\varphi_{ij} = 0$, (6.58.c) and (6.59)$\Rightarrow \frac{\partial f}{\partial \varphi_{ij}} - \lambda_i \geqslant 0$

whence the necessary conditions of optimality :

$$\frac{\partial f}{\partial \varphi_{ij}} - \lambda_i = 0 \qquad \text{if } \varphi_{ij} > 0 \quad \forall\, i,j \qquad (6.60.a)$$

$$\frac{\partial f}{\partial \varphi_{ij}} - \lambda_i \geqslant 0 \qquad \text{if } \varphi_{ij} = 0 \quad \forall\, i,j \qquad (6.60.b)$$

Interpretation : For every node i, the arcs (i,j) such that $\varphi_{ij} > 0$ have an identical marginal cost $\frac{\partial f}{\partial \varphi_{ij}}$ which is less than the marginal cost of the arcs such that $\varphi_{ij} = 0$.

Calculation of the marginal cost $\frac{\partial f}{\partial \varphi_{ij}}$

(1) If φ_{ij} increases by ε, the corresponding flow increase on the arc (i,j) is εt_i. This perturbation induces the cost variation $\varepsilon t_i \frac{\partial f_{ij}}{\partial x_{ij}}$ on (i,j). Its effect on the rest of the network from the node j can be considered as a perturbation εt_i on r_j whence :

$$\frac{\partial f}{\partial \varphi_{ij}} = t_i \left[f'_{ij} + \frac{\partial f}{\partial r_j} \right] \qquad (6.61)$$

with $\quad f'_{ij} = \frac{\partial f_{ij}}{\partial x_{ij}}$

(2) To calculate $\frac{\partial f}{\partial r_i}$, let us note that a perturbation ε on r_i causes a flow perturbation $\varepsilon \varphi_{ij}$ on the arc (i,j) $\forall j \in 0(i)$,

which can be considered as a perturbation $\varepsilon\,\varphi_{ij}$ on r_j from the node j, whence :

$$\frac{\partial f}{\partial r_i} = \sum_{j \in O(i)} \varphi_{ij} \left[f'_{ij} + \frac{\partial f}{\partial r_j} \right] \qquad (6.62)$$

This equation, which implicitly defines $\frac{\partial f}{\partial r_i}$ can be easily calculated as we shall see in the description of the algorithm. (6.61) can also be calculated. Moreover, it is possible to prove [GAL - 77]:

Theorem 6.2

If $f_{ij}\ (x_{ij}) \in C^1$ and if $\varphi \in \Phi^*$, the system of equations (6.61) has a unique solution. Furthermore $\frac{\partial f}{\partial \varphi_{ij}}$ and $\frac{\partial f}{\partial r_i}$ are continuous functions of r and φ .

Sufficient optimality conditions

Problem (6.56) is not convex because the functions $f_{ij}(\varphi)$ are not convex. Inflexion points may exist and the necessary optimality conditions are therefore unfortunately not sufficient. The difficulties stem from the term t_i in (6.61) because the necessary optimality conditions are automatically satisfied in i when t_i=0. In [GAL - 77] , Gallager shows on a small example the existance of a point of inflexion for f (φ).

According to the conditions (6.60) we can write :

$$\sum_{j \in O(i)} \varphi_{ij} \frac{\partial f}{\partial \varphi_{ij}} = \lambda_i \leqslant \frac{\partial f}{\partial \varphi_{ij}} \qquad (6.63)$$

which is satisfied with strict equality for $\varphi_{ij} > 0$.

If $t_i > 0$, (6.63) can be divided by t_i and we have :

$$\sum_{j \in O(i)} \varphi_{ij} \left[\frac{1}{t_i} \frac{\partial f}{\partial \varphi_{ij}} \right] \leqslant \frac{1}{t_i} \frac{\partial f}{\partial \varphi_{ij}} \qquad (6.64)$$

satisfied with strict equality for $\varphi_{ij} > 0$. Using (6.61) and (6.62) we can write :

$$\frac{\partial f}{\partial r_i} \leqslant f'_{ij} + \frac{\partial f}{\partial r_j} \qquad (6.65)$$

satisfied with strict equality for $\varphi_{ij} > 0$. We can show [GAL - 77] that this expression is a sufficient condition of optimality which remains valid in the case where t_i=0. These results are summarised in the following theorem :

Theorem 6.3

Let us suppose that $f_{ij}(x_{ij})$ is a convex function $\in C^1$ for $0 \leq x_{ij} \leq c_{ij}$. Let $\Psi = \{\varphi / x_{ij} < c_{ij} \quad \forall (i,j) \in L\}$. Then (6.60) are necessary conditions to minimise f on Ψ and (6.65) are sufficient conditions.

6.4.2.3. Calculation of marginal delays

The test of optimality consists of verifying whether the conditions (6.65) are satisfied $\forall i$, $\forall j \in 0(i)$. In this section we give details of how the terms $\frac{\partial f}{\partial r_i}$ can be calculated in order to do the optimality test. We shall show in the next section that if the initial routing policy is such that $\varphi^\circ \in \Phi^*$, the algorithm generates a sequence of policies $\varphi^\ell \in \Phi^*$.

Theorem 6.4

There exists an optimal policy $\hat{\varphi}$ for (6.54) such that $G^+ = (N, L^+)$ has no loop.

Proof : Let $\varphi \in \Phi^*$ be an optimal policy (6.54). According to the assumptions (6.5), if there is a loop in G^+ the flow on this loop must be zero, for if not a flow decrease would cause a decrease of the criterion. Since $\varphi \in \Phi^*$ there is at least one node l of the loop which has two immediate successors in G^+ one of which belongs to the loop (noted as m) and the other which does not (noted as i). (See figure 6.12). This policy φ is such that $\varphi_{lm} > 0$, $\varphi_{li} > 0$. By making $\varphi_{lm}=0$, $\varphi_{li}=1$, we define a new policy $\hat{\varphi}$ without a loop, which is optimal since $f(\hat{\varphi}) = f(\varphi)$. It is therefore always possible to find an optimal policy without a loop.

This theorem is fundamental as it is possible to restrict the search for an optimal routing policy to the set of policies $\varphi \in \Phi^*$ which have in addition the property of having no loop. Let us consider such a policy and let $G^+ = (N,L^+)$ be its representative graph. It is well known that a graph with no loops can be decomposed into levels. From every node $i \in N$, there exists in G^+ at least one path towards the destination n. Let us call the successor of i a node l which belongs to the paths from i to n. The node i is then called the predecessor of l.

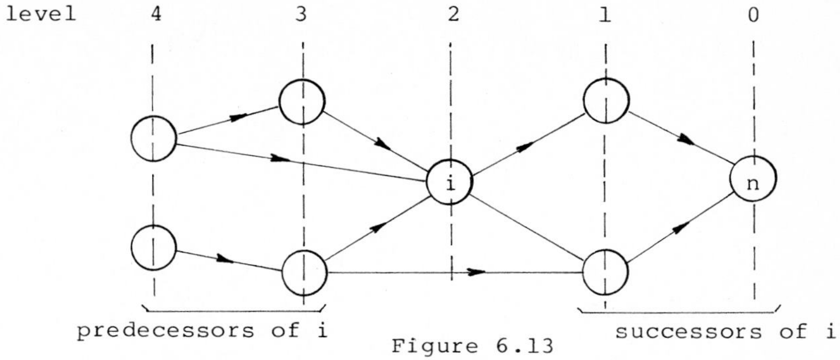

Figure 6.13

The node n is the only node of the level 0, the successors of a node i of level k belong to the levels 0, ... k-1. The relationship predecessor-successor forms a partial order on the set of nodes. Such a structure makes it possible to distribute the calculations of the marginal delays. Every node i can calculate $\frac{\partial f}{\partial r_i}$ by the following procedure :

(a) i waits until it has received $\frac{\partial f}{\partial r_j}$ from its immediate successors in G^+, ie. from the nodes $j \in 0(i)$ such that $\varphi_{ij} > 0$.

(b) i calculates $\frac{\partial f}{\partial r_i}$ using (6.62). It is supposed, in fact, that each node can estimate x_{ij} and $f'_{ij}(x_{ij})$, $\forall j \in 0(i)$. Moreover, we put $\frac{\partial f}{\partial r_n} = 0$.

(c) i transmits $\frac{\partial f}{\partial r_i}$ to its immediate predecessors.

Calculation of the marginal delays can only be distributed if G^+ has no loops. To guarantee this property we shall see that it is necessary to introduce the notion of set of blocked nodes $B(i) = \{ j \in 0(i) \; / \; \varphi_{ij} = 0$ and must remain zero at the following iteration$\}$ linked with every node $i \neq n$.

Interpretation of the optimality conditions

The condition (6.65) is satisfied with strict equality for $\varphi_{ij} > 0$. Consequently, for the optimal routing policy $\hat{\varphi}$, the marginal cost $\frac{\partial f}{\partial r_i}$ can be interpreted as the minimal distance from the node i to n for the metric $d = \nabla f(\hat{x}) = \{ f'_{ij}[x_{ij}(\hat{\varphi})] \}$ on the arcs of the network G. The graph G^+ relative to this policy $\hat{\varphi}$ is a set of minimal paths with bifurcations which converge towards n. We obviously find again the conventional result that every product uses paths whose marginal cost is identical to or less than the marginal cost of the unused paths.

6.4.2.4. Algorithm

From marginal delays we can verify in each node i whether the sufficient conditions of optimality are satisfied on the arcs (i,j), $j \in O(i)$. If this is not the case, from the present policy $\varphi \in \Phi^*$ without loops, the algorithm generates a new policy $\varphi^1 = F(\varphi)$ which has the same properties. The sufficient optimality conditions (6.65) suggest a decrease of the variables φ_{ij} for which the marginal cost :

$$\delta_{ij} = f'_{ij} + \frac{\partial f}{\partial r_j} \tag{6.66}$$

is important and an increase of those for which this marginal cost is low. The procedure is as follows :

(a) for $j \in B(i)$ (immediate successor of i blocked) do

$$\varphi^1_{ij} = 0 \tag{6.67}$$

(b) for $j \notin B(i)$ we define the difference of marginal cost by :

$$a_{ij} = \delta_{ij} - \min_{l \notin B_i} \delta_{il} \tag{6.68}$$

and we call j* the preferred successor, the node which guarantees the minimum in (6.68). We call (i,j*) the preferred arc. Namely :

$$v_{ij} = \min \left\{ \varphi_{ij},\ \eta\, a_{ij} / t_i \right\} \tag{6.69}$$

We do :

$$\varphi^1_{ij} = \begin{cases} \varphi_{ij} - v_{ij} & \text{for } j \neq j^* \\ \varphi_{ij} + \sum_{j \neq l} v_{ij} & \text{for } j = j^* \end{cases} \tag{6.70}$$

This algorithm reduces the flow on the non-optimal arcs proportionally to the difference of marginal cost in relation to the preferred arc (with the restriction that the flow must remain non-negative), and sends the totality of the negative flow variations on the preferred arc.

Note : For the nodes i such that $t_i=0$, it can easily be verified that the algorithm consists of doing $\varphi_{ij}=0$ for $j \neq j^*$, $\varphi_{ij^*}=1$.

Convergence of the algorithm depends on the parameter η . For small η convergence is guaranteed but can be slow. If η is too big convergence is no longer guaranteed. The choice of this parameter is the delicate point of the method.

Blocked nodes

This notion is fundamental to generate a series of routing policies $\varphi^\ell \in \Phi^*$. The set of blocked nodes relating to the node i is defined by :

$$B(i) = \{ j \in O(i) \,/\, \varphi_{ij} = 0 \text{ and in addition there exists a path from } j \text{ to } n \text{ which contains an improper arc } (l,m) \}.$$

It is said that an arc (l,m) is improper if the following two conditions are satisfied :

(a) $$\varphi_{lm} > 0 \text{ and } \frac{\partial f}{\partial r_l} \leqslant \frac{\partial f}{\partial r_m} \quad (6.71a)$$

(b) $$\varphi_{lm} \geqslant \eta \left[\delta_{lm} - \frac{\partial f}{\partial r_l} \right] / t_l \quad (6.71b)$$

The sets B(i) are introduced to avoid the creation of loops. Indeed, if φ_{ij} is increased from 0 and if there exists a path from j to n containing an improper arc, according to condition (6.71a) there is a risk that this path may pass through i, thus causing the appearance of a loop. Condition (6.71b) makes the notion of an improper arc less restrictive. Indeed, it is easy to see that φ_{lm} will necessarily decrease in an improper arc so that if φ_{lm} becomes zero this arc can no longer cause the existence of a loop. According to (6.69) it would be enough to impose $\varphi_{lm} \geqslant \eta\, a_{lm} / t_l$ as a condition. In fact the condition (6.71b) is a little more restrictive than this.

It can be noted that if the conditions (6.71) are satisfied there is not necessarily the creation of a loop because (l,m) may belong to a path from j to n which does not pass through i. On the other hand, if (6.71) is not satisfied φ^1 will have no loop.

The sets B(i) can be generated by tags linked to the nodes and calculated at the same time as $\frac{\partial f}{\partial r_i}$. Every node l, having calculated $\frac{\partial f}{\partial r_l}$, can verify whether the conditions (6.71) are satisfied.

If such is the case for an arc (l,m) a tag is set to signify that l is blocked for the destination n. This tag must be reflected back on all the predecessors of l. The tag is transmitted at the same time as $\frac{\partial f}{\partial r_l}$ is sent to the predecessors of l. Every node i which receives such a tag has no need to test the conditions (6.71). It is itself blocked for the destination n. It sets its tag and reflects it back on its predecessors with $\frac{\partial f}{\partial r_i}$. From these tags, calculation of B(i) sets is immediate.

Convergence of the algorithm

Convergence can be proved in the case where the coefficient η is sufficiently small [GAL - 77].

Theorem 6.5

On the assumptions 6.5 concerning $f_{ij}\ (x_{ij})$

$\forall\ f_o > 0,\quad \exists\ \eta$ for F such that if $f\ (\varphi^\circ) \leq f_o$ then

$$\lim_{l \to \infty} f\ (\varphi^l) = \min_{\varphi} f(\varphi)$$

with

$$\varphi^l = F[\varphi^{l-1}] \qquad \forall\ l \geq 1$$

6.4.3. Bertsekas method [BER - 78]

This can be seen as an extension of the Gallager's method. Like the latter, it is a feasible sequential method which works on the routing variables φ . Its originality stems from the fact that the optimisation technique is based on the method of the gradient projection of Goldstein, Lévitin, Polyak [LEV - 66]. The Gallager's method becomes a special case of the Bertsekas method.

6.4.3.1. Goldstein, Lévitin, Polyak's method of the gradient projection

Let the program be :

$$\begin{array}{l} \min \quad f(x) \\ \text{subject to} \\ Ax = b, \quad x \geq 0 \end{array} \tag{6.72}$$

with $f \in C^2$. Knowing a feasible initial solution x_o, the successive solutions are

$$x^{l+1} = \bar{x}^l$$

with $\bar{x}^l$ the solution of the auxiliary problem

$$\begin{array}{l} \min \quad \nabla f\ (x^l)\ \ (x-x^l) + \frac{1}{2}\ \ (x-x^l)'\ M^l\ (x-x^l) \\ \text{subject to} \\ Ax = b, \quad x \geq 0 \end{array} \tag{6.73}$$

where M^l is a symmetrical, positive definite matrix on $\mathcal{N}(A)$, the null space of A. For M^l sufficiently large it is possible to show that the algorithm is convergent.

Notes :

(1) If $M^l = \nabla^2 f(x^l)$ and supposing the Hessian to be positive definite on $\mathcal{N}(A)$, this method can be interpreted as a version of the Newton's method in the constrained case. It is often inconvenient to use the Hessian. An approximation of $\nabla^2 f(x^l)$ is used (for example a diagonal matrix).

(2) Let us put $g^l = \nabla f(x^l)$. If M^l has an inverse, the auxiliary problem (6.73) can be written as :

$$\min \frac{1}{2} \left[x - x^l + (M^l)^{-1} g^l\right]' M^l \left[x-x^l+(M^l)^{-1} g^l\right] \tag{6.74}$$

subject to

$$Ax = b, \quad x \geqslant 0$$

and $\bar{x}^l$ is therefore the projection of $x^l - (M^l)^{-1} g^l$ on the constraints with the norm linked to M^l.

6.4.3.2. Application to the routing problem

The problem of optimal routing is formulated as in Gallager's method. We recall that the problem is written as :

$$\min \ f\left[\varphi_1, \varphi_2, \ldots \varphi_{n-1}\right] \tag{6.56a}$$

subject to

$$\varphi_i \in \Phi_i \qquad \text{with } \Phi_i = \left\{\varphi_i \geqslant 0 \ / \sum_{j \in 0(i)} \varphi_{ij} = 1\right\}$$

(6.56b and c)

Here Φ is made to appear in decomposed form for reasons which will become obvious in the sequel.
In the auxiliary problem, if one chooses a block diagonal matrix M^l composed of square sub-matrices of dimension equal to the dimension of the vectors φ_i, the auxiliary problem can be decomposed into n-1 independent problems. The successive solutions will be given by :

$$\varphi_i^{l+1} = \bar{\varphi}_i^l \qquad i=1, \ldots n-1 \tag{6.75}$$

with φ_i^l solution of the auxiliary problem :

$$\min \ \nabla f_{\varphi_i} \ (\varphi_i - \varphi_i^l) + \frac{1}{2} \ (\varphi_i - \varphi_i^l) \ M_i^l \ (\varphi_i - \varphi_i^l) \tag{6.76}$$

subject to $\varphi_i \in \Phi_i$

As in Gallager's method, one must search only for routing policies without loops. If the algorithm is initialised with a policy $\varphi°$ having no loop to guarantee that successive policies have this property, the notion of blocked nodes B(i) from every node i for the destination n must be introduced. On the other hand, we know that the problem (6.56) is not convex and may present inflexion points when $t_i=0$, which implies $\nabla f_{\varphi_i} = 0$.
In order to avoid these difficulties Bertsekas proposes finding $\bar{\varphi}_i^l$, the solution of the following auxiliary problem :

$$\text{Min } \delta_i' \ (\varphi_i - \varphi_i^l) + \frac{1}{2} \ (\varphi_i - \varphi_i^l) \ M_i^l \ (\varphi_i - \varphi_i^l) \tag{6.77a}$$

subject to

$$\left|\begin{array}{l} \sum_{j \in 0(i)} \varphi_{ij} = 1 \\ \varphi_i \geqslant 0 \\ \varphi_{ij} = 0, \qquad \forall\ j \in B(i) \end{array}\right. \tag{6.77b}$$

with

$$\delta_i = \frac{1}{t_i} \nabla f'_{\varphi_i} = \left\{ \delta_{ij} = f'_{ij} + \frac{\partial f}{\partial r_j} \right\} \tag{6.78}$$

which is consistent with (6.66).

In addition :

. for i such that $t_i > 0$, M_i^l is a symmetrical positive definite matrix on the sub-space

$$\left\{ v_i \ / \sum_{j \in B(i)} v_{ij} = 0 \right\}.$$

. for i such that $t_i = 0$, $M_i^l = \mathbb{O}$.

This new auxiliary problem is identical to the preceding one, dividing the criterion by t_i and including t_i^{-1} in the matrix M_i^l. Its solution is unique for $t_i > 0$.

Blocked nodes. The definition is slightly different from that given by Gallager. $B(i) = B^1(i) \cup B^2(i)$ with :

$$B^1(i) = \left\{ j \in 0(i) \ / \ \varphi_{ij} = 0 \text{ and } \frac{\partial f}{\partial r_i} \leq \frac{\partial f}{\partial r_j} \right\}$$

$B^2(i) = \{ j \in 0(i) \ / \ \varphi_{ij} = 0$ and there exists a path from j to n which containts an improper arc (l,m) for the destination n$\}$.

An arc (l,m) is said to be improper for the destination n if

$$\varphi_{lm} > 0 \text{ and } \frac{\partial f}{\partial r_l} \leq \frac{\partial f}{\partial r_m}$$

$B^2(i)$ is identical to B(i) defined by Gallager without the condition (6.71b).

$B^1(i)$ was useless in Gallager's method because its algorithm reduces the flow on the improper arcs. In the Bertsekas method which is more general, this additional constraint has to be introduced because, for certain choices of M_i^l, it may happen that φ_{ij} increases on an arc such that

$$\frac{\partial f}{\partial r_i} \leq \frac{\partial f}{\partial r_j}$$

which could cause the appearance of a loop.

<u>Theorem 6.6</u>

(1) If φ^l has no loop, then φ^{l+1} has no loop

(2) If φ^l has no loop and resolves the auxiliary problem (6.77), then φ^l is optimal.

(3) If φ^l is optimal, then φ^{l+1} is also optimal

(4) If $\bar{\varphi}_i^l \neq \varphi_i^l$ for i such that $t_i > 0$, then $\exists\, \hat{\eta}_l > 0$ such that

$$f\left\{\varphi^l + \eta(\bar{\varphi}^l - \varphi^l)\right\} < f(\varphi^l) \quad \forall \eta \in (0, \hat{\eta}_l)$$

For proof of this theorem, see [BER - 78].
Choice of the M_i^1 is crucial because it has a bearing on the rate of convergence of the algorithm. A particular choice of M_i gives the Gallager's method.

6.4.3.3. Return to Gallager's method

Let there be $j_1, j_2, \ldots j_p \in O(i)$ such that $j. \notin B(i)$. The indices can always be rearranged so that :

$$\delta_{ij} \geqslant \delta_{ij_p} \qquad \forall j \in O(i),\ j \notin B(i) \tag{6.79}$$

We consider the matrix Z_i^1 (p,p-1) :

$$Z_i^1 = \begin{bmatrix} 1 & & & \mathbb{O} \\ & 1 & & \\ \mathbb{O} & & \ddots & \\ -1 & -1 & \cdots & 1 \end{bmatrix} \tag{6.80}$$

This is a linear operator which, to every vector $v_i = \left\{v_{ij},\ j=j_1,\ldots j_{p-1}\right\} \in R^{p-1}$ links a vector $\Delta\varphi_i = Z_i^1\ v_i \in \text{Im}\ (Z_i^1) \subset R^p$, where Im (Z_i^1) is the image space of Z_i^1. We can easily verify that :

$$\text{Im}\ (Z_i^1) = \left\{\Delta\varphi_i \in R^p \ /\ \sum_{j=j_1}^{jp} \Delta\varphi_{ij} = 0\right\} \tag{6.81}$$

We put $$M_i^1 = \frac{t_i}{\eta}\ Z_i^1 \left[(Z_i^1)'\ Z_i^1 \right]^{-2} (Z_i^1)' \quad , \eta > 0$$

By changing the variables $\varphi_i = \varphi_i^1 - Z_i^1\ v_i$, the auxiliary problem (6.77) becomes the following :

$$\min - a_i'\ v_i + \frac{t_i}{2\eta}\ v_i'\ v_i \tag{6.82a}$$

subject to the constraints

$$v_{ij} \leqslant \varphi_{ij}^1 \qquad j = j_1, j_2, \ldots j_{p-1} \tag{6.82b}$$

$$\sum_{j=j_1}^{j_{p-1}} v_{ij} \geqslant -\varphi_{ij_p}^1 \tag{6.82c}$$

with $a_i = \left\{a_{ij} = \delta_{ij} - \delta_{ij_p}\right\} \geqslant 0$ for $j = j_1, \ldots j_{p-1}$, a_{ij} is coherent with (6.68).
By introducing a Lagrange parameter for the equation (6.82c)

we can show that it is always zero for $a_i \geqslant 0$. The solution of the auxiliary problem is :

$$v_{ij} = \min\left\{\varphi^l_{ij} \; ; \; \frac{\eta \, a_{ij}}{t_i}\right\} \qquad j=j_1, j_2, \dots j_{p-1} \quad (6.83)$$

or, returning to the variables φ_i, we have the algorithm :

$$\varphi^{l+1}_{ij} = \begin{cases} \varphi^l_{ij} - v_{ij} & j=j_1, j_2, \dots j_{p-1} \\ \varphi^l_{ij} + \sum_{j=j_1}^{j_{p-1}} v_{ij} & j=j_p \end{cases} \quad (6.84)$$

which is precisely Gallager's algorithm (6.70). The choice of η is difficult if the variation range of r(n) can be wide. If it is very small it guarantees convergence but the rate of convergence is slow, if it is too large it can make the algorithm non-convergent.

6.4.3.4. Use of second derivatives for the choice of M_i^l

We are choosing M_i^l composed of the diagonal elements $\frac{1}{t_i} \frac{\partial^2 f}{\partial \varphi_{ij}^2}$ for $j= j_1, j_2, \dots j_p$. We must therefore calculate $\frac{\partial^2 f}{\partial \varphi_{ij}^2}$. Using (6.61) we can write :

$$\frac{\partial^2 f}{\partial \varphi_{ij}^2} = \frac{\partial}{\partial \varphi_{ij}} \left[t_i \left(f'_{ij} + \frac{\partial f}{\partial r_i}\right)\right]$$

But since we are working with flows with no loops, we have $\frac{\partial t_i}{\partial \varphi_{ij}} = 0$ since i is the predecessor of j; knowing that

$$\frac{\partial f'_{ij}}{\partial \varphi_{ij}} = \frac{\partial f'_{ij}}{\partial x_{ij}} \frac{\partial x_{ij}}{\partial \varphi_{ij}} = f''_{ij} \, t_i, \text{ (6.85) becomes :}$$

$$\frac{\partial^2 f}{\partial \varphi_{ij}^2} = t_i^2 \, f''_{ij} + t_i \frac{\partial^2 f}{\partial \varphi_{ij} \, \partial r_j} \quad (6.86)$$

Using (6.61) we deduce :

$$\frac{\partial^2 f}{\partial \varphi_{ij} \, \partial r_j} = \frac{\partial}{\partial r_j} \left\{ t_i \left(f'_{ij} + \frac{\partial f}{\partial r_j}\right)\right\}$$

Since the flow has no loops, with j the successor to i, we have

$$\frac{\partial t_i}{\partial r_j} = 0, \quad \frac{\partial f'_{ij}}{\partial r_j} = 0, \text{ whence } \frac{\partial^2 f}{\partial \varphi_{ij}^2} = t_i^2 \left(f'_{ij} + \frac{\partial^2 f}{\partial r_j^2}\right) \quad (6.87)$$

Calculation of $\frac{\partial^2 f}{\partial r_j^2}$. Using (6.62) we can write :

$$\frac{\partial^2 f}{\partial r_j^2} = \frac{\partial}{\partial r_j}\left\{\sum_m \varphi_{jm}\,(f'_{jm} + \frac{\partial f}{\partial r_m})\right\}$$

Noting that $\frac{\partial \varphi_{jm}}{\partial r_j} = 0$ and $\frac{\partial f'_{jm}}{\partial r_j} = f'_{jm}\,\varphi_{jm}$ since φ has no loops, we have :

$$\frac{\partial^2 f}{\partial r_j^2} = \sum_m \varphi_{jm}\left\{\varphi_{jm}\, f''_{jm} + \frac{\partial^2 f}{\partial r_j\,\partial r_m}\right\} \tag{6.88}$$

Similarly

$$\frac{\partial^2 f}{\partial r_j\,\partial r_m} = \frac{\partial}{\partial r_m}\left\{\sum_n \varphi_{jn}\,(f'_{jn} + \frac{\partial f}{\partial r_n})\right\} = \sum_n \varphi_{jn}\,\frac{\partial^2 f}{\partial r_m \partial r_n}$$

whence, finally

$$\frac{\partial^2 f}{\partial r_j^2} = \sum_m \varphi_{jm}^2\, f''_{jm} + \sum_{m,n} \varphi_{jm}\,\varphi_{jn}\,\frac{\partial^2 f}{\partial r_m \partial r_n} \tag{6.89}$$

We can show that f is convex in r for given φ. We therefore have $\frac{\partial^2 f}{\partial r_m\,\partial r_n} \geq 0$, $\forall m,n$ and a lower limit for (6.89) is given by :

$$\underline{R}_j = \sum_m \varphi_{jm}^2\,(f''_{jm} + \frac{\partial^2 f}{\partial r_m^2}) \tag{6.90}$$

Since f is convex in r, the Hessian is positive semi-definite, so that :

$$\frac{\partial^2 f}{\partial r_m\,\partial r_n} \leq \left(\frac{\partial^2 f}{\partial r_m^2}\cdot\frac{\partial^2 f}{\partial r_n^2}\right)^{1/2}$$

whence the higher limit for (6.89) :

$$\bar{R}_j = \sum_m \varphi^2_{jm}\quad f''_{jm} + \left\{\sum_m \varphi_{jm}\,\left(\frac{\partial^2 f}{\partial r_m^2}\right)^{1/2}\right\}^2 \tag{6.91}$$

Using (6.90) and (6.91) we can write :

$$\underline{R}_j \leq \frac{\partial^2 f}{\partial r_j^2} \leq \bar{R}_j \tag{6.92}$$

with

$$\left|\begin{array}{l} \underline{R}_j = \sum_m \varphi_{jm}^2 \, (f''_{jm} + \underline{R}_m) \\ \overline{R}_j = \sum_m \varphi_{jm}^2 \, f''_{jm} + \left(\sum_m \varphi_{jm} \, \overline{R}_m^{1/2}\right)^2 \\ \underline{R}_k = \overline{R}_k = 0 \end{array}\right. \tag{6.93}$$

$\underline{R}_j$ and $\overline{R}_j$ are calculated and transmitted in $G^+ = (N, L^+)$ at the same time as $\frac{\partial f}{\partial r_j}$. Using (6.87) we can easily deduce the upper and lower limits for $\frac{\partial^2 f}{\partial \varphi_{ij}^2}$

$$\begin{array}{l} \underline{F}_{ij} = t_i^2 \quad (f''_{ij} + \underline{R}_j) \\ \overline{F}_{ij} = t_i^2 \quad (f''_{ij} + \overline{R}_j) \end{array} \tag{6.94}$$

Note : If we eliminate in G^+ all the nodes which do not have convergent arcs and if the sub-graph obtained is an oriented tree, then

$$\underline{F}_{ij} = \overline{F}_{ij} = \frac{\partial^2 f}{\partial \varphi_{ij}^2} \qquad \text{for all the arcs of } G^+.$$

6.4.3.5. Bertsekas algorithm based on the second derivatives

For $t_i \neq 0$ the auxiliary problem (6.77) becomes

$$\min \sum_j \left\{ \delta_{ij} \, (\varphi_{ij} - \varphi_{ij}^1) + \frac{\overline{F}_{ij}}{2\alpha t_i} (\varphi_{ij} - \varphi_{ij}^1)^2 \right\} \tag{6.95a}$$

Under the constraints $\varphi_{ij} \geq 0$, $\sum_{j \in O(i)} \varphi_{ij} = 1$, $\varphi_{ij} = 0 \quad \forall j \in B(i)$ (6.95b)

ie. M_i^1 has been replaced by a diagonal matrix with elements $\overline{F}_{ij} / 2\alpha t_i$ where α is a scale factor which guarantees convergence if it is sufficiently small. In fact $\alpha = 1$ should give good rate of convergence for a wide range of variation in r.
As for (6.77) we say $M_i = 0$ if $t_i = 0$.
Using a Lagrange parameter λ for the constraint $\sum_j \varphi_{ij} = 1$, we can easily find the solution of the auxiliary problem (6.95).

$$\overline{\varphi}_{ij}^1 = \max \left\{ 0, \varphi_{ij}^1 - \frac{\alpha \, (\delta_{ij} - \lambda)}{t_i \, (f''_{ij} + \overline{R}_j)} \right\} \tag{6.96}$$

where λ is the solution of the piece-wise linear equation :

$$\sum_{j \notin B_i} \max \left\{ 0, \varphi^1_{ij} - \frac{\alpha (\delta_{ij} - \lambda)}{t_i (f''_{ij} + \bar{R}_j)} \right\} = 1 \qquad (6.97)$$

Note :

(1) Bertsekas shows that if $r > 0$ and α is sufficiently small, the algorithm converges towards the optimal solution. In numerical applications, even with components $r_i = 0$, no difficulty appeared in convergence.
(2) The use of second derivatives generates good search directions. The correct values of α (not too small for the convergence to be rapid, not too big for convergence to be guaranteed) are near to 1 and relatively independent of r as in the Newton's method, which is a considerable advantage for a quasi-static algorithm. It is only in pathological cases that $\alpha < 0{,}5$ is necessary.
(3) As we indicated at the begining, this method is applied sequentially to the products, ie. each destination plays sequentially the role of the node n. We could consider making the routing variables of the various products vary simultaneously. The choice of α to guarantee convergence appears to be more difficult.

6.4.4. Distributed calculation and distributed algorithms [LEL - 79] ,[COU - 80]

In the Gallager and Bertsekas methods we introduced the notion of distributed calculation. We showed, in fact, that if a calculation unit is linked with each node of the network the information which is used to update the routing variables can be estimated at each node provided that the nearby calculation units exchange certain data in a defined order. We shall now develop these aspects.

6.4.4.1. Distributed systems

Let us consider a set of geographically distributed processors. Each processor carries out a program and defines a process (totally scheduled set of events) which can be seen as a sequential machine. The set of processes constitute a distributed system if they must co-operate to carry out certain operations, called activities. Since the processors are a long way from each other, the only method they can use to interact is to communicate by sending messages.
At the instant t the state of the process i can be defined by the state vector s (i,t) which must include all the information necessary for the process i to be able to continue to evolve.
The state of the distributed system at the instant t is the vector $\{ \dots s(i,t) \dots \}$ $i \in N$.
The control of a distributed system is a set of decisions and the way of modifying the state of each process in order to make them co-operate so that they carry out the activities for which the system was conceived. Two types of controls can be conside-

red :
(1) a centralised control in which a particular process, called the controller, forces all the other processes to have the same vision of the state of the system. The methods described in section 6.3 assume a centralised (or hierarchical) control as the calculations can only be made from knowledge of the global state of the system.
(2) a distributed control where there is no controller in the system. Each processor executes an algorithm which works with a partial vision of the state of the system or with an approximation of that state. The methods examined in sections 6.4 and 6.5 can be put into operation by a distributed control.

It is obvious that in order to reach a given objective co-operation between the processors requires a certain type of synchronisation. A distributed control imposes an asynchronous functioning of the processes ie. physical time must not be used to carry out this synchronisation.

Synchronisation is guaranteed by the dispatch of synchronisation messages which create an order relation between the events of the different processes. Every machine passes through a succession of active states and waiting states.
(1) a process i becomes active when it has received all the messages of synchronisation necessary to enable the continuation of its evolution. Let H(i) be the set of neighbours of i and let $I^+(i) \subset H(i)$ be the set of its neighbours from which it is waiting for a message of activation. The process i only becomes active when it has received a synchronisation message from every $j \in I^+(i)$.
(2) During its active state the process i sends a synchronisation message to a sub-set $O^+(i) \subset H(i)$ of its neighbours, then goes back to a waiting state until the activation conditions are once again fulfilled.

This synchronisation mechanism is independent of the operating speed of the processors. The activation messages which guarantee synchronisation are generally accompanied by data messages as the aim of the co-operation of machines is to calculate optimal routing variables by successive iterations. In our problem the messages constitute the activation orders and carry the information for the calculation when in the general case these two functions could be disassociated.

6.4.4.2. Distributed algorithms for optimisation

In this section we are introducing informally the notion of distributed algorithms for optimisation. Every calculation system can be seen as a piece of architecture on several levels where the execution of a program, comprising a series of instructions, corresponds to one level of observation. The word "instruction" is used in the broad sense, ie. it represents the smallest set of operation with a physical sense at the observation level of the system. It is not permitted to observe the system during the execution of an instruction, which gives a more or less global view of the system according to the definition of the instruction. We propose to structure a calculation system of optimal control policies through a three hierarchical level architecture.

(1) The global level. This one has the most abstract vision of the system. This level is not dependent on the material on which the optimisation algorithm is executed and is only concerned with the succession of iterations. The algorithm is considered as a quadruple (Q, I, Ω, f) whose arguments represent respectively the set of calculation states, the starting sub-sets ($I \subset Q$) and solution sub-sets ($\Omega \subset Q$) and the law of calculation. For every input $x^\circ \in I$, the algorithm defines a sequence $x^\circ, x^1, \ldots$ with $x^{k+1} = f(x^k)$ for $k \geq 0$ which stops as soon as $x^k \in \Omega$. We say that x^k is the solution which corresponds to x°. The global level is concerned simply with the sequence $x^\circ, x^1, \ldots$ and is not permitted to observe the way in which we pass from one solution to the next, ie. the instruction consists of an iteration of the global optimisation algorithm.

(2) The co-operation level is concerned with the way in which x^{k+1} is obtained from x^k. In the system which we are studying, where there exist several entities of calculation which co-operate in order to get x^{k+1}, we must analyse the way in which the machines interact by exchanging messages to calculate x^{k+1} from x^k. At this level we shall say that a distributed calculation occurs. Each process passes through a series of successive states which constitute the observable points of the co-operation level.
An instruction is the set of operations which modify the state of a process.

(3) The elementary level is the one where we have a detailed vision of the functioning of each processor but where the level of abstraction is the weakest, ie. we have no perception of the general functioning of the system. Every process which changes state starts calculation programs and message dispatching programs for the transmission of information to its neighbours. At this level the instructions are understood in the conventional sense of a programming language.

Therefore our calculation system can be seen as a hierarchy of three levels of abstraction where at each level the instructions keep the system in a legitimate state, ie. a state which has a meaning for the level under consideration or for the outside world. At the global level we must define an optimisation algorithm without thinking about its implementation. At the co-operation level we must define the method of implementing an algorithm by co-ordinating the machines which work in parallel to achieve a common objective. We must now formally define a distributed algorithm.

6.4.4.3. Distributed calculation protocols [MER - 79]

Distributed calculation protocols is the set of rules governing exchanges of messages between processors. The protocol may be so complex that it is necessary to use a specification technique based on a formal model in order to define it fully and unambiguously. Moreover, it is necessary to use a validation technique to prove that the protocol is correct, ie. that it achieves the objective for which it was conceived. There is no universal method to model and validate a protocol. Before giving a brief description of modelling and validation techniques we must define the characteristics of a protocol.

Characteristics of a protocol.

A protocol can be considered to have two characteristics :

(1) Party characteristic. The part of the protocol residing in a processor is called the party characteristic of the protocol. It is the set of all the possible sequences of input-output pairs of messages. We remember that every machine i enters its active state i when it has received a message from a sub-set $I^+(i)$ of its neighbours and that it then sends a message to a sub-set $O^+(i)$ of its neighbours. A machine decides to act in terms of the messages received and perhaps of the order of receipt. The party characteristic is the set of all the possible operations.
(2) Topology characteristic. The topology of a protocol is the graph whose nodes are the elements of the protocol and whose arcs represent the interactions between these elements. The topology characteristic is the set of topologies for which the protocol can act (pair of elements, elements in ring form, in star form...). Some protocols are capable of acting with an evolutive topology ie. the topology can be modified during operation as in computer networks where new entities can come into action or conversely disconnect themselves from the network.

Let us note that in the problem which we are studying there is identity between the topology of the network of processors and the topology of the protocol since a processor must exchange messages of synchronisation with all its neighbours. We shall work on a non-evolutive network topology.

Each of the two protocol characteristics can be simple or complex. Early studies concentrated on protocols with simple characteristics (protocols of communication between pairs of entities). At present there are a lot of works devoted to protocols with complex element characteristics applicable to simple topologies. For our own study we are dealing with protocols with simple element characteristics and complex topology characteristics (any type of network with non-evolutive structure). The complexity of the protocol is a result of its topology characteristic.

Modelling techniques.

A model can be used to specify a protocol formally if it is capable of representing the party and topology characteristics at the same time. Three essential models can be used :

(1) finite state machines essentially applicable to protocols with simple element characteristics.

(2) Petri nets - the applicability of which is higher than that of the finite state machines. They are better adapted to represent parallelism.
(3) High level programming languages particularly adapted to the representation of data, they do not seem adequate for the representation of complex control structures.

For the calculation protocols that we are studying the finite state machines (used in section 6.4.5.) and the Petri nets (used in section 6.5.4.) are the best adapted.

Validation techniques.

In order to prove that a protocol is functioning correctly we generally validate a certain number of properties (the system does not get blocked; does not loop; is capable of returning to a normal state after going into an abnormal state). Every validation method uses one of the four following techniques :
(1) Global state generation used particularly with modelling by Petri nets (token machine). It is applied to protocols with simple topology characteristics.
(2) Assertion proving where the protocol is considered as the execution of a parallel program. It is used with modelling by high level language but can be applied to the other models. It consists of attaching a predicate to the values of variables at certain points of the program and proving that this predicate is always true when those points are reached.
(3) Induction over the topology, which consists of showing that certain conditions are propagated in the topology. With this technique it is possible to prove that properties are satisfied or that events appear in protocols with simple elements and complex topologies (like those we are studying).
(4) Adherence to sufficient conditions, where the protocol is conceived to satisfy conditions at each stage of its elaboration which are sufficient to satisfy the desired properties. The protocol is thus correct by construction and it is therefore not necessary to validate it a posteriori.

6.4.5. Distributed calculation protocol for the Gallager's algorithm

In this section we are proposing a calculation protocol to implement the Gallager's algorithm. The specification is made by representation with coupled state machines and validation, which goes beyond the context of this study, could be made by using techniques of induction on the topology and assertion proving. Let us note that an identical protocol could be proposed for the Bertsekas algorithm. The structure for exchanging messages remaining the same, it would be sufficient to modify the calculation program when a processor is activated.

We know that at every iteration of the Gallager's algorithm the processors can be classified in levels because the routing policy is loopless. The destination node n is the root of the partial graph $G^+ = (N,L^+)$, which gives it a special role for

the initialisation of the distributed algorithm. Each processor carries out the same algorithm and proceeds in two phases :
(1) during the first phase the root initialises the calculation of the dual variables. Each processor i calculates $\frac{\partial f}{\partial r_i}$ in terms of the information transmitted by the lower levels together with an indicator e_i which will be used for calculation of the blocked nodes. One wave of calculation moves up the levels of the graph G^+ and the activation order of the processors is consistent with the partial order induced by classification into levels.

(2) during the second phase the primal variables (flows) are calculated by sending messages and activating the processors in an order consistent with the new classification into levels (from the new highest level towards the root). When n has received all the information from its neighbours the distributed algorithm stops and can be started again by the root in order to carry out a new iteration of the optimisation method.

Protocol specification

We model this by a finite state machine (FSM). Each processor other than the root n can be represented by a machine with three states noted as E1, E2, E3 and carries out successively the transitions T_{12}, T_{23}, T_{31}.

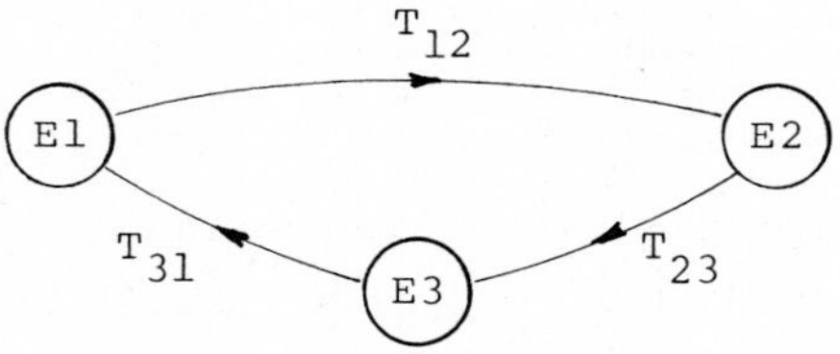

Figure 6.14

For a transition to take place a certain "condition" must be satisfied, then an "action" is undertaken. In the formal description of the algorithm we calculate a routing policy φ^1 from a policy φ . We shall use the following notations :

$$O^+(i) = \{ j \in O(i) \,/\, \varphi_{ij} > 0 \}$$
$$O_1^+(i) = \{ j \in O(i) \,/\, \varphi^1_{ij} > 0 \}$$

$\langle i, \frac{\partial f}{\partial r_i}, e_i \rangle$ is a synchronisation message from node i which transmits the information $\frac{\partial f}{\partial r_i}$ and the indicator e_i which will be used for calculation of the blocked nodes.

$\langle i, x^1_{ij} \rangle$ is a message from node i to destination j which transmits the flow on (i,j) resulting from application of the new policy.

B(i) is a vector linked to node i for calculation of the blocked nodes. Initially set at zero, its elements are progressively calculated. $B_j(i) = 1$ indicates that from i the node j is blocked for destination n.

Algorithm for a node $i \neq n$

Transition T_{12} : Calculation of the variables $\frac{\partial f}{\partial r_i}$ and e_i

Condition : To have received a message $\langle j, \frac{\partial f}{\partial r_j}, e_j \rangle$ from every $j \in O^+(i)$.

Action :

(a) Calculate $\frac{\partial f}{\partial r_i}$ using (6.62). Calculation is possible as soon as a message has been received from every $j \in O^+(i)$.

(b) Calculate :

- if $B_j(i) = 1$ for at least one element $j \in O^+(i)$, then $e_i=1$
- else calculate whether there exists an improper arc from i (only the arcs (i,j) for $j \in O^+(i)$ can be improper). If such is the case, then $e_i=1$, else $e_i=0$.

(c) Send a message $\langle i, \frac{\partial f}{\partial r_i}, e_i \rangle$ to every $j \in I(i)$.

Transition T_{23} : Calculation of the variables φ^1_{ij} and $O^+_1(i)$

Condition : To have received a message $\langle j, \frac{\partial f}{\partial r_j}, e_j \rangle$ from every $j \in O(i)$.

Action :

(a) Calculate φ^1_{ij} for $j \in O(i)$ using (6.66) to (6.70). Calculation is possible as B(i) has been totally determined.

(b) Calculate $O^+_1(i)$

(c) send a message $\langle i, x^1_{ij} = 0 \rangle$ to every $j \in O(i)$, $j \notin O^+_1(i)$

Transition T_{31} : Calculation of the variables x^1_{ij} and t^1_i

Condition : To have received a message $\langle j, x^1_{ji} \rangle$ from every $j \in I(i)$

Action :

(a) calculate $t^1_i = \sum_{j \in I(i)} x^1_{ji} + r_i$

(b) calculate $x^1_{ij} = \varphi^1_{ij} t^1_i$ for $j \in O^+_1(i)$

(c) do $B(i) = 0$

(d) send a message $\langle i, x^1_{ij} \rangle$ to every $j \in O^+_1(i)$

Receipt of a message $\langle j, \frac{\partial f}{\partial r_j}, e_j \rangle$ for a node i

(a) do $B_j(i) = e_j$

(b) execute finite state machine

Receipt of a message $\langle j, x^1_{ji} \rangle$ by a node i

execute finite state machine.

Algorithm for the root n
The root can be modelled by a two state machine

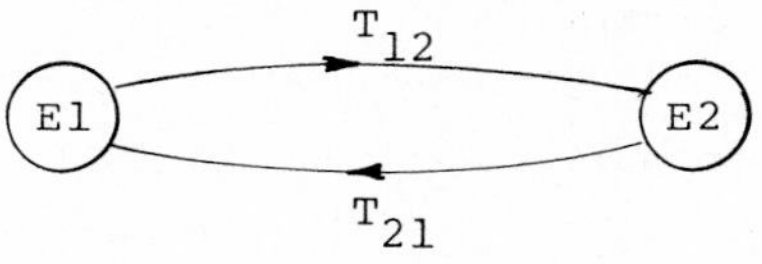

Figure 6.15

Transition T_{12}: Initialisation of the distributed algorithm

Condition : none

Action : send a message $\langle n, \frac{\partial f}{\partial r_n} = 0, e_n=0 \rangle$ to every $j \in I(n)$

Transition T_{21} : Stopping of the distributed algorithm

Condition : to have received a message $\langle j, x^1_{ji} \rangle$ from every $j \in I(n)$

Action : increase the iteration number of the optimisation method.

Validation of the protocol.

This goes beyond the context of our study and is centred around three properties which we shall not prove here. We say that the protocol is in its resting state if all the processors are in state E1, if there is no message in circulation in the network and if the graph G^+ defined from the present routing policy has no loops.

Property 1. If the distributed algorithm (the protocol) is in its resting state and if it is started by the destination, its execution stops within finite time. (there is no deadlock).

Property 2. After each execution of the distributed algorithm the flow generated has no loop.

Property 3. For a static resources vector, after a sufficient number of executions of the distributed algorithm, the routing policy minimises the criterion.

6.4.6. Conclusion.

The first optimal routing distributed algorithms attributed to Gallager and Bertsekas still fall within the area of research and have not been tested on real computer networks. The distribution of calculations must nevertheless make their application possible for obtaining optimal routing policies, when the methods used at present are based on the search for minimal paths which do not generate the optimal policy. Parallel calculation gives rise to new problems, particularly the specification of a calculation protocol which must guarantee synchronisation of the processors by making them co-operate in the search for the optimal policy. The validation of such protocols is non trivial; it is a research subject in the area of data processing. In the last section we propose an original optimisation method which can be easily distributed.

6.5 A dual method for optimal routing

6.5.1. Formulation of the problem

The method proposed in this section to calculate the optimal routing policy is a sequential method like the Gallager's and Bertsekas methods examined in the previous paragraph. We are therefore considering one product and, without losing generality, we are considering the problem of optimal routing of the product destined for n. To simplify the notation, the index n is omitted in the variables of the problem. In relation to the methods we have just examined, the two essential differences are :

(1) a formulation of the problem in x instead of formulation in φ .
(2) resolution of the dual problem instead of resolving the primal.

Let us remember that the primal problem can be written in the form :

$$\min \sum_{(i,j)\in L} f_{ij}\ (x_{ij}) \tag{6.98a}$$

subject to

$$Ax = r, \quad x \geqslant 0 \tag{6.98b}$$

We are using the Lagrangian formulation of the dual of problem (6.98). By dualising in relation to the constraints Ax = r and by introducing the vector p(1,n) of dual variables associated with these constraints, the Lagrangian is written as :

$$L(x,p) = \sum_{(i,j)\in L} f_{ij}(x_{ij}) + p\ (r-Ax)$$

$$= \sum_{(i,j)\in L} \left\{ f_{ij}(x_{ij}) - p\ A^{ij}\ x_{ij} \right\} + pr \tag{6.99}$$

Knowing that $A^{ij} = I^i - I^j$, (6.99) is written in the form :

$$L(x,p) = \sum_{(i,j)\in L} L_{ij}\ (x_{ij},p) + pr \tag{6.100}$$

with

$$L_{ij}\ (x_{ij},p) = f_{ij}\ (x_{ij}) - (p^i - p^j)\ x_{ij} \tag{6.101}$$

For a value p of the Lagrange parameters, the dual function is defined by

$$\psi(p) = \min_{x \geq 0} L(x,p)$$

which, using (6.100) is written as :

$$\psi(p) = \sum_{(i,j) \in L} \left\{ \min_{x_{ij} \geq 0} L_{ij}(x_{ij},p) \right\} + pr \qquad (6.102)$$

As the Lagrangian is decomposable by arc for a given value p, calculation of the dual function $\psi(p)$ consists of m independent minimisation problems, one problem being associated with each arc of the network.
According to the assuptions 6.5 on the cost functions $f_{ij}(x_{ij})$, the elementary Lagrangian (6.101) is a convex function in x_{ij} for $x_{ij} \geq 0$. Its minimum for $x_{ij} \geq 0$ exists $\forall\, p \in R^n$ and consequently $\psi(p)$ exists $\forall\, p \in R^n$.
The dual problem of (6.98) is written as :

$$\max_{p \in R^n} \psi(p) \qquad (6.103)$$

Note : p^i is a variable associated with the node i, called the potential of i. In the elementary Lagrangian (6.101), it is only the difference of potential $p^i - p^j$ which appears. This property results from the redundancy of system $Ax = r$ ($\rho(A)=n-1$). It is therefore possible to choose arbitrarily the potential of a node; we shall do, for example $p^n = 0$.
According to the weak duality theorem [LAS - 70] we know that

$$\psi(p) \leq \sum_{(i,j) \in L} f_{ij}(x_{ij}) \quad \forall\, p \in R^n, \quad \forall x \geq 0 \text{ such that } Ax=r.$$

Consequently, if we can find $\hat{p}$ and $\hat{x}$ such that :

$$\psi(\hat{p}) = \sum_{(i,j) \in L} f_{ij}(\hat{x}_{ij}) \qquad (6.104)$$

where $\hat{x}$ is a feasible solution of the primal, then $\hat{p}$ is the solution of the dual problem and $\hat{x}$ is the solution of the primal problem.

On the other hand, from the theory of Lagrangian duality we know that if the primal problem is convex and has at least one feasible solution, then there is no duality gap, ie. one can find $\hat{p}$ and $\hat{x}$ which verify (6.104).
As we shall see in the following section, the resolution of the dual problem is easier than the resolution of the primal problem, as dualisation gives rise to a decomposition of the Lagrangian by arcs.

6.5.2. Resolution of the dual problem by decomposition

6.5.2.1. Principle of the method

The dual function $\psi(p)$ has two useful properties :

(1) $\psi(p)$ is a concave function

(2) $\nabla \psi(p) = r - A\, x^*(p)$ where $x^*(p)$ results from minimisa-

tion of the Lagrangian (6.100) for the value p of the dual variables.

These properties suggest the use of a gradient type method for
the resolution of the dual problem, the principle of which is
the following :

Step 1. Initialisation. Choose an initial arbitrary vector of
dual variables (for example $p(0) = 0$)

Step 2. Minimisation of the Lagrangian. For the vector $p = p(\ell)$
of the dual variables, calculate x* which minimises the Lagrangian (6.100). The minimisation is decomposable by arc.

Step 3. Maximisation of the dual function. Calculate $\nabla \psi(p) =$
$r - Ax*$

- If $|\nabla \psi(p)| < \varepsilon$, stop. $\hat{p}=p$ is the solution of the dual and
  $\hat{x} = x*$ is the solution of the primal.

- Else, choose a direction $\pi = \pi(\ell)$ such that a movement in this
  direction from $p(\ell)$ increases $\psi(p)$. Let $\lambda*$ be the solution
  of $\underset{\lambda \geq 0}{max}\ \psi[p(\ell) + \lambda \pi(\ell)]$

  Do $p(\ell+1) = p(\ell) + \lambda*\pi(\ell)$; $\ell=\ell+1$ and return to step 2.

This method has three main advantages :

(1) it is not necessary to know an initial feasible solution of
the problem as for all the other methods described in this study.
The initial potential can be chosen arbitrarily.

(2) The dual problem is a maximisation of a concave function
without constraints. It is by nature simpler than the primal
problem.

(3) Often the disadvantage of a dual method lies in the calculations which are relatively complex to get a point of the dual
function (minimisation of the Lagrangian). In this problem,
these calculations are immediate.

6.5.2.2. Minimisation of the elementary Lagrangian L_{ij}

We recall that the Lagrangian associated with the arc (i,j) is :

$$L_{ij}(x_{ij},p) = f_{ij}(x_{ij}) - (p^i-p^j)\ x_{ij} \qquad (6.101)$$

Let us put $g_{ij}(x_{ij})= f'_{ij}(x_{ij})$. According to the assumptions
6.5, this function is convex, increasing for $x_{ij} \geq 0$. The form
of such a function could be that of figure 6.16.

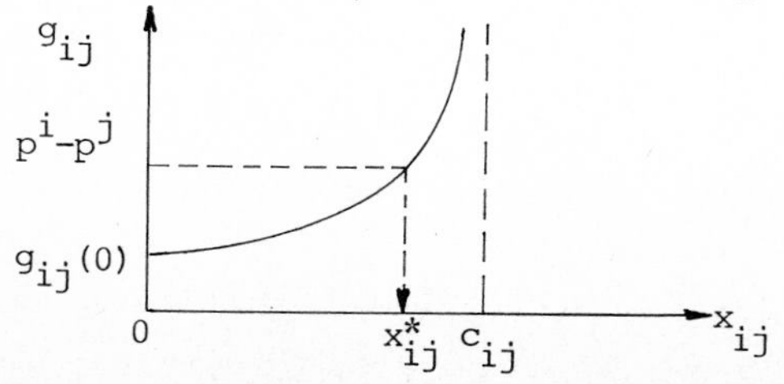

Figure 6.16

For the minimisation of L_{ij}, there are two cases :

(1) L_{ij} has a minimum for $x_{ij} > 0$. It is obtained for x_{ij}^* such that :

$$\frac{\partial L_{ij}}{\partial x_{ij}} = g_{ij}(x_{ij}^*) - (p^i - p^j) = 0, \text{ namely, } x_{ij}^* = g_{ij}^{-1}(p^i - p^j).$$

This situation is obtained for $p^i - p^j > g_{ij}(0)$.

(2) L_{ij} is an increasing function for $x_{ij} > 0$. Its minimum is obtained for $x_{ij}^* = 0$. This is the case where $p^i - p^j \leqslant g_{ij}(0)$.

In conclusion for a given p, the primal variables x* which minimise the Lagrangian are given by the analytical relation :

$$x_{ij}^* = \begin{cases} 0 & \text{if } p^i - p^j \leqslant g_{ij}(0) \\ g_{ij}^{-1}(p^i - p^j) & \text{if } p^i - p^j > g_{ij}(0) \end{cases} \qquad (6.105)$$

for every $(i,j) \in L$. This result is very important since step 2 of the dual method is immediate.

Note. We can compare the relationships (6.105) with the sufficient optimality conditions (6.65). If we put $p^i = \frac{\partial f}{\partial r_i}$, the relationships are equivalent. In the primal methods these relationships are only satisfied at the optimum, when in the dual method the optimality conditions are always satisfied, but on the other hand x is a feasible solution only at the optimum of the dual problem. For a given p, minimisation of the Lagrangian generates the optimal solution of the primal problem which would have $r^* = Ax^*$. In addition we have $p^i = \frac{\partial f}{\partial r_i^*}$ which is a well known result of the duality theory.

6.5.2.3. Maximisation of the dual function

Instead of using a standard gradient method, we used the Fletcher and Reeves conjugate gradient method [LUE - 73] :

Initialisation : Do $y_1 = p(o)$; calculate $x^*(y_1)$ and deduce from it $\pi_1 = \nabla \psi(y_1)$.

Do $\ell = 0$, $j=1$

Step 1. If $\|\nabla \psi(y_1)\| < \varepsilon$, stop. Else, let λ_j be the solution of

$$\max_{\lambda \geq 0} \psi[y_j + \lambda \pi_j]$$

Do $y_{j+1} = y_j + \lambda_j \pi_j$. If $j < n$ go on to step 2, else go to step 3.

Step 2. Calculate $x^* (y_{j+1})$ and deduce from it $\nabla\Psi(y_{j+1})$

Do $\pi_{j+1} = \nabla\Psi(y_{j+1}) + \alpha_j \pi_j$ with $\alpha_j = \dfrac{\| \nabla \Psi (y_{j+1}) \|^2}{\| \nabla\Psi(y_j) \|^2}$

Do $j = j+1$ and return to step 1.

Step 3. Let $y_1 = y_{n+1}$ and let $\pi_1 = \nabla\Psi(y_1)$

Let $j = 1$, $\ell = \ell+1$ and return to step 1.

6.5.2.4. An example

We are considering a problem dealt with in[SCH - 76]. The network is represented in figure 6.3, where 2 information units per second enter node 1 and leave at node 4. The graph $G = (N,L)$ is composed of the ordered sets :

$$N = \{1, 2, 3, 4\}$$
$$L = \{(1,2),(1,3),(2,4),(3,4),(3,2),(2,3)\}$$

Its incidence matrix is given by (6.1) and the resources vector is $r = [2\ 0\ 0\ -2]'$. For each arc the transmission cost function is in the form :

$$f_{ij}(x_{ij}) = \frac{x_{ij}}{c_{ij} - x_{ij}} + b_{ij}\, x_{ij} \qquad (6.106)$$

with the parameters given in the table below :

arc (i,j)	c_{ij}	b_{ij}
1,2	3	0
1,3	2.5	0.5
2,4	3	0
3,4	2	0.5
3,2	2	0.5
2,3	2.5	0.5

(6.107)

The function (6.106) satisfies the assumptions 6.5 with the choice of parameters in the table (6.107). This is a single commodity flow problem (only one product circulates from nodes 1 to 4) for which $p(o) = [0\ 0\ 0\ 0]$ is chosen as initial vector of the dual variables.
The results are shown in the following table for the first 4 iterations.
By way of comparison one can refer to the results of [SCH - 76] pp. 452 obtained by application of the projected gradient method. An identical precision on the primal variables is obtained after 7 iterations of the method, initialised with the vector

Table (6.108)

Iteration	$\|\nabla\Psi(p)\|$	p^1	p^2	p^3	p^4	x_{12}^*	x_{13}^*	x_{24}^*	x_{34}^*	x_{32}^*	x_{23}^*
1	0.3789	1.2665	0		-1.2665	1.4609	0.6940	1.4609	0.3847	0	0
2	0.0623	1.2328	0	0.0875	-1.3203	1.4401	0.5319	1.4926	0.5157	0	0
3	0.0112	1.2493	-0.0349	0.1037	-1.3182	1.4716	0.5322	1.4711	0.5271	0	0
4	0.0004	1.2484	-0.0349	0.1051	-1.3187	1.4710	0.5287	1.4713	0.5286	0	0

Notes :

(1) For the unidirectional maximisation of step 1, if π is the search direction and $\nabla\Psi$ the gradient of the dual function, the stopping criterion is $\nabla\Psi . \pi \leq \|\nabla\Psi\| \|\pi\| /100$.

(2) If the potential of node 4 is chosen as reference, the maximum of the dual function is obtained for $p = [2.5671 \quad 1.2838 \quad 1.4238 \quad 0]$.

$x^{\circ} = [1\ 1\ 1\ 1\ 0\ 0]'$.

6.5.3. Distribution of calculations

The dual method presented in the preceding section can be adapted to a distribution of the calculations. In this section we show the modifications to be made to it and in section 6.5.4. we shall propose a protocol of distributed calculation for maximisation of the dual function.

6.5.3.1. Sequential distributed method

The calculation entities are located at the nodes of the network. Distribution of the dual method consists of proposing an algorithm for each calculation entity using information transmitted by its neighbours to obtain the optimal global solution.

We know that the minimisation of the Lagrangian is decomposable by arc. Minimisation of L_{ij} can be made at i or at j but it is more advantageous to carry out this calculation at the same time at i and j. In fact the calculation of x^*_{ij} is immediate because it results from the analytical expression (6.105) and can be carried out at i and j, which avoids having to transfer the variables x^*_{ij} towards j if it was decided to calculate them at i. For this, every node i must receive the potential of its neighbours (which supposes that when (i,j) exists, there is also the possibility of sending the potential p^j towards the node i). Then node i can calculate the primal variables x^*_{ij} and $x^*_{\ell i}$ if the set of functions $g^{-1}_{ij}(.)$ and $g^{-1}_{\ell i}(.)$ are stored at i. For a given vector p, calculation of the primal variables can therefore be distributed.

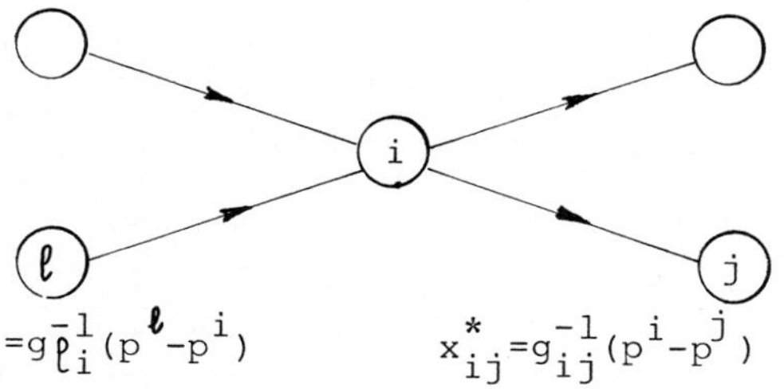

Figure 6.17

At node i it is also possible to calculate the component i of the gradient of the dual function. Indeed, $[\nabla\Psi(p)]_i = r_i - A_i x^*(p)$ with A_i the co-cycle of node i. This equations of conservation can be calculated locally after minimisation of the Lagrangian on the incident arcs of i. $\nabla\Psi(p)$ defines a direction in which Ψ can be increased but the search for the maximum in this direction requires a centralisation of the information relating to the whole network. Since only $[\nabla\Psi(p)]_i$ is known,

at the node i, the maximum of $\Psi(p)$ can be found in the direction i by fixing the potential of all the nodes around i. The maximum is obtained for the potential p^i which guarantees satisfaction of the equation of conservation at i, namely
$r_i - A_i\ x^*(p^i) = 0$.
Since $\Psi(p)$ is a function of class C^1, a sequential maximisation in the different directions achieves the maximum of Ψ. Then the calculation entities are sequentialy activated each in turn adjusting its potential. A calculation protocol for the implementation of this algorithm must create an order in the activation of the processors respected at each calculation cycle.

6.5.3.2. Improvement of the sequential method

The method proposed, which consists of making the dual variables vary sequentially in the independent space directions of $\Psi(p)$, can be improved because of the particular structure of the problem. Let us consider, for example, the network in figure 6.21 composed of 4 processors.

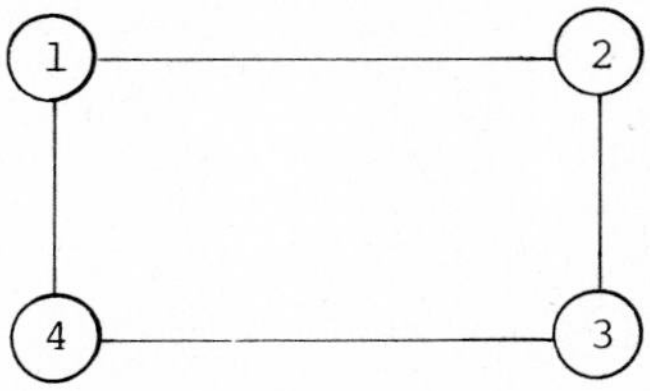

Figure 6.18

When p^1 is changed, the links between the entities 2-3 and 4-3 are not. If p^3 is then changed, the links 1-2 and 1-4 are not affected. The search for the maximum of Ψ by changing p^1 then p^3 is therefore identical to a search where p^1 and p^3 can change simultaneously. For the example in figure 6.18 the nodes can be divided into two groups $C_1 = \{1,3\}$ and $C_2 = \{2,4\}$. The distributed algorithm will consists of changing the potentials of the same group simultaneously by establishing a sequence on the activation of the groups. The potentials of the nodes of the same group define independent directions.
In general one must search for a partition of the set of nodes N into groups C_i, $N = \{C_1, C_2, \ldots C_\gamma\}$ so that two nodes are not found to be in the same group if they are adjacent.
The partition which generates the smallest value of γ is called chromatic decomposition of the graph and γ is called the chromatic number of the graph. When the potentials of the same group change the potentials of the neighbours of the elements of this group are fixed. So the maximum of Ψ is determined in several directions simultaneously. In the network in figure 6.19 the processors can be divided into two groups (the nodes are marked by the group to which they belong).

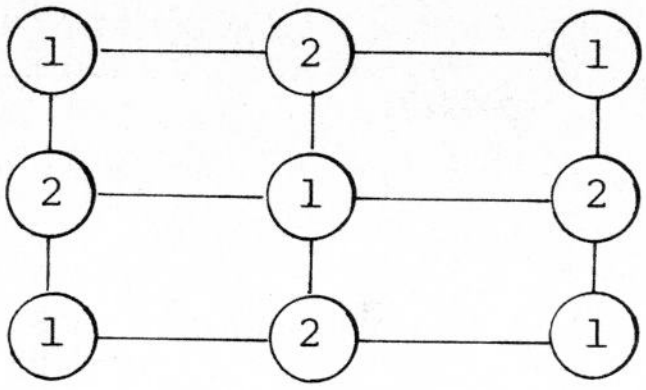

Figure 6.19

6.5.3.3. Maximisation of $\Psi(p^i)$ by a Newton's method

When the processor i is activated to calculate its potential p^i, all its neighbours are in a waiting state. To find the potential p^i which maximises Ψ it is possible to use the second derivative $\frac{\partial^2 \Psi}{\partial p^{i^2}}$ in a Newton type method. When p^i varies from $-\infty$ to $+\infty$, the critical points of this potential can be defined.

Critical points of p^i. A critical point of p^i is defined by :

$$\gamma_j^i = \begin{cases} p^j - g_{ji}(o) & \text{if } j \in I(i) \quad (6.109a) \\ p^j + g_{ij}(o) & \text{if } j \in O(i) \quad (6.109b) \end{cases}$$

For $j \in I(i)$
$$\begin{cases} p^i < \gamma_j^i \Rightarrow x_{ji}^* > 0 & (6.110a) \\ p^i > \gamma_j^i \Rightarrow x_{ji}^* = 0 & (6.110b) \end{cases}$$

For $j \in O(i)$
$$\begin{cases} p^i > \gamma_j^i \Rightarrow x_{ij}^* > 0 & (6.110c) \\ p^i < \gamma_j^i \Rightarrow x_{ij}^* = 0 & (6.110d) \end{cases}$$

If $p^i = \gamma_j^i$ the arc (i,j) or (j,i) is in its critical state, ie. a perturbation of potential makes it used or non-used.

Continuity of $\frac{\partial \Psi}{\partial p^i}$. We know that Ψ is differentiable and that :

$$\frac{\partial \Psi}{\partial p^i} = [\nabla \Psi(p)]_i = r_i - A_i\, x\,(p) \qquad (6.111)$$

According to the assumptions 6.5, $x^*(p)$ given from the expressions (6.105) is unique for arbitrary p. Consequently $\frac{\partial \Psi}{\partial p^i}$ is a continuous function.

Discontinuity of $\frac{\partial^2 \Psi}{\partial p^{i^2}}$. Let us suppose $p^i \neq \gamma_j^i$ for $j \in H(i)$ and let us put :

$$O^*(i) = \{ j \in O(i) / \; p^i > \gamma_j^i \}$$

$$I^*(i) = \{ j \in I(i) / \; p^i < \gamma_j^i \}$$

The expression (6.111) can be re-written in the form :

$$\frac{\partial \psi}{\partial p^i} = r_i + \sum_{j \in I^*(i)} x^*_{ji} - \sum_{j \in O^*(i)} x^*_{ij} \qquad (6.112)$$

By noting that

$$\frac{\partial x^*_{ji}}{\partial p^i} = - \frac{1}{g'_{ji}(x^*_{ji})} = - \frac{1}{f''_{ji}(x^*_{ji})} \quad \text{for } j \in I^+(i) \text{ and}$$

that $\quad \dfrac{\partial x^*_{ij}}{\partial p^i} = \dfrac{1}{g'_{ij}(x^*_{ij})} = \dfrac{1}{f''_{ij}(x^*_{ij})}$, we have :

$$\frac{\partial^2 \psi}{\partial p^{i^2}} = - \sum_{j \in I^*(i)} \frac{1}{f''_{ji}} - \sum_{j \in O^*(i)} \frac{1}{f''_{ij}} \qquad (6.113)$$

In the intervals defined by the critical points $\dfrac{\partial^2 \psi}{\partial p^{i^2}}$ is defined in a unique way. On the other hand, at a critical point there is a discontinuity of $\dfrac{\partial^2 \psi}{\partial p^{i^2}}$ since $I^*(i)$ or $O^*(i)$ is modified. Let us put :

$$I^+(i) = I*(i) \ , \quad \left\{ I^-(i) = j \in I(i) \ / \ p^i \leq \gamma^i_j \right\}$$

$$O^-(i) = O^*(i) \ , \quad \left\{ O^+(i) = j \in O(i) \ / \ p^i \geq \gamma^i_j \right\}$$

As for (6.113) it is easy to show that the second derivative on the left and right of a critical point are respectively :

$$\left[\frac{\partial^2 \psi}{\partial p^{i^2}} \right]^- = - \sum_{j \in I^-(i)} \frac{1}{f''_{ji}} - \sum_{j \in O^-(i)} \frac{1}{f''_{ij}} \qquad (6.114)$$

$$\left[\frac{\partial^2 \psi}{\partial p^{i^2}} \right]^+ = - \sum_{j \in I^+(i)} \frac{1}{f''_{ji}} - \sum_{j \in O^+(i)} \frac{1}{f''_{ij}} \qquad (6.115)$$

Maximisation of $\psi(p^i)$ in two steps

Step 1. Search for the interval in which the maximum of $\psi(p^i)$ is found. The critical points define intervals. As $\dfrac{\partial \psi}{\partial p^i}$ is a decreasing, monotone function, we can easily determine the two adjacent critical points

$\left[p^i \right]^-$ and $\left[p^i \right]^+$ such that $\left[\dfrac{\partial \psi}{\partial p^i} \right]_{\left[p^i \right]^-} > 0$, $\left[\dfrac{\partial \psi}{\partial p^i} \right]_{\left[p^i \right]^+} < 0$. If $\dfrac{\partial \psi}{\partial p^i} = 0$

at a critical point, p^i optimal is determined in step 1.
Step 2. Search for optimal p^i in the interval $[p^i]^-$, $[p^i]^+$.
In an interval, Ψ is a function of class C^2. The terms $\frac{\partial \Psi}{\partial p^i}$ and $\frac{\partial^2 \Psi}{\partial p^{i2}}$ are easily calculable in i with the help of the relationships (6.112) and (6.113). Then it is possible to use the Newton's method, the convergence of which is rapid near the optimum. We shall therefore do :

$$p^i = p^i - \frac{\partial \Psi / \partial p^i}{\partial^2 \Psi / \partial p^{i2}} \qquad (6.116)$$

until $\left| \frac{\partial \Psi}{\partial p^i} \right| < \varepsilon$ fixed.

6.5.4. Distributed calculation protocol

We now propose a distributed algorithm which is the implementation of the dual method in section 6.5.3.. We use two types of modelling for the distributed algorithm :
(a) by Petri nets, to make the process of synchronisation of the processors appear clearly on a given example
(b) by Finite state machines to formalise the protocol in the general case of an arbitrary topology.

6.5.4.1. Modelling by Petri nets

Here we are considering the example dealt with in section 6.5.2. on which we explain the development of the distributed algorithm. In a Petri net a condition is represented by a circle called "place" and an event is represented by a bar called "transition". Satisfaction of a condition is symbolised by the presence of a token in the corresponding place. The input places of a transition are linked to this transition by oriented arcs; oriented arcs link this transition to its output places. A transition fires when each input place has a token. The transition fire consists of taking a token from each input place and putting a token in each output place of this transition. The distributed calculation protocol for the example in figure 6.3 can be represented by the Petri net in figure 6.20.

Transitions are related to the nodes of the graph in figure 6.3 and have the same number. The transition fire is the activation of the corresponding processor.
The places represent the receipt of activation messages. Transition 1 can fire (processor 1 can start the calculation of its potential) as soon as the potential of the neighbouring processors has been received, ie. as soon as a token is present in places A and B. When transition 1 fires the potential of node 1 is calculated and sent to the neighbouring nodes, ie. a token is put in places C and F.

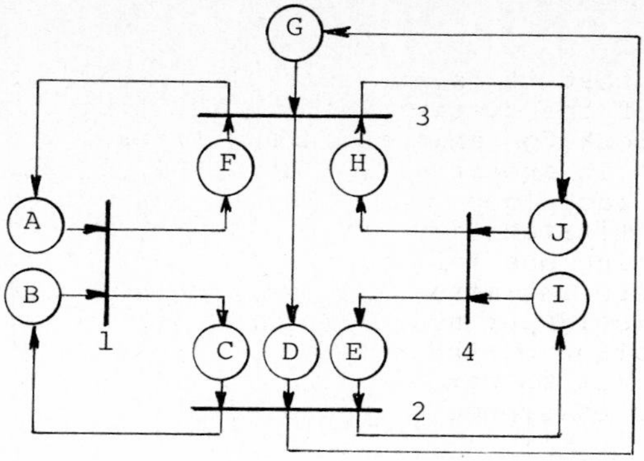

Figure 6.20

A state is a distribution of tokens in the places of the network. It defines the global state of the network. From an initial state, all the accessible states and the transitions to pass from one to the other define a token machine. For the example under consideration, the choice of initial state ABGIJ defines the token machine in figure 6.21.

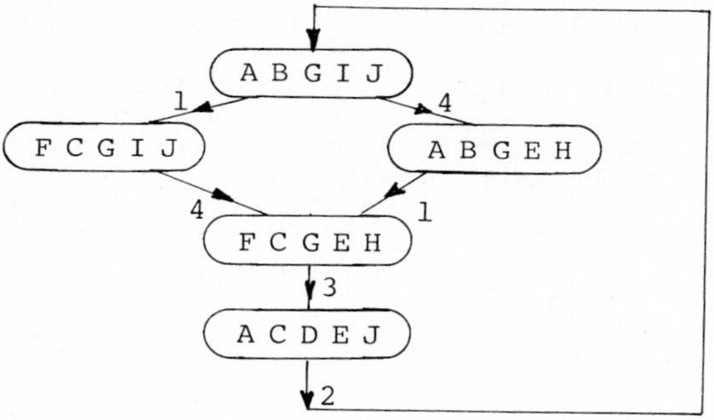

Figure 6.21

In a Petri net the transition can fire from the moment when it has been sensitized (all its input places have a token) but in our case we suppose that the processor associated with a transition is activated as soon as the latter is sensitized, the fire only taking place when the calculation of potential is terminated. In the token machine it appears a periodic functionning of the distributed algorithm. When the net enters into the global state ABGIJ, the processors 1 and 4 are simultaneously activated, but depending on the order of fire of the transitions 1 and 4, one or the other of the two branches of the token machine can be followed. We can state that in any case the potentials p^1 and p^4 are calculated in parallel and are followed by the calculation of p^3, then of p^2 before reactivating processors 1 and 4.

It is obvious that it is the initial state which determines the functionning of the system. Bad initial state, such as ABIJ, causes a deadlock for only the transitions 1 and 4 can fire. In figure 6.20 it appears five loops : A1F3, B1C2, I4E2, J4H3 and G3D2.. In each loop there are two adjacent processors which must not be activated simultaneously; such a condition can be met by allocating one token per loop. As each loop comprises two places, if there was more than one token per place, the fire of a transition would put a second token in one place, ie. a second information on the potential of the neighbouring processor and the first potential (first token) would no longer have any reason for existence.

Search for an initial state.

The processors are divided in groups C_1, C_2, ... C_γ. Each group is assigned an index which represents its activation rank in the cycle (the groups must be activated sequentially).

. Place a token on the input places of the transitions of group C_1, then fire these transitions.

. Place a token on the input places of the transitions of group C_2 which are empty and fire these transitions.

. Continue the procedure as far as group C_γ . The fire of the transitions of groupe C_γ gives an initial marking which will induce a sequential activation of the processors C_1, C_2 ... C_γ.

6.5.4.2. Modelling by finite state machines

In this algorithm all the processors have identical roles. Their function can be modelled by a machine with two states noted as E1 and E2.

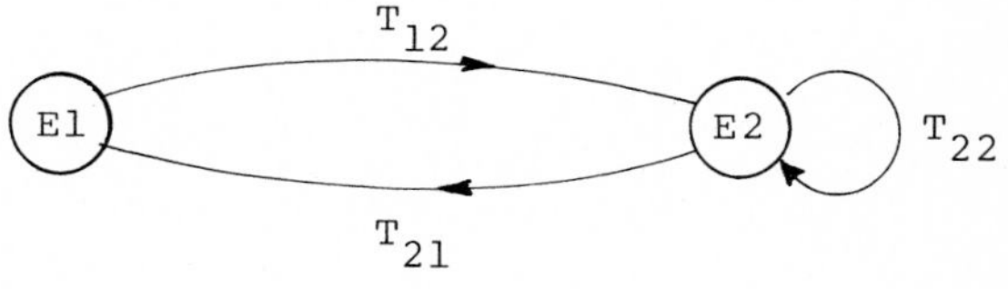

Figure 6.22

where E1 is a waiting state and E2 a calculation state. The machines are synchronised by sending messages. We are using the following notations :

$H(i) = I(i) \cup O(i)$ is the set of neighbours of node i

$\langle i, p^i \rangle$ is a message from node i which carries the potential p^i.

<u>Transition T_{12}</u>

Condition : To have received a message $\langle j, p^j \rangle$ from every $j \in H(i)$

Action : Calculate x^*_{ij} and x^*_{ji} for every $j \in H(i)$ from information received p^j and from its own potential p^i using (6.105). Deduce from this $r_i - A_i x^*$.

Transition T_{22}

Condition : $r_i - A_i x^* \neq 0$

Action : Update p^i. Recalculate x^*_{ij}, x^*_{ji} for $j \in H(i)$ and deduce $r_i - A_i x^*$.

Transition T_{21}

Condition : $r_i - A_i x^* = 0$

Action : (a) Calculate $\varphi_{ij} = x^*_{ij} / \sum_{j \in O(i)} x^*_{ij}$ for every $j \in O(i)$

(b) Send a message $< i, p^i >$ to every $j \in H(i)$

Receiptof a message $< j, p^j >$ by node i

Comment : node i is necessarily in state E1

Action : the new potential p^j replaces the preceding one. Execute the finite state machine.

6.5.4.3. Example of application

Here we take up again the example dealt with in section 6.5.2.4. by the global method. The chromatic number of the corresponding graph (figure 6.3) is $\gamma = 3$. We divide the nodes into three groups with $C_1 = \{1,4\}$, $C_2 = \{3\}$ and $C_3 = \{2\}$. The initial potentials are given by $p(o) = [0\ 0\ 0\ 0]$ and the distributed algorithm is initialised by supposing that processors 1 and 4 have received a message from 2 and 3 and that processor 3 has received a message from 2. This initialisation is the initial marking of section 6.5.4.1.. The algorithm then operates cyclically with sequential activation of classes C_1, C_2, C_3.

The results of the optimisation are summarised in table (6.117). In a node i the maximum of $\psi(p^i)$ is found by using the Newton's method up to $|r_i - A_i x^*(p^i)| < 10^{-4}$. It will be noted that apart from the first calculation of p^1 and p^2 the number of iterations in the line search is very low. At the third iterations of the distributed algorithm, the variables x^*_{ij} are a feasible solution of the primal problem and if the potential p^4 is chosen as reference, the optimal solution corresponds to the potential $p = [2.5673 \quad 1.2836 \quad 1.4239 \quad 0]$. Comparing these results with those of the global method (table 6.108), we note that identical precision was obtained after 4 iterations of the dual method with a maximisation of $\psi(p)$ by the Fletcher-Reeves method. The advantage of the distributed algorithm lies in the simplicity of the calculations, in addition to the interest contributed by the possibilites of parallel calculation.

Table 6.117

Iteration	Group	p^1	p^2	p^3	p^4	x^*_{12}	x^*_{13}	x^*_{24}	x^*_{34}	x^*_{32}	x^*_{23}	$\mathcal{N}^{(+)}$
1	C_1	1.1857			-1.3671	1.4094	0.5906	1.5186	0.4813			7;6
	C_2			0.0473			0.5211		0.5211	0	0	2
	C_3		-0.0907			1.4669		1.4669		0	0	2
2	C_1	1.1913			-1.3756	1.4703	0.5297	1.4720	0.5279			1;1
	C_2			0.0481			0.5285		0.5285	0	0	1
	C_3		-0.0922			1.4711		1.4711		0	0	1
3	C_1	1.1915			-1.3758	1.4712	0.5288	1.4713	0.5287			1;1
	C_2			0.0481			0.5288		0.5287	0	0	0
	C_3		-0.0922			1.4712		1.4713		0	0	0

$\mathcal{N}^{(+)}$ is the number of iterations for the search for the potential p^i which maximises $\psi(p^i)$ by the Newton's method.
In group C_1 the first number is in relation to processor 1 and the second to processor 4.

The major disadvantage of a dual method comes from the fact that one does not have a primal feasible solution as long as the optimum of the dual function has not been reached. In the case of routing problems this disadvantage may disappear under certain conditions. In fact after an iteration of the dual method, one has a non-feasible vector x* if the optimum ψ (p) is not yet reached, but the control variables of a routing problem are the components of the routing vector φ. From x* one can determine locally the components of φ as indicated by action (a) of the transition T_{21} (section 6.5.4.2.). One can in fact calculate

$$\varphi_{ij} = x^*_{ij} \Big/ \sum_{j \in O(i)} x^*_{ij} \quad \text{for every } j \in O(i).$$

and the resulting vector φ can define a routing policy in the sense of the definition 6.11. In the example dealt with a set of independent routing variables is given by $\{\varphi_{12}, \varphi_{23}, \varphi_{32}\}$.

In table (6.118) we give the value of these variables after each iteration of the distributed algorithm together with the primal feasible variables which would result from the application of this policy.

Table 6.118

Iteration	Routing variables			Primal feasible variables					
	φ_{12}	φ_{23}	φ_{32}	x_{12}	x_{13}	x_{24}	x_{34}	x_{32}	x_{23}
1	0.7047	0	0	1.4094	0.5906	1.4094	0.5906	0	0
2	0.7351	0	0	1.4703	0.5297	1.4703	0.5297	0	0
3	0.7356	0	0	1.4712	0.5288	1.4712	0.5288	0	0

At the first iteration the vector φ deduced from x* is a routing policy which is already very near to the optimal policy. It could be immediately applied before being refined.

6.5.5. Implementation on a microcomputer network

It has been possible to test the distributed algorithm which we have just proposed on a microcomputer network with variable topology. Each of the 6 microcomputers is a complete system constructed around a Motorola M 6800 microprocessor, equipped with memory and 8 parallel input-output ports which enable it to be connected to a maximum of 4 neighbours (one must provide for the possibility of having 2 transmission lines of opposite directions between 2 adjacent processors). Links are used to interconnect the microcomputers representing a network of arbitrary topology. It is to be noted that in a real network the connections between geographically distributed computers are of the series type for reasons of economy and reliability of transmission. In the near future we expect to replace the present parallel connections by connections through optic fibres. Exchange of messages is carried out in interruptible mode.

The manipulation proceeds in two stages :
(1) A conversational program implanted in each microcomputer is used to enter all the data of the network relating to this processor and to the connections with its neighbours, together with its initial marking. All this information is then summarised in a table.
(2) The distributed algorithm is then started on all the processors of group 1. Calculations are made in floating point format and at each iteration every processor visualises the number of the iteration, its potential and the flow on each of its incident arcs.

The example in section 6.5.4.3. was dealt with on this network and the results conform to table 6.117 which results from a program passed on an IBM 370 computer. The satisfactory functioning of this manipulation is a partial validation of the calculation protocol which is correct at least for the topology tested.

6.5.6. Conclusion

The dual method presented in this section can be regarded in two different ways :

(1) as a centralised method of resolution of convex flow or multicommodity flow problems. Its principal advantage is its simplicity and its programming facility. When the potential approaches the optimum it is possible to link to it a routing policy, which is very advantageous since in dual methods it is necessary to wait for the optimum to be obtained in order to have a primal feasible solution.

(2) this method which works essentially with the potentials, is well suited for a distribution of calculations. The distributed calculation protocol is simple, easy to put into operation, as has been proved on a microcomputer network. This protocol has original properties like the distribution of processors in groups coming from a chromatic decomposition of the network. The degree of parallelism is higher than in the distributed algorithms of section 6.4 which, on the other hand, has the interesting property of having a means of control over the satisfactory functioning of the algorithm by the priviledged role which the destination plays. These algorithms have the disadvantage of having to be initialised from a primal feasible solution when the dual algorithm can be initialised from an arbitrary potential.

We believe that studies should turn towards the specification of protocols with evolutive topologies (connection or disconnection of processors in real time) when here we have only considered algorithms functioning on arbitrary but fixed topologies. Such protocols have already been proposed for the search for minimal paths, but this search is still only beginning with respect to optimal flow problems.

REFERENCES

AUT - 79 AUTHIE G., "Recherche d'un flot minimisant une fonction de coût linéaire. Méthode primale", Note Interne LAAS n° 18, Mai 1979.

BER - 78 BERTSEKAS D.P., "Algorithms for Optimal Routing of Flow in Networks", Coordinated Science Laboratory Working Paper, University of Illinois at Champain-Urbana, June 1978.

CAN - 74 CANTOR D.G. and M. GERLA, "Optimal Routing in a Packet Switched Computer Network", IEEE Trans. on Computers, vol. C-23, no. 10, October 1974, pp. 1062 - 1069.

COU - 80 COURCOUBETIS C. and P. VARAIYA, "A Preliminary Model for Distributed Algorithms", 2nd IFAC Symposium on Large Scale Systems : Theory and Applications, Toulouse, France, June 24-26, 1980.

FRA - 73 FRATTA L., M. GERLA and L. KLEINROCK, "The Flow Deviation Method : An Approach to Store and Forward Communication Network Design", Networks vol. 3, 1971, pp. 97-133.

GAL - 77 GALLAGER R., "A Minimum Delay Routing Algorithm Using Distributed Computation", IEEE Trans. on Communications, vol. COM-25, no. 1, January 1977, pp. 73-85.

GAN - 59 GANTMACHER F.R., "Theory of Matrices", Vol. 2, Chelsea, 1959

GRE - 80 GREEN P.E., "An Introduction to Network Architectures and Protocols", IEEE Trans. on Communications, vol. COM-28, no. 4, April 1980, pp. 413-424.

LAS - 70 LASDON L.S., "Optimization Theory for Large Scale Systems", Macmillan, 1970.

LEL - 79 LE LAN G., "An Analysis of Different Approaches to Distributed Computing", 1st Int. Conf. on Distributed Computing Systems, Huntsville, USA, October 1979.

LEV - 66 LEVITIN E.S. and POLYAK B.T., "Constrained Minimization Problems", USSR Comput. Math. Math. Phys., vol. 6, 1966, pp. 1-50.

LUE - 73 LUENBERGER D.G., "Introduction to Linear and Nonlinear Programming", Addison-Wesley, 1973.

MER - 79 MERLIN P.M., "Specification and Validation of Protocols", IEEE Trans. on Communications, Vol. COM-27, no. 11, November 1979, pp. 1671-1680.

SCH - 76 SCHWARTZ M. and C.K. CHEUNG, "The Gradient Projection for Multiple Routing in Message-switched Networks", IEEE Trans. on Communications, vol. COM-24, April 1976, pp. 449-456.

SCH - 80 SCHWARTZ M. and T.E. STERN, "Routing Techniques Used in Computer Communication Networks", IEEE Trans. on communications, vol. COM-28, no. 4, April 1980, pp. 539-552.

SEG - 79 SEGALL A., "Optimal Distributed Routing for Virtual Line-Switched Data Networks", IEEE Trans. on Communications, vol. COM-27, no. 1, January 1979, pp. 201-209.

ZIM - 80 ZIMMERMANN H., "OSI Reference Model - The ISO Model of Architecture for Open Systems Interconnection", IEEE Trans. on Communications, vol. COM-28, no. 4, April 1980, pp. 425-432.

SUBJECT INDEX